Indigenous Peoples, Consent and Benefit Sharing

Rachel Wynberg · Doris Schroeder
Roger Chennells
Editors

Indigenous Peoples, Consent and Benefit Sharing: Lessons from the San-Hoodia Case

Editors
Dr. Rachel Wynberg
Environmental Evaluation Unit
University of Cape Town
Environmental & Geographical
  Science Building
South Lane, Upper Campus
Private Bag X3
Rondebosch 7701
South Africa
rachel@iafrica.com

Roger Chennells
Chennells Albertyn
Attorneys, Notaries & Conveyancers
44 Alexander Street
Stellenbosch 7600
South Africa
scarlin@iafrica.com

Prof. Doris Schroeder
Centre for Professional Ethics
University of Central Lancashire
Preston
Lancashire
PR1 2HE
United Kingdom
dschroeder@uclan.ac.uk
and
Centre for Applied Philosophy
  and Public Ethics
The University of Melbourne
Victoria 3010
Australia

Front Cover: © *Louise Gubb/Corbis*
Back Cover: Hoodia gordonii with young, unopened flowers, Ceres-Karoo, South Africa
Photographer: David Newton

Additional material to this book can be downloaded from http://extra.springer.com.

ISBN 978-90-481-3122-8   e-ISBN 978-90-481-3123-5
DOI 10.1007/978-90-481-3123-5
Springer Dordrecht Heidelberg London New York

Library of Congress Control Number: 2009930524

© Springer Science+Business Media B.V. 2009
No part of this work may be reproduced, stored in a retrieval system, or transmitted in any form or by any means, electronic, mechanical, photocopying, microfilming, recording or otherwise, without written permission from the Publisher, with the exception of any material supplied specifically for the purpose of being entered and executed on a computer system, for exclusive use by the purchaser of the work.

Printed on acid-free paper

Springer is part of Springer Science+Business Media (www.springer.com)

*For Our Children
Rebecca, Guy, Oliver,
Clara and Sebastian Chennells
Art and Mia Wynberg van der Lingen*

Preface

When one of our leaders informed an academic conference in 1997, 'The San will no longer be researched,' he spoke for us all. Our leaders had decided that we would never again be guinea pigs or objects of research, no matter how well meaning.

This study, however, is a collaborative project that is of clear benefit to the San. We were approached by the Universities of Central Lancashire and Cape Town to participate in a joint project funded by the Wellcome Trust to research and analyse the *Hoodia* case, with a special focus on benefit sharing and decision-making. It was clear to us that an objective view of the entire case, comparing it with other experiences elsewhere, would be very useful.

When the San challenged the CSIR on their patent in 2001, we were ignorant about our rights to traditional knowledge, and about intellectual property and international law. Not surprisingly, mistakes were made as we negotiated and concluded two benefit-sharing agreements over the following years, using the best knowledge available to us at the time.

Indigenous peoples elsewhere in the world supported us, and wanted to know more about how the *Hoodia* case was progressing. 'What about benefit sharing?' they asked. 'How are your decisions being made? What do the San feel about the *Hoodia* agreements?' And many other questions, to which we did not know all the answers.

We hope that this book will be useful, and that it answers many of these questions.

Collin Louw
Chairperson
Working Group of Indigenous Minorities in
Southern Africa (WIMSA)

Acknowledgements

This book would not have been possible without the support and help of a number of big-hearted organisations and individuals. Our sincere gratitude goes to the Wellcome Trust, whose research funding enabled us to undertake an in-depth analysis of the *Hoodia* case. Thanks in particular to Bella Starling from the trust, who discussed funding options with us, and to Tony Woods, Liz Shaw, Jackie Titley and Paul Woodgate for their support later.

The fairy godfather of this book is Fritz Schmuhl, our Springer editor, who combines a profound passion for books with speed and efficiency. Most importantly, though, his magic with Springer's sales department meant that the book became available in an affordable paperback version right at the outset. Thanks, Fritz!

Close collaboration with San representatives and San institutions made it possible to obtain commentary on our progress throughout the research project and avoid the one-sidedness that led Maori Linda Smith to declare that ' "research" ... is probably one of the dirtiest words in the indigenous world's vocabulary' (Smith 1999). On the contrary, the San adopted our research project as their own and helped generate and disseminate research results for and with us. Particular mention needs to be made of Andries Steenkamp, the chair of the South African San Council, whose vision, energy and humility supported the project throughout. Other San leaders who were closely involved and to whom we are most grateful are Collin Louw, Anna Festus, Mario Mahongo, Zeka Shiwarra, Jason Marende, Dawid Kruiper and, of course, the two colleagues who co-authored chapters of this book, Victoria Haraseb and Mathambo Ngakaeaja. Before research began, the Working Group of Indigenous Minorities in Southern Africa (WIMSA) approved its aims, objectives and fieldwork details, as did the Botswana government for research undertaken in that country.

Special thanks are due to all the people who helped Dr. Saskia Vermeylen with her fieldwork in Namibia and South Africa for their kindness, generosity of spirit, dedication and great sense of humour. Sincere thanks also to Richard Wicksteed, the creative mind behind the *Hoodia* DVD; Paul Wise, the most gifted and meticulous copy editor of them all; Meryl-Joy Wildschut and Grace Humphreys from the South African San Institute (SASI) for organising meetings in South Africa; Paula Watson for her efficient accounting of project resources; Samantha Williams and Roger Chennells for their fluent interpretation between English and Afrikaans on various occasions; Miltos Ladikas for helping secure research funding and for his advice on empirical research questions; Eileen Martin, dean of the Faculty of Health at the University of Central Lancashire, for letting Doris travel to the Kalahari rather than stay at home to teach; the Centre for Applied Philosophy and Public Ethics (CAPPE) for Doris' very generous research time Abduraghman Isaacs for very reliable logistical services in Cape Town; Albert Schroeder for racing Graham Dutfield and Julie Cook Lucas to our COP 9 dissemination event at the CBD meeting in Bonn; Samantha Williams for reporting on the Molopo Lodge event and assistance with fieldwork; Jack Beetson for wise words on benefit sharing, as always; Carolina Lasén Díaz and Pamela Andanda for early input into the project; Cyril Lombard of PhytoTrade Africa for ongoing insights and discussions about the *Hoodia* industry; David Newton of TRAFFIC for assisting with *Hoodia* policy work; and Carl van der Lingen, Armin Schmidt and Judy Beaumont (our partners) for serving as sounding boards for ideas and keeping home fires burning.

Thanks also to all authors, in particular those who compared the situation in their own countries with the San *Hoodia* case. Participants attending San-!Khoba workshops at Kalk Bay, Molopo Lodge and Upington, especially those who travelled for days at a time, are gratefully acknowledged for their vital inputs. Nanette Fleming assisted a team of talented amateur actors in putting on an unforgettable performance at the Molopo workshop. Axel Thoma, former coordinator of WIMSA, was an active supporter of the San struggle to secure rights to *Hoodia*, and continues to assist where possible behind the scenes.

Many individuals involved in trading or regulating *Hoodia* were interviewed for this research, often giving generously of their time and knowledge. It is impossible to name all of them, but in particular we would like to acknowledge the following for the open way in which they shared information. In South Africa, Elsabé Swart of the Northern Cape Department of Tourism, Environment and Conservation; Paul Geldenhuys, Melanie Simpson and Kas Hamman of CapeNature; Conrad Strauss of the Tankwa Karoo National Park; Helena Heystek and Marthinus Horak of the Council for Scientific and Industrial Research (CSIR); George Bowes; Robbie Gass of the Southern African *Hoodia* Growers Association; Rikus Muller of the Grassroots Group; Kobus Engelbrecht; Kersten Paulsen of the trading company BZH; and Danie Nel of Afriplex. The input of Kevin Povey of Unilever and Simon MacWilliam ('Mac') of Phytopharm is also gratefully acknowledged.

In Namibia we are grateful to the following people for explaining the intricacies of the region's 'green diamonds' and the difficulties of regulation and enforcement: Jörn and Adria Miller of the *Hoodia* Growers Association of Namibia (HOGRAN); Dougal and Kirk Bassingthwaighte; Bianca Braun and members of the Kalahari *Hoodia* Growers; Nico Visser of Hardap; Martin Cloete of Blouberg; Steve Carr and Gillian Maggs of the Namibian National Botanical Research Institute; Johnson Ndokosho of the Ministry of Environment and Tourism; Pierre du Plessis and Dave Cole of the Centre for Research Information Action in Africa – Southern African Development and Consulting (CRIAA SA-DC); and Charles Musiyaleka, chief control warden for nature conservation in Keetmanshoop. The drunken tales of two *Hoodia* smugglers, one an ex-policeman, provided light entertainment in Mariental.

Acknowledgements

At the Environmental Evaluation Unit, University of Cape Town, Associate Professor Merle Sowman showed constant enthusiasm for the project and was an invaluable sounding board; Fahdelah Hartley managed logistics with remarkable efficiency and goodwill; Quinton Williams, Penny-Jane Cooke and Paula Cardoso provided excellent research support; Shanaaz Saban kept the books in good shape; and Ntombovuyo Madlokazi helped with organisation of the Upington workshop. Additional financial support was provided to Rachel Wynberg by South Africa's National Research Foundation (NRF) for her ongoing research on the San-*Hoodia* case, although any opinion, findings and conclusions or recommendations expressed in this book are those of the authors and the NRF does not accept any liability for them.

Finally a note about a potential conflict of roles. One of the three editors, Roger Chennells, was personally and directly involved in the *Hoodia* case, as the main attorney representing the San in negotiations with the patent holder. To avoid any perception of bias or misrepresentation, he was not involved in reviewing or changing those sections that commented on the negotiations and related matters.

Wellcome Trust

The Wellcome Trust is an independent charity funding research to improve human and animal health. Established in 1936 and with an endowment of around £13 billion, it is the United Kingdom's largest non-governmental source of funds for biomedical research. This project was funded by the Medical Humanities stream. All opinions, findings, conclusions or recommendations expressed in this book are those of the authors and not of the Wellcome Trust.

Reference

Smith LT. Decolonizing methodologies: research and indigenous peoples. London: Zed Books; 1999. p. 1.

Contents

Part I Community Consent and Benefit Sharing: The Context

1 Introduction 3
   Rachel Wynberg, Doris Schroeder, and Roger Chennells

2 Justice and Benefit Sharing 11
   Doris Schroeder

3 Informed Consent: From Medical Research to Traditional
   Knowledge 27
   Doris Schroeder

4 Protecting the Rights of Indigenous Peoples:
   Can Prior Informed Consent Help? 53
   Graham Dutfield

5 Bioprospecting, Access and Benefit Sharing:
   Revisiting the 'Grand Bargain' 69
   Rachel Wynberg and Sarah Laird

Part II Learning from the San

6 Green Diamonds of the South: An Overview
   of the San-*Hoodia* Case 89
   Rachel Wynberg and Roger Chennells

7 Policies for Sharing Benefits from *Hoodia* 127
   Rachel Wynberg

8 The Struggle for Indigenous Peoples' Land Rights:
   The Case of Namibia 143
   Saskia Vermeylen

9 **Speaking for the San: Challenges for Representative Institutions** 165
Roger Chennells, Victoria Haraseb, and Mathambo Ngakaeaja

10 **Trading Traditional Knowledge: San Perspectives from South Africa, Namibia and Botswana** 193
Saskia Vermeylen

11 **Putting Intellectual Property Rights into Practice: Experiences from the San** 211
Roger Chennells

12 **Sharing Benefits Fairly: Decision-Making and Governance** 231
Rachel Wynberg, Doris Schroeder, Samantha Williams, and Saskia Vermeylen

Part III Reflections

13 **The Role of Scientists and the State in Benefit Sharing: Comparing Institutional Support for the San and Kani** 261
Sachin Chaturvedi

14 **The Law is Not Enough: Protecting Indigenous Peoples' Rights Against Mining Interests in the Philippines** 271
Rosa Cordillera A. Castillo and Fatima Alvarez-Castillo

15 **Benefit Sharing is No Solution to Development: Experiences from Mining on Aboriginal Land in Australia** 285
Jon Altman

16 **Human Research Ethics Guidelines as a Basis for Consent and Benefit Sharing: A Canadian Perspective** 303
Kelly Bannister

17 **The Limitations of Good Intent: Problems of Representation and Informed Consent in the Maya ICBG Project in Chiapas, Mexico** 315
Dafna Feinholz-Klip, Luis García Barrios, and Julie Cook Lucas

Part IV Conclusions and Recommendations

18 **Conclusions and Recommendations: Towards Best Practice for Community Consent and Benefit Sharing** 335
Rachel Wynberg, Roger Chennells, and Doris Schroeder

Index 351

List of Figures

Fig. 2.1	Domains of Justice	16
Fig. 3.1	Informed Consent Roles: Ideal Scenario	30
Fig. 3.2	Prior Informed Consent Roles: Simplified Scenario.....................	31
Fig. 6.1	Licence and Benefit-Sharing Agreements Developed Between the San, CSIR, Phytopharm and Unilever...	97
Fig. 6.2	The Distribution of *Hoodia* Species and Occurrence of the San in Southern Africa..	105
Fig. 6.3	Benefit Sharing and Value-Adding Under the San-CSIR-Phytopharm-Unilever Agreements.............................	108
Fig. 6.4	Benefit Sharing through SAHGA and the *Hoodia* Value Chain Based on Trade of Raw Material..	111
Fig. 6.5	Flowering *Hoodia gordonii*, Ceres (Karoo), Western Cape, South Africa...	120
Fig. 7.1	Process Prescribed by the ABS Regulations to Obtain a Bioprospecting Permit or Bioprospecting Export Permit	135
Fig. 9.1	An Organogram of the Kuru Family of Organisations	175
Fig. 9.2	An Organogram of the South African San Council	180
Fig. 12.1	Decision-Making in the Allocation of Funds from the San-*Hoodia* Trust	250

List of Tables

Table 3.1	International Guidelines Requiring Informed Consent	30
Table 3.2	International Guidelines Requiring (Free) Prior Informed Consent from Indigenous Communities	32
Table 3.3	Process for Obtaining Community Permission to Develop a Vaccine Study Site	44
Table 6.1	Chronology of the Commercial Development of *Hoodia*	98
Table 6.2	Benefit-Sharing Payments to the San-*Hoodia* Trust from the CSIR, Paid into the Trust Bank Account on 11 May, 2005	110
Table 7.1	Key Laws and Policies Pertaining to ABS and the Use, Trade and Conservation of *Hoodia* in South Africa, Namibia and Botswana	130
Table 10.1	Responses to Commodification of Medicinal Knowledge: Breakdown by Gender, Community, Country and Income	199
Table 12.1	Democratic and Traditional Indigenous Decision-Making: A Comparison	238
Table 12.2	Expenditure of Milestone Payments Received from the CSIR by the San-*Hoodia* Trust	245
Table 13.1	Comparison of San and Kani ABS Agreements between Stakeholders	269
Table 15.1	Schematic Representation of Development Challenges for Indigenous Beneficiaries of Agreements	297

List of Boxes

Box 3.1	Human Guinea Pigs	34
Box 3.2	Consent and Cultural Differences	35
Box 5.1	Regulating the Protection and Commercial Use of Traditional Knowledge	75
Box 6.1	What is *Hoodia*?	119
Box 7.1	The Legal and Institutional Framework for Bioprospecting and ABS in Southern Africa	133
Box 12.1	Conflict Resolution among the San	236
Box 12.2	The Molopo (San-!Khoba) Declaration	252
Box 14.1	The Alangan People	273

Contributors

Jon Altman Jon Altman is the inaugural director of the Centre for Aboriginal Economic Policy Research established at the Australian National University (ANU), Canberra, in 1990. He has an academic background in economics (University of Auckland) and anthropology (ANU) and has been researching indigenous economic development and policy issues since 1977. Of particular relevance to his contribution here are the book *Aborigines and Mining Royalties in the Northern Territory* (1983), a review of the Aboriginal Benefits Trust Account that he chaired for the Australian government (1984), his participation in the Review of Native Title Representative Bodies (1995) and his role as independent expert for the Kakadu Region Social Impact Study (1996 and 1997).

In the 1990s, he reviewed a number of Aboriginal royalty associations including the Nabarlek Traditional Owners Association (1994), the Gagudju Association (1996) and the Ngurratjuta Association (1998). More recently, he participated in the review of the Century Mine Agreement (2002) and between 2002 and 2007 headed an Australian Research Council project, 'Indigenous Community Organisations and Miners: Partnering Sustainable Regional Development?' Since 2006, he has participated in the international comparative study 'Identity, Power and Rights: The State, International Institutions and Indigenous Peoples' sponsored by the United Nations Research Institute for Social Development.

Kelly Bannister Kelly Bannister, M.Sc., Ph.D., is director of the POLIS Project on Ecological Governance and an adjunct professor in the Studies in Policy and Practice Program, Faculty of Human and Social Development, at the University of Victoria in British Columbia, Canada. An ethnobotanist by training, her main research interests are ethics and indigenous intellectual property rights in research involving biodiversity and traditional knowledge. She is particularly interested in institutional policy development for collaborative research between universities and Aboriginal communities. Her current research explores the ethics of community-based research and community protocols as a tool for facilitating equitable research practices. She has authored several book chapters, journal articles and reports on ethical and legal issues in ethnobotanical research. She is chair of the Ethics Committee of the International Society of Ethnobiology (http://ethnobiology.net/) and contributes to national policy development for biodiversity and for research ethics involving

Aboriginal peoples. She is involved in several federally funded research projects, including the Project for the Protection and Repatriation of First Nation Cultural Heritage (www.law.ualberta.ca/research/aboriginalculturalheritage/) and the Intellectual Property Issues in Cultural Heritage Project (http://cgi.sfu.ca/~ipinch/cgi-bin/).

Fatima Alvarez-Castillo Fatima Alvarez-Castillo is a professor of politics, research and ethics at the University of the Philippines, Manila. Apart from teaching, she undertakes research, training and advocacy. Her work over the past three decades has focused on the human rights of the marginalized and powerless. These include those who are discriminated against by reason of their gender, ethnicity or social class. The approach she uses for teaching, training, research and advocacy has always been participatory and enabling: she is inspired by Paolo Freire's empowering pedagogy.

Rosa Cordillera A. Castillo Rosa Cordillera A. Castillo teaches anthropology at the University of Manila, Philippines. She graduated with a MA in anthropology from the University of the Philippines, Diliman, following a degree awarded with academic honours. She temporarily stopped teaching at the university to complete ethnographic field work in a community on a small Philippine island. She is also actively engaged in campaigning for indigenous peoples' rights, starting as an undergraduate student of anthropology and later as a technical staff member of Anthropology Watch, Inc., a non-governmental organization (NGO) that helps indigenous peoples in the Philippines attain their rights. Her engagement with this NGO enabled her to visit several indigenous communities, which has given her valuable exposure to the variety of human rights issues faced by marginalized indigenous communities in the Philippines. She is currently a board member of the Anthropological Association of the Philippines, the only professional organization of anthropologists in the country. Her area of interest is the anthropology of human rights.

Sachin Chaturvedi Dr. Sachin Chaturvedi is a fellow at the Research and Information System for Developing Countries, a think tank with the Ministry of External Affairs in New Delhi, India. He is working on trade facilitation and World Trade Organization negotiations with the United Nations Economic and Social Commission for Asia and the Pacific in a project comparing seven countries. His areas of specialization include trade and economic issues related to technology and national innovation systems, and linkages with frontier technologies such as biotechnology.

He is the author of two books and has published several research articles in various prestigious journals. He has worked at the University of Amsterdam in a project supported by the Dutch Ministry of External Affairs on international development cooperation and biotechnology for developing countries, has been a member of the Independent Group on South Asian Cooperation Committee of Experts to evolve a framework of cooperation for the conservation of biodiversity in the South Asian

Association for Regional Cooperation region, and is on the editorial board of the *Biotechnology and Development Monitor*, the Netherlands, and the *Asian Biotechnology and Development Review*, India.

Roger Chennells Roger Chennells has been an attorney in private practice in South Africa since 1980, first in Durban and since 1984 in Stellenbosch. Over this period he has worked in various legal fields, ranging from commercial and constitutional law issues to labour, land, environmental and human rights law, with an emphasis on public interest law affecting rural communities.

Early cases revolved around police and state brutality, workplace discrimination and the rights of farmworkers and those living in informal settlements to housing security. Prior to 1990, most of his legal work involved representing and protecting those who opposed the apartheid state. During this period he became an active practitioner of alternative dispute resolution as a means of achieving fair outcomes to legal problems. After the emergence of a democratic state in the early 1990s, he began to represent indigenous peoples, initially in their struggle for the restitution of land and heritage rights from the state. He represented the ≠Khomani San community in their claim for land in and near the Kgalagadi Transfrontier Park in the Kalahari. During this time he began to assist the San peoples in the region with the formation of a regional organization that would represent their rights in Botswana, Namibia and South Africa. It was clear to the San that rights over land were closely associated with their entire heritage, and advocacy for San rights began to focus on issues of culture, heritage and intellectual property rights.

During the United Nations International Decade of the World's Indigenous People (1995–2004) he became an advocate for the emerging intellectual property and heritage rights of indigenous peoples, at times representing indigenous groupings elsewhere in Africa, as well as in Australia and Jamaica.

During 2001 the San became aware that their traditional knowledge had been used in the patenting of active constituents of *Hoodia* by South Africa's Council for Scientific and Industrial Research, and Roger Chennells was requested to assist them in challenging the patent. The San had not been consulted nor did they stand to benefit from the patent. During the ensuing years the San achieved various milestones in the protection of their intellectual property rights, some of which form the subject matter of this book.

Julie Cook Lucas Julie Cook Lucas has a special interest in issues around user representation and involvement in the development and delivery of health services, and the relationships between academia and activism. She developed national women's health initiatives at the UK National Health Service Health Education Authority in the 1980s and ran award-winning campaigns for the NGO Women's Environmental Network in the 1990s. She has worked for a range of NGOs, most recently providing women's mental health services, and as a community health representative. Her postgraduate research in the Philosophy Department at Lancaster University (1995–1998) examined the tensions between activism and academia, theory and practice. She joined the Centre for Professional Ethics as a researcher in 2005.

Graham Dutfield Graham Dutfield is professor of International Governance at the University of Leeds, United Kingdom. He has authored or co-authored five books on intellectual property, genetic resources, traditional knowledge and the life science industries, and edited two others. His latest publication is *Global Intellectual Property Law* (Edward Elgar). A second edition of *Intellectual Property Rights and the Life Science Industries* will be available during 2009 (World Scientific Publishing). His current research interests include innovation and creativity in law, economics, anthropology and history; the politics of intellectual property; history of patent law and the life science industries; intellectual property and genetic resources, traditional knowledge and folklore; and the TRIPS (Trade-Related Aspects of Intellectual Property Rights) Agreement and public health. He has served as a consultant or commissioned report author for several governments, international organizations, United Nations agencies and non-governmental organizations. He has a D.Phil. from the University of Oxford.

Dafna Feinholz-Klip Dafna Feinholz-Klip, with a B.A. and an M.A. in Psychology and a Ph.D. in Research Psychology from the Universidad Iberoamericana, Mexico, also has a master's in Bioethics from the Universidad Complutense, Madrid. She has studied at the Kennedy Institute of Ethics and the Harvard School of Public Health. She is currently executive director of the National Commission of Bioethics in Mexico and a member of the National Research System.

She has wide experience as a member of research ethics committees in Mexico and internationally. She was a member of the international team that wrote the World Health Organization's *Operational Guidelines for Ethics Committees that Review Biomedical Research* (2000) and was responsible for its translation into Spanish. She is founder and former chairperson of the Latin American Forum for Ethics Committees for Health Research (FLACEIS).

In 2005 she represented Mexico at UNESCO (the United Nations Educational, Scientific and Cultural Organization) intergovernmental meetings of experts to finalize the Universal Declaration of Bioethics and Human Rights. Since 2008 she has been Mexico's representative at the Steering Committee on Bioethics (CBDI) of the Council of Europe. As well as writing and teaching about bioethics, she serves as a peer reviewer and a member of the advisory board for prominent bioethics journals.

Luis García Barrios Dr Luis García Barrios has worked for 23 years at ECOSUR-San Cristóbal (El Colegio de la Frontera Sur, San Cristóbal de las Casas), a multi-disciplinary research centre in the highlands of Chiapas, Mexico, where he has worked extensively as an agroecologist with indigenous and mestizo farmers in the tropical mountains of Mexico. He has published in the fields of ethnobotany, the ecology of intercropping, agroforestry, silvopastoral systems, functional biodiversity in agriculture, rural development, socio-environmental systems, the dynamics and self-organization of complex systems, participatory agent-based modelling and sustainability evaluation. He actively participated in the International Cooperative Biodiversity Group (ICBG) Maya project from 1998 to 2000 as co-principal investigator of the community development subprogramme.

Victoria Haraseb Victoria Haraseb was born in the Outjo district of Namibia, on the way to the ancestral land of the Hai//om San peoples in and around the Etosha National Park. In 1997 she was appointed by the Hai//om Traditional Authority as their secretary, and she has been a community activist ever since. She played an important role in research that was done in the Hai//om community on cultural resource mapping and also contributed to a booklet, *Voices of the San*. She has attended several training workshops through the Working Group of Indigenous Minorities in Southern Africa (WIMSA) and has represented the San in several international forums concerning human rights, traditional knowledge and environmental issues. She has also participated in United Nations processes on indigenous peoples.

She has worked at WIMSA for the past 10 years as the regional education assistant, promoting San access to education and the position of San women. She has also served on various boards, including the Outjo Development Trust, and as vice-chair to WIMSA and the Community Empowerment and Development Association. She is currently enrolled for a degree in Business Administration through the Management College of Southern Africa.

Timothy Hodges Timothy Hodges is co-chair of the Working Group on Access and Benefit Sharing (ABS) of Genetic Resources under the United Nations Convention on Biological Diversity (CBD). Elected to this position in March 2006, he is co-chairing negotiations on an international ABS regime aimed at concluding in 2010.

Prior to assuming this position, he was Canada's National ABS Focal Point. In this capacity, he led Canada's delegations to the CBD's ABS working group meetings over several years and also had overall responsibility for the Canadian government's national ABS policy initiative. As such he chaired the federal government's ABS committee and co-chaired the Federal-Provincial-Territorial ABS Working Group.

Mr Hodges is a career diplomat. Much of his professional career has focused on bilateral and multilateral environmental, economic and trade negotiations. Over the past 25 years, he has been involved in a wide range of international files relating to technology transfer, science policy, biotechnology, intellectual property rights, biodiversity, indigenous issues, climate and global change and circumpolar affairs, negotiated under the United Nations, the G8, the World Trade Organization, Asia-Pacific Economic Cooperation, the Organization of American States, the Organisation for Economic Co-operation and Development and numerous other international instruments and forums.

Sarah Laird Sarah Laird is the director of People and Plants International, working in the field of forest and biodiversity conservation. In part her work has focused on building equity into the genetic resources trade and developing policies to guide access and benefit-sharing under the Convention on Biological Diversity. Publications in this field include *Biodiversity and Traditional Knowledge: Equitable Partnerships in Practice* (Earthscan, 2002) and *The Commercial Use of Biodiversity: Access to Genetic Resources and Benefit-Sharing* (with K. ten Kate, Earthscan, 1999). She also undertakes and manages research and applied projects on non-timber forest products, primarily in Africa, and since 1997 has also undertaken ethnobiological research around Mount Cameroon with Bakweri and other groups living in the area.

Mathambo Ngakaeaja Mathambo Ngakaeaja was born of a Tswana father and San mother at D'kar, a farm in the western Kalahari of Botswana. Since graduating with a B.Sc. in Geology from the University of Botswana, he has worked continuously for the San peoples, first as manager for the Dqae Qare Game Farm, part of the Kuru organizations, until 1998, and subsequently as coordinator for the Botswana chapter of the Working Group of Indigenous Minorities in Southern Africa (WIMSA) until 2007. He has represented the San in numerous international forums, and has engaged extensively over the past decade with the international indigenous peoples' movement as well as with United Nations processes on a broad range of issues including indigenous peoples' human, environmental and intellectual property rights.

He has served as a board member of Kuru Savings and Loans, San Arts and Crafts, the Botswana Khwedom Council, the !Khwa ttu San Education Centre and the Southern African *Hoodia* Growers Association.

Doris Schroeder Doris Schroeder is professor of Moral Philosophy in the Centre for Professional Ethics at the University of Central Lancashire and a professorial fellow in the Centre for Applied Philosophy and Public Ethics at the University of Melbourne. Her background is in philosophy, politics and economics. Prior to joining academia, she worked as a strategic planner for Time Warner. Her main areas of interest are international justice, human rights, benefit sharing and global bioethics. She was the principal grant-holder for the San-!Khoba research project.

Saskia Vermeylen Dr. Saskia Vermeylen is a lecturer at the Lancaster Environment Centre, Lancaster University, UK. Her research is focused on studying the cultural property rights of indigenous peoples from a socio-legal perspective. She has conducted fieldwork in southern Africa, specifically engaging with the commodification of cultural practices and the protection of both the tangible and the intangible heritage of indigenous peoples. Her work is inspired by critical legal studies, legal anthropology and discourses of post-colonialism. In her most recent work on museum practices, she explores the concept of giving a voice to indigenous peoples through new media art and 'cybermuseology' as alternative curatorial practices for indigenous peoples. Her research on intellectual property rights and traditional knowledge and indigenous museum practices has been published in internationally acclaimed journals.

Samantha Williams Samantha Williams is a full-time student reading for a doctorate at the University of Cape Town. She completed her undergraduate and master's degrees at the University of the Western Cape. In 2005 she was appointed as a junior researcher at the Environmental Evaluation Unit, University of Cape Town, and participated as a research assistant in the San-!Khoba project. Samantha's research interests include issues related to coastal access, sustainable rural livelihoods and small-scale fisheries.

Rachel Wynberg Dr. Rachel Wynberg is a senior researcher and deputy director at the Environmental Evaluation Unit, University of Cape Town. Trained as both a natural and a social scientist, she has worked widely on environmental policy and strategy, specializing in the commercialization and trade of natural products, and the integration of social justice into biodiversity concerns.

Over the past 15 years she has worked closely with several governments, international organizations and NGOs to formulate appropriate policy frameworks for biotrade, access and benefit-sharing; intellectual property rights and traditional knowledge; and community-based natural resource management. She is actively involved in civil society movements, and is trustee and founding member of two South African non-governmental organizations, the Environmental Monitoring Group and Biowatch South Africa. She holds two master's degrees from the University of Cape Town and a PhD from the University of Strathclyde.

In 1997 she came across the *Hoodia* patent filed by South Africa's Council for Scientific and Industrial Research, did research to uncover the traditional use of the plant and began a campaign through Biowatch to alert the media and the San to the exploitative use of this knowledge. She has been involved in research relating to *Hoodia* and its commercialization for the past 10 years.

Foreword

This book arrives at a critical juncture in the history of genetic resource use and policymaking – not only in southern Africa, but across that continent and, indeed, around the world.

The volume's arrival also coincides with the growing awareness and concern over the loss of biological diversity and what this loss means for the health of the planet and survival of the human species.

At first glance, such global and momentous concerns might appear remote in relation to the 'San-*Hoodia*' story. But, in fact, in this unfolding drama of *Hoodia* and its many embroiled stakeholders, the book's contributors depict a microcosm of the global debate over genetic resources, traditional knowledge, bioprospecting and economic and social development.

This book will prove highly instructive to the providers and users of genetic resources and associated traditional knowledge around the world who share the frustrations and disappointments, as well as the expectations and desires, of the *Hoodia* stakeholders. The significance of this book, however, goes beyond the myriad lessons it has to offer to those involved in similar cases in other parts of the world.

Importantly, this book serves as a timely and substantive reminder to those negotiating a new international regime on Access and Benefit Sharing (ABS) under the United Nations Convention on Biological Diversity (CBD). ABS, and all its attendant issues, is about real people – in local and indigenous communities, on farms, in public research laboratories and business boardrooms – struggling to feed families, fuel economies, cure diseases, conserve biodiversity, address injustices and account to shareholders. This book brings readers' feet to the ground and we are reminded by many of the volume's highly respected contributors that ABS is about the well-being of communities, the universal struggle for just societies and the desire for fair deals.

The San-*Hoodia* story is far from over and, indeed, the nature of its ending is far from clear. With the opportunity that this book brings to share the complexities and importance of the San-*Hoodia* case with the world, the stakes have never been higher for the indigenous communities, farmers, governments and firms involved. The question remains: Is it practicable to develop and implement a fair and equitable ABS model involving a range of stakeholders in a multijurisdictional context? We anxiously await an answer.

And in the complex international ABS regime talks an equally challenging question may be posed. How will the Regime assist in generating the mutually acceptable outcomes sought in cases such as San-*Hoodia*? The answer, of course, lies with the ABS regime negotiators and they would do well to study closely the pages of this insightful and provocative volume.

Timothy J. Hodges
Co-Chair
Working Group on Access and Benefit Sharing
United Nations Convention on Biological Diversity

List of Acronyms

ABS	Access and Benefit Sharing
ABTA	Aboriginals Benefit Trust Account
AICRPE	All India Coordinated Research Project on Ethnobiology
ALAMIN	Alyansa Laban sa Mina or Alliance Against Mining
AVP	Arya Vaidya Pharmacy
BMC	Business Management Committee
CAH	Consejo Aguaruna Huambisa
CBD	Convention on Biological Diversity
CEGA	Cape Ethno-Botanical Growers Association
CIHR	Canadian Institutes of Health Research
CIOMS	Council for International Organizations of Medical Sciences
CIPR	Commission on Intellectual Property Rights
CITES	Convention on International Trade in Endangered Species of Wild Fauna and Flora
COMPITCH	*Consejo Estatel de Organizaciones de Médicos y Parteras Indígenas Tradicionales de Chiapas*
CONAP	Confederación de Nacionalidades Amazónicas del Perú
CPA	Community Property Association
CRIAA SA-DC	Centre for Research Information Action in Africa – Southern African Development and Consulting
CSIR	Council for Scientific and Industrial Research
ECOSUR	El Colegio de la Frontera Sur
FDA	Food and Drug Administration
FPIC	Free and Prior Informed Consent
FPK	First People of the Kalahari
FTC	Federal Trade Commission
GMP	Good Manufacturing Practice
ICBG	International Cooperative Biodiversity Groups
ICESCR	International Covenant on Economic, Social and Cultural Rights
IGC	Intergovernmental Committee on Traditional Knowledge, Genetic Resources and Folklore
IKS	Indigenous Knowledge Systems
ILO	International Labour Organisation

IPACC	Indigenous Peoples of Africa Coordinating Committee
IPHR	Indigenous Peoples and Human Rights
IPR	Intellectual Property Rights
IPRA	Indigenous Peoples Rights Act
ISE	International Society of Ethnobiology
IWGIA	International Workgroup for Indigenous Affairs
JBDF	Ju/wa Bushman Development Foundation
KAMTI	Kaisahan Mangyan Tadyawan Inc
KFO	Kuru Family of Organisations
MCA	Minerals Council of Australia
MNL	Molecular Nature Limited
NBAC	National Bioethics Advisory Commission
NCIP	National Commission of Indigenous Peoples
NGO	Non-Governmental Organization
NIH	National Institutes of Health
NSERC	Natural Sciences and Engineering Research Council of Canada
OCCAAM	Organización Central de Comunidades Aguarunas del Alto Marañon
OMIECH	Organización de Médicos Indígenas del Estado de Chiapas
PIC	Prior Informed Consent
PRECIS	Pretoria Computerised Information System
PROCOMITH	Programa de Colaboración sobre Medicina Indígena Tradicional y Herbolaria (Collaborative Programme in Traditional Indigenous Herbal Medicine)
PROMAYA	Promotion of Intellectual Property Rights of the Highland Maya of Chiapas
R&D	Research and Development
RAFI	Rural Advancement Foundation International
RRL	Regional Research Laboratory
SAHG	South African *Hoodia* Growers (Pty) Limited
SAHGA	South African *Hoodia* Growers Association
SAHRC	South African Human Rights Commission
SANAMA	Samahan ng Nagkakaisang Mangyan Alangan
SASI	South African San Institute
SEMARNAP	*Secretaría de Medio Ambiente, Recursos Naturales y Pesca*
SHDC	Sustainably Harvested Devil's Claw Project
SSHRC	Social Sciences and Humanities Research Council of Canada
SWAPO	South West Africa People's Organization
TBGRI	Tropical Botanic Garden and Research Institute
TRIPS	Trade-Related Aspects of Intellectual Property Rights
UCLAN	University of Central Lancashire
UCT	University of Cape Town
UNDP	United Nations Development Programme
UNESCO	United Nations Educational, Scientific and Cultural Organization
WHO	World Health Organization

List of Acronyms

WIMSA	Working Group of Indigenous Minorities in Southern Africa
WIPO	World Intellectual Property Organization
WMA	World Medical Association
WSSD	World Summit on Sustainable Development
WTO	World Trade Organization

Part I
Community Consent and Benefit Sharing: The Context

Chapter 1
Introduction

Rachel Wynberg, Doris Schroeder, and Roger Chennells

The story of *Hoodia* has captured the world's imagination. A plant used by the San to quench thirst and possibly hunger for centuries suddenly enters world markets as an appetite suppressant. Pictures from the Kalahari of poverty-induced thinness mingle with pictures of obese Westerners. A showcase for the Convention on Biological Diversity (CBD) in terms of the conservation of biodiversity, sustainable use and fair and equitable benefit sharing? Alas, not quite. But it is a showcase for the challenges that indigenous communities, national and international policymakers, and industry face in realizing the letter and the spirit of the CBD. This book explains why.

Few other bioprospecting cases have started as dramatically as the *Hoodia* case did, with a leading article in a British newspaper citing the perceived extinction of the San, and few have gone through as many ups and downs. The world's largest pharmaceutical company, Pfizer, undertook to bring *Hoodia* to market, then withdrew from the task. Next, Unilever, one of the largest multinational food manufacturers, aimed to add *Hoodia* to its slimming range, yet also withdrew after 4 years of research and an investment of more than €20 million. Meanwhile, natural *Hoodia* habitats were ravaged to supply material for a booming market while commercial growers committed themselves to sharing some of their profits with the San. Hidden behind the hype of this case are highly valuable lessons applicable beyond southern Africa.

R. Wynberg (✉)
Environmental Evaluation Unit, University of Cape Town, Private Bag X3, Rondebosch 7701, Cape Town, South Africa
e-mail: rachel@iafrica.com

D. Schroeder
UCLAN, Centre for Professional Ethics, Brook 317, Preston PR1 2HE, United Kingdom
e-mail: dschroeder@uclan.ac.uk

R. Chennells
Chennells Albertyn: Attorneys Notaries and Conveyancers, 44 Alexander Street, Stellenbosch, South Africa
e-mail: scarlin@iafrica.com

This book, the result of a 2.5-year project funded by the Wellcome Trust, and its accompanying DVD, presents the first in-depth account of the *Hoodia* benefit-sharing case. It is unique in bringing together disciplines that to date have never engaged collectively on the dilemmas of just how prior informed consent and benefit sharing are effected in practice. This has included the academic fraternity of philosophers applying its mind to questions of justice in the CBD; those in legal disciplines interrogating the use of intellectual property rights to protect traditional knowledge; environmental scientists analysing the extent to which the case reflects the intent of the CBD and national policies; anthropologists grappling with questions of how and whether knowledge should be commodified; and, uniquely, those with knowledge of other benefit-sharing arrangements throughout the world bringing their collective expertise to compare and contrast their experiences with those of the San.

The book is divided into three main parts. Part 1 contains articles of an overarching nature, which describe the setting and the challenges in the brave new world of business between indigenous peoples and the bioprospecting industry. Part 2 contains articles specifically focused on the San-*Hoodia* benefit-sharing case. The rich findings articulated in the first two sections are discussed and debated by a range of experts in Part 3, to tease out the similarities and differences between the San-*Hoodia* case and others. The book concludes with a synthesis of main points and specific recommendations.

Following this introduction, Chapter 2 queries the ethical foundation of the CBD. Doris Schroeder is the first philosopher to ask how benefit sharing fits into philosophical debates of justice. Why should it be just to restrict the 'common heritage of humankind' rule by giving sovereignty over biological resources to national governments and requiring prior informed consent and benefit sharing? The chapter looks at questions of justice in exchange (e.g. traditional knowledge for royalty payments) and those of global distributive justice. It argues that the CBD is an example of a set of social rules designed to increase social utility. This imposition of rules, which adds a new bureaucratic layer to biodiversity access, is ethically justified as long as the international economic order is characterized by serious distributive injustices, reflected in the enormous poverty-related death toll in developing countries. Any ethical attempt to redress the balance in favour of the disadvantaged, as the CBD does, has to be welcomed. By legislating for a 'justice in exchange' system covering non-human biological resources and traditional knowledge in preference to the tacit 'common heritage of humankind' principle, the CBD provides a small step forward in redressing the balance. The author concludes that the convention presents just legislation sensitive to the international relations context of the twenty-first century. However, its implementation is enormously challenging.

One of the main implementation challenges relates to the requirement of consent. Obtaining informed consent has become an essential part of modern medical practice. Today, patients and research subjects are actively involved in medical decision-making and are no longer expected to defer responsibility to paternalistic, benevolent doctors. Since the early 1990s, the concept has also been employed systematically in connection with indigenous peoples' rights of self-determination.

The CBD, for instance, requires that prior informed consent be obtained from indigenous communities before their traditional knowledge, innovations and practices may be accessed. Chapters 3 and 4 by Doris Schroeder and Graham Dutfield respectively introduce the wider discourse on informed consent. Schroeder provides a useful overview focusing on a comparison between informed consent in medicine and that in the field of natural product development. After describing the necessary stages of concluding a consent process, Schroeder argues that the similarities between obtaining informed consent in the medical context and obtaining prior informed consent according to CBD requirements are strong enough to warrant mutual learning. Such learning is particularly appropriate when dealing with the inherent power imbalances between medical staff and research subjects, and the similar imbalances between bioprospectors and indigenous communities.

Following this overview chapter, Dutfield provides an analysis of what prior informed consent means in practice when accessing traditional knowledge and biological resources. Using a case study approach, his chapter shows why applying prior informed consent requirements in very diverse cultural settings and tense political contexts can be immensely challenging. Even with the best intentions and the most carefully drawn-up plans, things can go wrong, as Dutfield illustrates convincingly with a case from Peru. He also shows that prior informed consent may not be a requirement in many cases because a great deal of knowledge and resources are already in free circulation and can no longer be attributed to a single originator community or country. This should not, he argues, lead to the conclusion that there can be no moral obligations even in the absence of legal ones. While prior informed consent may not resolve biopiracy satisfactorily in all cases, it can nevertheless be a useful concept. Effective, culturally appropriate, transparent and flexible prior informed consent procedures should be seen as a necessary but not a sufficient requirement for the establishment of more equitable bioprospecting arrangements.

The challenges of obtaining prior informed consent are replicated when negotiating benefit-sharing agreements. In Chapter 5 Rachel Wynberg and Sarah Laird set the wider international context of bioprospecting, access and benefit sharing, and describe the fraught policy process that has evolved since the adoption of the CBD in 1992. Notwithstanding the abundance of new policies and laws to control access to genetic resources and ensure fair benefit sharing, their effectiveness has been questionable. The complexity and diversity of bioprospecting activities and commercial players are often poorly recognized, and policy has lagged behind the practice of biprospecting. Moreover, the vast range of issues involved – from trade to conservation, intellectual property, biotechnology and traditional knowledge – has resulted in the policy process becoming a forum for much wider concerns dealing with globalization, corporate behaviour and the disparities between rich and poor. Some of the key issues that remain unresolved in the run-up to finalizing an international regime on access and benefit sharing revolve around compliance, and whether or not patent holders should be obliged to disclose the origin of biological resources and knowledge in patent applications; the scope of the agreement, and whether or not it should go beyond the CBD to address biochemicals and derivatives; and even its purpose.

Part 2 takes the book to its main focus: the San, their cultures and institutions, their use of *Hoodia* and, importantly, the benefit-sharing agreements they have entered into with the South African-based Council for Scientific and Industrial Research (CSIR) and the Southern African *Hoodia* Growers Association (SAHGA). In a comprehensive analysis, the historical detail of the San-*Hoodia* case is reported in Chapter 6 by Rachel Wynberg and Roger Chennells, who introduce the San and chart the history of *Hoodia* development, its patenting by the CSIR and the flurry of activity after the infamous comment quoted in the British *Observer* that the San were extinct. The process of developing a benefit-sharing agreement between the San, the CSIR and the SAHGA is described in detail, along with the elaborate processes that have been followed to secure San representation, develop a representative trust and set in place mechanisms to distribute resources fairly. As Wynberg and Chennells note, the challenges of implementation are substantial, and are exacerbated by regional differences in benefit-sharing policies and highly unstable *Hoodia* markets, more especially in light of Unilever's decision to terminate its involvement.

The complexities of access and benefit sharing and their interface with government regulation, conservation and compliance are well illustrated in Rachel Wynberg's Chapter 7 on policy frameworks for *Hoodia*. *Hoodia* is a biological resource that is shared across national political boundaries, in particular those separating Namibia, South Africa and Botswana, and knowledge of the plant is similarly shared by communities straddling these boundaries. Yet each country is involved in diverse initiatives to commercialize the plant and has different policy approaches to prior informed consent, commercialization, benefit sharing, conservation and the recognition of indigenous peoples. Regional strategies to control illegal trade, develop benefit-sharing approaches, obtain prior informed consent from communities and cooperate on value-adding and marketing are vital constituents of a viable industry, especially in the face of increasing international competition.

One of the policies that differ between countries with San populations is support for land claims. While South Africa is broadly supportive of San land rights and has transferred six Kalahari farms to the San as part of its land reform programme, the San continue to be dispossessed of land in Botswana and encounter difficulties in realizing their land claims in Namibia. In fact, the San are Namibia's poorest, most vulnerable group, living as scattered itinerant labourers, often on the outskirts of cities or settlements. Yet, as Saskia Vermeylen shows in Chapter 8, indigenous peoples often explicitly link rights over knowledge, culture, natural resources and land. Traditional knowledge is seen as closely tied to land and its resources; in fact, such knowledge encapsulates spiritual experience and deep relationships with the land.

In order to claim rights, be they land rights or rights over cultural heritage, indigenous peoples must become organized and empower themselves. Drawing on their considerable experience of working with the San, Roger Chennells, Victoria Haraseb and Mathambo Ngakaeaja show in Chapter 9 that strong institutions are essential to realizing rights in practice. Chennells, Haraseb and Ngakaeaja examine the status of the San as the poorest and most dispossessed peoples in southern Africa and raise the question: why have they collectively been unable to compete in the modern world? The legacy of a hunter-gatherer world view, pervasive poverty and landlessness, and collective trauma are examined as potential sources of societal problems. Against this background, the chapter describes institutions that are trying to assist and guide the San in Namibia, Botswana and South Africa. These include non-governmental organizations (NGOs) such as the Kuru Development Trust in Botswana and San representative organizations such as the Working Group of Indigenous Minorities in Southern Africa (WIMSA), which has represented the San in the *Hoodia* benefit-sharing negotiations.

One of the most controversial aspects of the access and benefit-sharing debate relates to the way in which traditional knowledge is used and commercialized. Many critics have pointed out the inherent contradictions between traditional knowledge systems, which are typically collective, based on sharing and characterized by their non-barter nature, and Western approaches to knowledge protection such as patenting, which by contrast are monopolistic and individualistic. Few, if any, empirical studies have documented the relationship between these systems and community perceptions of the so-called commodification of traditional knowledge. Based on fieldwork conducted in South Africa, Namibia and Botswana, Vermeylen presents in Chapter 10 a compelling account of how these issues are perceived by San communities and the inherent linkages between knowledge, land and economic status. While indigenous peoples are often portrayed in the literature as homogenous groups voicing uniform opinions, Vermeylen's scenario surveys clearly indicate that within the communities studied, there were many different opinions on whether or not to commodify traditional knowledge. This diversity of voices is not surprising when one takes into account the local context or the current and historical socio-economic and political circumstances of individuals and communities. Although there was widespread acceptance of commodification in principle, it is important to be aware of the cultural and symbolic, as well as economic, value of a commodity. At the same time, the scenario surveys showed that many respondents wanted to keep control of their knowledge rather than part with it for economic benefit (royalties) only. Notably, a gender divide could be observed, with women more likely to settle for royalties – to finance, for instance, their children's education – and men more likely to either reject all commodification or opt to be co-holders of patents.

These findings are developed by Roger Chennells through practical demonstration of the actions taken by the San to protect their traditional knowledge. After outlining the basics of the intellectual property rights system in Chapter 11, Chennells summarizes the growing discontent of indigenous peoples about the failure of the system to prevent the misappropriation of their knowledge and culture. The theft of music, folk law, traditional art and innovations shows that the current system is inadequate to secure the full protection of indigenous rights. Yet the system leaves room for flexible, local initiatives driven by indigenous peoples to remedy the situation. One example is the 'research and media contract' drafted by a San NGO and now used widely, which requires aspirant researchers not only to provide full details of the applicant and of the nature, content and purpose of the research, but also to negotiate terms with an appointed San leader. Chennells shows that there are practical methods for regaining control over traditional knowledge and heritage, but that indigenous peoples need to be proactive in asserting their own rights and using the existing laws and tools at their disposal.

Understanding how decisions were made by the San in the *Hoodia* case and how decision-making and governance structures vary between bioprospectors and indigenous communities is essential for the implementation of effective benefit sharing. Drawing on academic literature as well as interviews undertaken in South Africa, Wynberg, Schroeder, Samantha Williams and Vermeylen show in Chapter 12 that decision-making processes in benefit-sharing negotiations vary significantly from party to party. In corporate hierarchies, decision-making is usually centred around a small number of individuals and does not involve the wider consultation of stakeholders. Decisions are made on a routine basis by highly educated personnel in positions of power who are well versed in the legalities and implications of their decisions. By contrast, decision-making in traditional indigenous communities such as the San often involves a large number of community members. Discussions are seldom limited to a single event, but rather emerge over time during conversations among friends, relatives and neighbours. In the case of the San, decisions are taken by consensus, which is reached when significant opposition no longer exists. These differences in decision-making practice place an obvious burden on negotiations, with one party requiring fast decisions to satisfy shareholders while the other needs significant time to allow meaningful community consultation and digest the implications of different options. This clash over decision-making procedures and speed often turns out to be detrimental to traditional knowledge holders, whose decision-making abilities are compromised by the commercial partners' need for urgent resolution. Wynberg, Schroeder, Williams and Vermeylen point to a possible solution embraced by the South African Biodiversity Act, which now locates support for consultation firmly with the government to ensure that negotiations are on an equal footing when benefit-sharing agreements are negotiated. However, practical implementation of this requirement remains hampered by constraints of capacity, resource and knowledge.

The challenges of the San-*Hoodia* case described in Part 2 provide the basis for the commentaries from around the world presented in Part 3, in which knowledge from similar cases is used to stimulate debate and mutual learning. One of the oldest benefit-sharing trusts was established on behalf of the Kani tribe in Kerala, India, in 1997. Since then, the trust has received a constant income that has funded a number of projects. Chapter 13, by Sachin Chaturvedi, provides a fascinating comparison of access and benefit-sharing arrangements between the San and the Kani communities in order to identify potential keys to success. Drawing from the Kani experience, Chaturvedi suggests that strong institutional frameworks and committed staff (the bioprospecting researchers in the Kani case) are essential ingredients for effective benefit sharing. In this case, the Indian-based research institute funded a lawyer to draft the trust deed for the Kani community and also provided skills training in the cultivation and harvesting of *Trichopus zeylanicus travancoricus*, so that income from benefit sharing could be supplemented with income from growing. The institute also decided to share royalties from the patent licensee equally with the Kani. By contrast, the San had little institutional support from the research institute concerned and none from the government. Instead, they were supported by several NGOs. Chaturvedi commends the San and their support teams, however, for the longer-standing relationships and benefit-sharing agreements established with research institutes as well as with *Hoodia* growers and traders.

The CBD is a relatively recent development in international law, but benefit sharing with indigenous communities is not without precedent. Particularly relevant are the agreements that have developed between mining companies and indigenous peoples, and subsequent negotiations for compensation when mining takes place on indigenous land. Chapters from the Philippines and Australia deal with this matter and compare experiences there with those of the San. In a rich narrative, Rosa Cordillera Castillo and Fatima Alvarez-Castillo illuminate in Chapter 14 the difficulties of obtaining genuine consent in the Philippines. While the Philippine legal situation is unambiguous and gives indigenous peoples clear protection against the involuntary intrusion of mining operations, the authors argue that the law is not enough. Despite legal protection of indigenous peoples' right of autonomous decision-making regarding the use of their lands and resources, inadequacies in the implementation of the law and the complicity of state agencies in circumventing its requirements are among the major problems. Given this situation, Philippine indigenous peoples and advocates have resorted to direct political action to assert their right to autonomous decision-making over their lands. Comparing the situation in the Philippines with the San case, the authors point to the importance of collective, participatory action and the delicate role that legal advocates can play.

Another experience of benefit sharing from mining is related in an incisive chapter by Jon Altman. In Chapter 15, Altman draws on experiences from Australia, where indigenous people have negotiated with multinational corporations engaged in mineral exploration on their lands. The chapter gives a brief background of the situation of indigenous peoples in Australia and their relationship with miners as mediated by the state. Altman asks, among other things, how relatively powerless groups can gain leverage for commercial negotiations and to whom payments should be distributed under benefit-sharing agreements. He also points to the inevitable tension between the interests of those directly affected by resource extraction and wider indigenous community interests. He argues that this tension needs to be addressed productively and that benefit-sharing incomes must not detract from government responsibilities and lead to cost-shifting. Looking back on decades of agreements negotiated in Australia, Altman concludes that benefit sharing is no easy solution to the development problems faced by indigenous peoples. While the capital generated by benefit-sharing agreements should help to ameliorate these problems, it is important to acknowledge that no single agreement can provide more than a partial solution. Managing expectations while sustainably implementing agreements is clearly a challenge, and Altman argues that early investment in capacity-building is one of the most important answers.

Canada is another highly interesting country for comparative purposes. While no access and benefit-sharing policy is yet in place in Canada, consent, benefit sharing and other issues relevant to bioprospecting and biodiversity research are important points of discussion at national as well as institutional and community levels. Interestingly, as Kelly Bannister shows in Chapter 16, most of the impetus for discussions comes from new national human research ethics guidelines for research involving indigenous communities. In its development, Canadian biodiversity policy will likely look to policies that protect research subjects when issuing guidance on good ethical practice and also take into account significant foreign cases, such as San-*Hoodia*. This case is believed to set an important precedent by enabling indigenous communities to share in benefits even when the traditional knowledge used is based on literature that is already in the public domain.

The final chapter in this section compares one of the best-documented cases in the history of benefit-sharing agreements, that emerging from the Maya International Cooperative Biodiversity Group (ICBG) in Chiapas, Mexico, with the *Hoodia* case. The researchers involved in the Mexican case were particularly keen for the indigenous Maya people to contribute to and benefit from research. However, gaps in the way local communities were included became a focus for resistance to the project, which was eventually abandoned. Dafna Feinholz-Klip, Luis García Barrios and Julie Cook Lucas argue persuasively in Chapter 17 that no single actor should bear the total responsibility for what happened to the Chiapas project, but none is devoid of such responsibility. Through a comparison with the San-*Hoodia* case, the authors discuss how parties had conflicting assumptions about how and to what extent different groups of people should benefit from the potential royalties and who should make these decisions. Like the San, the Maya stood to receive a very small proportion of any profit that might come from the development of commercial products. Both cases played out in a domestic legal and policy vacuum. Questions about the legitimacy of processes and decisions emerged as fundamental points of conflict. Indeed, the Chiapas case failed largely due to the lack of an appropriate prior informed consent process built on trust and adequate representation.

We hope that the book and our conclusions summarized in the last chapter will serve as a valuable resource for indigenous communities, policymakers, the range of industry sectors involved in bioprospecting, NGOs and academics as they move forward in realizing the principles and the ambitions of the CBD.

Chapter 2
Justice and Benefit Sharing

Doris Schroeder

Abstract Benefit sharing as envisaged by the Convention on Biological Diversity (CBD) is a relatively new idea in international law. In the context of non-human biological resources, it aims to guarantee the conservation of biodiversity and its sustainable use by ensuring that its custodians are adequately rewarded for its preservation.

Prior to the adoption of the CBD, biological resources were regarded as the common heritage of humankind. Bioprospectors were able to take resources out of their natural habitat and develop commercial products without sharing benefits with states or local communities. Since 1992, prior informed consent has been required from states for access to biological resources, and users of these resources are required to develop fair and equitable benefit-sharing agreements on mutually agreed terms.

This chapter asks how benefit sharing fits into debates on justice. It argues that the CBD is an example of a set of social rules designed to increase social utility. It also proposes that the common heritage of humankind principle would be preferable to assigning bureaucratic property rights to biological resources. However, as long as the international economic order is characterized by serious distributive injustices, reflected in the enormous poverty-related death toll in developing countries, any ethical attempt to redress the balance in favour of the disadvantaged has to be welcomed. By legislating for a justice in exchange system covering non-human biological resources and traditional knowledge in preference to the tacit common heritage of humankind principle, the CBD has provided a small step forward in redressing the balance. It therefore presents just legislation sensitive to the international relations context in the twenty-first century.

Keywords benefit sharing • Convention on Biological Diversity • cosmopolitanism • international justice • traditional knowledge

D. Schroeder (✉)
UCLAN, Centre for Professional Ethics, Brook 317, Preston PR1 2HE, UK
e-mail: dschroeder@uclan.ac.uk

George moved to Texas and bought a donkey from an old farmer for 100 dollars. The farmer agreed to deliver the donkey the next day.
In the morning, the farmer drove up and said:

> 'Sorry, but I got some bad news. The donkey died.'
> 'Well then, just give me my money back.'
> 'Can't do that. I went and spent it already.'
> 'OK then, just unload the donkey.'
> 'What are you gonna do with him?'
> 'I'm gonna raffle him off.'
> 'You can't raffle off a dead donkey!'
> 'Sure, I can. Watch me. I just won't tell anyone he's dead.'
> A month later the farmer met up with George and asked,
> 'What happened with the dead donkey?'
> 'I raffled him off. I sold 500 tickets at two dollars apiece and made a profit of 898 dollars.'
> 'Didn't anyone complain?'
> 'Just the guy who won. So I gave him his two dollars back.

Did George act unjustly? It was not his fault that the donkey was dead. He had lost 100 dollars, after all. And besides, nobody was made worse off through the raffle. With one exception, those who took part did not win the donkey and 'the guy who won' was given a refund. Well, this is one perspective on the justice issues posed by George's conundrum. Philosophers have been debating matters of justice for millennia, but instead of calming down, the discussions are heating up. One of the newer debates surrounds justice and benefit sharing.

'Benefit sharing' is a technical term, which was popularized by the Convention on Biological Diversity (CBD) adopted at the 1992 Earth Summit in Rio de Janeiro, Brazil. This global convention aims to achieve three objectives (CBD 1992):

1. The conservation of biological diversity
2. The sustainable use of its components
3. The fair and equitable sharing of benefits from the use of genetic resources

The CBD was the first international treaty to recognize that the conservation of biodiversity is a 'common concern of humankind' (CBD 1992). Today its 190 parties cooperate to stop the destruction of biodiversity by attempting to ensure its sustainable use, and by requiring users of this natural wealth to share the benefits with those who provide knowledge of and access to genetic resources.

The custodians of biodiversity are often traditional or local communities in developing countries whose rights were strengthened by CBD parties in 2000 when it was decided that access to traditional knowledge should be subject to formal prior informed consent from the holders of such knowledge (COP 2000).

This chapter will situate the relatively new concept of benefit sharing within long-standing debates of justice. First, I shall explain how genetic resources can be viewed as the common heritage of humankind or as 'property' falling under the sovereignty of states. Second, I shall outline the difference between distributive justice and justice in exchange. Third, drawing on the above, I shall present answers to the following two questions:

- What type of justice does the CBD demand with its principles?
- Can the CBD be regarded as *just* legislation?

2.1 Common Heritage of Humankind Versus National Sovereignty

Who legally owns biological resources? Some answers to this question are straightforward, others are not. Farmers and smallholders own the crops, vegetables and flowers they grow in their fields and gardens. They can take them to a market and sell them for cash. Private landowners, companies or traditional communities with secure rights over their ancestral land often own the gold, oil or minerals found on their property. Where they do not, the state owns all mineral rights within its territory. Land that is not registered in anybody's name usually belongs to the state, and if it contains minerals or other valuable resources those can be exploited by the state. The same applies to forests not in private hands, whose timber can be sold by the state. But what about resources found on the seabed, or even resources found on the moon?

It is in this context that the idea of the common heritage of humankind was made explicit in the late twentieth century with the United Nations Agreement Governing the Activities of States on the Moon and Other Celestial Bodies (UN 1979) and the Convention on the Law of the Sea (UN 1982). The conventions declared that the seabed, the ocean floor and the subsoil thereof, as well as the surface and the subsurface of the moon, should not become the property of any state, organization or individual. Instead, the exploitation of their possible resources must be carried out so as to benefit humankind as a whole.[1] This principle was also expressed in the non-binding Statement on Benefit Sharing by the Human Genome Project's Ethics Committee recognizing that 'the human genome is part of the common heritage of humanity' (HUGO 2000).

Does the above cover all eventualities that might mandate property ascriptions? We have renewable resources (e.g. crops, timber), non-renewable resources (e.g. oil, minerals) and potential finds from the seabed and the moon, as well as human genetic resources. This sounds like a comprehensive list, but it is not. And it is the missing item that has been subject to contentious debates before, throughout and beyond CBD negotiations. As Shiva (1991:257) a prominent Indian environmentalist, writes:

---

[1] The administration of US president Bill Clinton managed to undermine the common heritage principle through a superseding agreement in 1994 that opened seabed resources to the first comer while eliminating the benefit-sharing provisions of the original text (Pogge 2008:131-2).

> [T]he North has always used Third World germplasm as a freely available resource and treated it as valueless. The advanced capitalist nations wish to retain free access to the developing world's storehouse of genetic diversity, while the South would like to have the proprietory varieties of the North's industry declared a similarly 'public' good.

Germplasm is the genetic material of an organism, which carries in it all inherited qualities, usually in one entity. For instance, the seeds of an *Artemisia* plant would be called the plant's germplasm. Through its germplasm, the plant itself can be recreated or its properties can be used, for instance, in malaria medication. In her criticism, Shiva refers to the fact that the world of bioprospecting was a free-for-all until the CBD was adopted. For hundreds of years, Northern plant specialists travelled to the South and took non-human genetic resources without asking permission or sharing potential benefits with states or local communities.

An example: After obtaining a patent, Merck Pharmaceuticals started marketing a treatment for glaucoma derived from a bush (jaborandi) found exclusively in the Amazon region. The plants' leaves are harvested by Indians in Brazil and transported to Germany, where its relevant parts (alkaloids) are refined and transformed into eye drops. A Brazilian wanting to use the eye drops would have to buy them at German-set prices, while any Brazilian company wanting to produce a generic version of the treatment would have to pay royalties to Merck for the period of the patent. 'Northern biotechnology companies see this as a right to earnings on their investments. Southern nations see this as more of the all-too-familiar exploitation' (Rolston 1995).

Before the CBD was adopted, renewable resources were assumed to belong to the common heritage of humankind (Srinivas 2008). And in contrast to the provisions made through the above-mentioned UN conventions on the moon and the seabed, they were not protected through an explicit demand that all of humanity must benefit from their exploitation. Following the adoption of the CBD, such resources are no longer regarded as a public good, available to all. According to the preamble of the CBD, biological resources fall under the national sovereignty of states. This move, it was assumed by CBD proponents, would help facilitate their sustainable use and preservation much more than the common heritage paradigm had done in the past. It would also contribute to combating incidents of exploitation, as noted by Shiva above, by imposing restrictions on access and by setting requirements to share benefits with the providers of resources.

But what about knowledge? Does it fall under the common heritage of humankind or the national sovereignty of states? A subcategory of knowledge consists of creations of the brain such as music, literature, paintings, inventions, designs and recipes. And it is those that are of interest for this chapter. Is Wagner's opera *Tristan und Isolde* the common heritage of humankind? If not, to whom does it belong? To resolve such questions, most states adhere to the idea of intellectual property that is entitled to legal protection through tools such as copyrights, trademarks and patents. Laws are put in place to give exclusive property rights over creations of the mind to their creators. Accordingly, Wagner and his descendants own *Tristan und Isolde*, Merck owns the glaucoma drops, sparkling wines can only be called 'champagne' if they were produced in a certain region in France, Philip Roth controls and profits from his novels and the Mercedes-Benz three-pointed star must not be used by anyone else. To reward and thereby encourage such creations and creativity, social rules were *created* to protect designers, inventors and others.

Extending this practice and rationale, the parties to the CBD have agreed that access to traditional knowledge must be subject to formal prior informed consent, and if such consent is forthcoming, benefits arising from its exploitation must be shared equitably. Today traditional knowledge is therefore removed from the common heritage of humankind, as is the case with plants, animals and micro-organisms. The following section outlines basic ideas on justice to enable me to explain how benefit sharing for non-human genetic resources and traditional knowledge can be viewed from within a justice framework.

2.2 Different Concepts of Justice

What exactly is justice? One cannot point at it and say: 'Look, there is justice!' as one can point to a castle or a person. Justice is a concept, which can be attributed to certain entities, namely agents, actions, social rules/institutions and states of affairs (Pogge 2006). These four entities are not mutually exclusive, and in fact one can draw simplified links between them as follows.

George, in raffling off a dead donkey, used deception for personal gain. Leaving aside the complication that the old farmer should have returned the 100 dollars to him, one could say that George acted unjustly. It is simply unjust to profit through deception. This makes George an unjust agent. At the same time, the story describes an unjust act (profiting through deception). But what about the bigger picture? In the case of George, 499 people were deceived – possibly the population of an entire village in deep Texas. If he continued to do so with repeated raffles, he might create a state of affairs in which he gained wealth significantly, while others unjustly lost some of their income. One would then talk about an unjust state of affairs.

Sometimes unjust states of affairs are due to unjust social rules. For instance, in a country where only those who could afford to hire an attorney were allowed access to the courts, the poor would not be able to achieve redress if, for instance, they were deceived by George in a business dealing. Such a rule would unjustly favour the rich over the poor. Ideally, social rules enable the peaceful flourishing of citizens under the authority of those rules, avoiding injustices as far as possible.

The following diagram shows a simplified version of the domains of justice and how they influence one another (Fig. 2.1).

Let me apply the above categorization to Merck's glaucoma treatment as an example. The tacit social rules prior to the adoption of the CBD were that wild plants and germplasm belonged to the public domain and formed part of the common heritage of humankind. This rule enabled Merck (the agent) to obtain valuable plant material in the Amazon and market a profitable product without obtaining consent for access and without sharing benefits (the action). At the same time this tacit social rule led to a state of affairs which Shiva (1991:257) described as exploitative and

Fig. 2.1 Domains of Justice

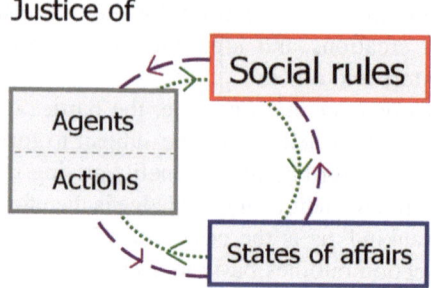

unjust. As a result of lobbying by developing countries, the tacit rule was abandoned and an explicit international legal rule put in its place. Since 1992, wild plants and germplasm have fallen under the sovereignty of individual states – assuming they have signed the CBD – and are thereby subject to regulations governing access and benefit sharing. One could say that bioprospectors who nowadays neglect to respect the CBD are unjust agents, committing unjust actions, as they violate a legitimate social rule set up to prevent exploitation. Before the CBD was signed, one could not make the same claim.

It was always true that the physical plant belonged either to the Brazilian state or to local landowners. But prior to the adoption of the CBD, the plant genetic material and its biochemical properties could have been regarded as the common heritage of humankind (or, at least, Brazil had no recourse against someone patenting a molecule derived from the plant). It is not immediately obvious why the plant genetic material should belong to Brazil or other countries which host the species, rather than to humankind. This claim requires further substantiation and an excursion into the justification of social rules.

2.2.1 *Natural Rights Versus Social Utility*

Human beings live with rules: moral rules ('do not kill'), etiquette ('do not speak with your mouth full'), prudential rules ('avoid sugar if you suffer from diabetes'), legal rules ('do not run a red light') and so on. Simplified, some rules are intuitively plausible to human beings, while others require a justification. For instance, if you arrived in a foreign country and a native told you that they ate their dead instead of burying them (see the example of Schroeder, Chapter 3), you would probably be stunned and possibly horrified. It would not seem immediately plausible that doing such a thing could be justified. But you might be given certain reasons by insiders or outsiders, e.g. 'Our Gods demand this of us,' 'This rule developed during a famine,' or 'We believe that the spirits of our ancestors would perish if buried in soil.' At the same time, the native could say that they have a rule against killing except in self-defence. This would presumably seem a lot more plausible to you than the first rule.

In this context, philosophers and political theorists distinguish so-called natural rights from rights based upon laws or beliefs. A natural right is universal and applies to all human beings, irrespective of laws, traditions or culture (Brown 1960:1f). The right not to be killed, with certain provisos (e.g. self-defence), is considered such a right (Finnis 1980:281). It applies everywhere, for all human beings, whether it is enshrined in law or not (D'Entrèves 1970:22-36). Natural rights cannot be overruled by legislation. Even if a law said that it was acceptable to kill those belonging to a particular ethnic group, it would still be wrong to do so.

Many rights and obligations that inform human interaction are not based on natural law. Some, for instance, are specifically *created* by human beings in order to improve human flourishing. They require reference to social utility to be acceptable. For instance, the obligation to stop at red traffic lights is based on a rule that can be justified with reference to social utility. Traffic lights ease the flow of traffic at crossings and avoid human deaths. Rules that depend on their social utility for justification are not permanent and unchangeable. On an island that once had 5,000 cars, but now has 5, it might make sense to abolish all traffic lights to save maintenance costs and energy. The social utility of traffic lights would have changed. Such potential changes need to be monitored in order to avoid a once beneficial rule becoming problematic, to test whether social rules are indeed improving human flourishing or to assure ourselves that there is no alternative set of rules that would do even better (Pogge 2007).

Could the claim that the genetic make-up of a wild plant growing in the Amazon belongs to the Brazilian state rather than, for instance, humankind be based on natural law? Is there a natural right that requires states to be assigned ownership of plant species? No, because justifying such a rule would face obvious and grave difficulties. For instance, why should a state own the resources of its people when many governments today are military dictatorships for whom the human flourishing of their citizens is the least pressing concern? Why should human flourishing be hampered through property rights that limit benefits for humankind? For instance, if the earth were an island with plentiful resources for its small number of egalitarian and affluent citizens, a land where milk and honey flowed, it would not make sense to restrict access to wild plants. Nobody would object to a particularly inventive chap taking a plant and extracting its active ingredients in order to create an antidiabetes drink, even if he charged for the end-product.[2] The CBD rule that wild plants and other resources fall under the sovereignty of states is a rule whose justification must appeal to social utility. And the legitimacy of such rules depends on context, in particular on the international economic order and distributions of wealth. One can therefore note that CBD rules, not being mandated by natural rights, are open to discussion about their social utility (relative to alternative regulations), which needs to be tested in the context of today's economic order.

---

[2] Of course, people might object if he simultaneously demanded monopoly powers over his antidiabetes drink for more than a decade.

Of course, it is important to note that decisions about social utility are not always easily made. When we examine a particular issue, such as access to plant species, and then compare alternative sets of rules with one another and with the option of leaving the issue unregulated, we realize how complex such matters are. In some comparisons, we will find that one set of rules is unambiguously worse than another: everybody is worse off with the former rules than with the latter. However, even if obviously unattractive proposals are eliminated, several contenders will normally remain. When we look at any two such proposals, we will find that one of them is better than the other for some people, yet worse for others. It is at this juncture that it is hard to decide which of the two should be favoured in terms of their social utility.

The justified selection of certain social rules over others falls within the domain of distributive justice and involves weighing the respective gains and losses of separate groups against one another. However, although this sounds straightforward and almost mathematical, it is not an easy task. Robin Hood would say that a rule permitting theft from the rich to feed the starving has high social utility. He would argue that, from a distributive justice perspective, it favours a disadvantaged group by 'only' harming an advantaged group where the life of the former is at stake. He might add that the rich are exploiting the poor as serfs on their land and therefore have contributed to the starvation. By contrast, the Sheriff of Nottingham and English law impose the same rules against theft upon poor and rich alike. They could argue that human beings have a natural right to property, which must not be violated – or that the social utility of a blanket prohibition on theft is sufficiently high even if this might lead to some individuals starving.

This Robin Hood example illuminates two potential justice considerations. First, should anybody ever starve in the presence of the rich, if such starvation could be avoided? This is a distributive justice issue in a world of scarce resources that is characterized by vast inequalities in wealth. Second, should serfs ever starve when they are working on a landlord's property? In other words, do they not deserve a decent reward for their labour that lifts them at least to the subsistence level? This is a justice in exchange issue.

I need to introduce both concepts in order to situate CBD-style benefit sharing within established justice debates. Let me first clarify what the terms 'distributive justice' and 'justice in exchange' mean.

2.2.2 Distributive Justice and Justice in Exchange

Essentially, justice in exchange regulates the justice of giving one thing and receiving an appropriate return, while distributive justice deals with the division of existing resources among a group of qualifying recipients.[3]

---

[3] It would go beyond the scope of this chapter to explain the other two main concepts of justice, namely corrective and retributive justice (Pogge 2006).

Justice in exchange mainly establishes the fairness of transactions. For instance, is the rent charged for a particular flat in central London appropriate – or *just*?[4] An interaction is considered just if all parties in the exchange receive an appropriate return for their contribution.

Distributive justice deals with access to scarce resources – from the division of an apple pie among friends to the structure of an economic order that regulates access to raw materials and the distribution of the jointly created social product. The further one moves away from individual actions (e.g. sharing an apple pie) towards actions impacting on large groups (e.g. all those requiring tuberculosis treatment), the more complex social rules come into play.

Philosophers have been debating issues of distributive justice for a long time. I will introduce the briefest of answers to the essential question within the sphere of distributive justice, namely: 'Who deserves what from whom?'

In the mid-twentieth century, there was some consensus on the 'Who deserves what from whom?' question, at least among European welfare-focused politicians and theorists (henceforth 'welfare liberals'). Simplified, those who live legitimately within a state ('*who*'), qualify for the receipt of income support at subsistence level plus other services to cover their basic needs ('*what*') from the state ('*from whom*') (Beveridge 1942).

Later in the twentieth century, however, the proviso that the distributive justice realm aligns with national borders was questioned on the basis that all human beings are equal, independent of their country of birth. It is now increasingly argued that distributive justice demands a universal, cosmopolitan response (Cole 2001; Pogge 2008). This understanding also seems to align with the Universal Declaration of Human Rights and its article 25(1), which reads (UN 1948):

> Everyone ['*who*'] has the right to a standard of living adequate for the health and wellbeing of himself and of his family, including food, clothing, housing and medical care and necessary social services, and the right to security in the event of unemployment, sickness, disability, widowhood, old age or other lack of livelihood in circumstances beyond his control ['*what*'].

One might argue that there is no disagreement between the welfare liberal and the cosmopolitan human rights answer to my distributive justice question. In response to the '*who*' question, the cosmopolitan says that everyone has entitlements within the realm of distributive justice, while the welfare liberal says that everyone who lives within a state has such entitlements. Over recent decades, the two realms have aligned: everybody is born into a state. Hence, the answer to the '*who*' question is identical for all practical purposes. At the same time, both the welfare liberal and the cosmopolitan answer the '*what*' question with reference to basic needs fulfilment, one of the most prominent distributive justice positions.

---

[4] I am not using the understanding of 'justice in exchange' based on Roman law, which only requires that two competent adults have voluntarily agreed on a price. Instead, I am referring to the Aristotelian notion of 'justice in exchange', which requires that a price and a good are proportionate requitals, i.e. the intrinsic worth of a good is mirrored in a monetary sum. On this understanding, a landlord who overcharges a tenant could thereby be violating justice in exchange even if the tenant agrees (see Aristotle 1934:279-82).

This position argues that no human being should starve, freeze to death due to lack of shelter, die prematurely from easily curable diseases or suffer violent aggression because they lack support (e.g. shelter, police) (Raz 1986; Frankfurt 1987).

But the divergence of the welfare liberal and the cosmopolitan view becomes apparent when one looks at the last element of our question, *'from whom'*. Welfare liberals answer that the state is responsible for the satisfaction of the basic needs of only its citizens, while cosmopolitans typically argue that national borders make no difference to questions of distributive justice and impose duties on all states and their citizens to provide for those in need.

To summarize on distributive justice: a tentative consensus has been expressed through the Universal Declaration of Human Rights and ongoing work by philosophers and political scientists that every human being (*'who'*) deserves to have their basic needs satisfied (*'what'*) through their state and others, if necessary (*'from whom'*). How can these two concepts of justice be applied to benefit sharing?

2.3 Benefit Sharing: Distributive Justice or Justice in Exchange?

Let us remind ourselves of the main principles expressed in the CBD and how they relate to benefit sharing. First, benefit sharing aims to improve the conservation of biological diversity. It is one thing to look after a resource for the benefit of humankind, and quite another to look after it when its flourishing benefits one directly. Hence, by including custodians of this natural wealth in the receiving of benefits, the loss of biodiversity, a 'common concern of humankind', can hopefully be addressed. Second, benefit sharing aims to enable access to biodiversity for sustainable use, with the emphasis on *use*. In the context of increasing criticism from developing countries regarding the exploitation of their biological resources (Shiva 1991), it is much more likely that access for use will be granted if the exploitation critique has been addressed satisfactorily, through access and benefit-sharing agreements, than if it is left unresolved. And hence the third principle of the CBD – the fair and equitable sharing of benefits from the use of genetic resources – is instrumental in achieving the first two principles.

With the clarifications above on common heritage and national sovereignty, social utility and natural rights, as well as distributive justice and justice in exchange, one can now situate benefit sharing within justice frameworks by answering the following two questions.

- What type of justice does the CBD demand with its principles?
- Can the CBD be regarded as *just* legislation, i.e. as a just set of social rules?

Let me begin with the first question: what type of justice does the CBD demand with its principles? From the preceding, it is obvious that the CBD has created requirements of justice in exchange. Let me take the famous neem tree case to make this point clearer. The neem tree's medicinal properties have been known for thousands of years, in particular in India, Sri Lanka and Burma. Yet a patent was taken out by an international agrochemical business ignoring this prior art and aiming for monopoly control. Led by Vandana Shiva, an international lobbying movement managed to have the patent revoked after a battle that took almost 10 years (Sheridan 2005). In this case, an international agrochemical business used a resource from a region outside its own country for shareholder profit without rewarding local people for their knowledge contribution. With CBD legislation in place, such appropriation by outsiders of plants, animals, micro-organisms or traditional knowledge without the consent of local people and without compensation is no longer legal for signatories to the CBD. The CBD has declared that such resources fall under the sovereignty of states, and anybody wishing to access and use them therefore has to fulfil the demands of justice in exchange (Schroeder 2007). If you give something, you need to receive an appropriate return, a fair and equitable share of the benefits. Resource providers need to be given something in exchange for their resources or their traditional knowledge, according to the CBD.

Yet some academics such as De Jonge and Korthals (2006) maintain that benefit sharing

> should not merely be seen as an instrument of compensation. ... Instead, and in the face of the harsh reality that more than 800 million people are undernourished, benefit sharing should also ... be a tool to improve food security.

Similarly, Castle and Gold (2006) argue that

> [b]enefit sharing ... is an obligation owed to all peoples regardless of their ability to provide traditional knowledge. The obligation to share benefits derives from one's status as a person and not from one's control over certain knowledge.

One's first reaction to the above quotations might be that to throw two injustices into the same basket does not necessarily resolve them fairly. Imagine two communities that are undernourished and lack safe drinking water, adequate sanitation and access to essential medicines. One group holds traditional knowledge that is being used by a pharmaceutical company and leads to a patent, the other one does not. The abject poverty of both communities is one justice issue. Cosmopolitan philosophers would argue that countries in the North and their citizens have a duty to contribute to solutions. However, the pharmaceutical company that uses traditional knowledge surely has the added duty of compensating one of the two communities for the contributions they made to a patented product. Guarding this traditional knowledge requires effort, such as teaching the next generation the knowledge passed on from one's ancestors and conserving the biological resource itself.

It seems, therefore, that one should not mix up justice in exchange with distributive justice issues. However, a closer look reveals that the two are linked with regard to the CBD, and we shall see this more clearly when responding to the second question: can the CBD be justified as *just* legislation – that is, as a just set of social rules? The answer to this question is more complex.

As I have argued above, benefit sharing is not mandated by natural law. There is no natural right to, for instance, ownership of germplasm. Whether it belongs to individuals, on a first-come, first-served basis, or to local communities or states or humankind as a whole is an open question; a question that must be settled with reference to social utility. Likewise, there is no natural right to veto or restrict an outsider's use of traditional knowledge, which is often in the public domain.

For instance, traditional German knowledge has it that cold, soured, curdled milk (quark) helps to reduce fever if applied onto the lower legs inside a wrap. (Sounds obscure, but it works!) Yet it would seem ludicrous to suggest that Germans have a natural right to stop the French, for example, from applying this fever remedy.

Why then, does the CBD create property rights where there were none before, such as in traditional health knowledge? And are these new rights just? I noted earlier that any attempt to create new social rules, such as the CBD, must make reference to context, in particular to the international economic order. Social utility is measured in human flourishing, and if one relies on the slowly evolving consensus around the Universal Declaration of Human Rights, one will understand social utility as involving the fulfilment of basic human needs first and foremost.

The 'harsh reality' that De Jonge and Korthals (2006) point out above is not restricted to food security. While distributive justice as basic needs fulfilment has almost been achieved in European welfare states, the situation in other parts of the world is desperate: 800 million human beings are undernourished, 1,085 million lack access to safe drinking water, 2,600 million lack adequate sanitation and 1,577 million have no electricity (UNDP 2007). About 2,000 million lack access to essential medicines (FIC 2003), 1,000 million lack adequate shelter and 774 million adults are illiterate (UIS, n.d.). Each year, some 18 million or fully one-third of all human deaths are due to avoidable poverty-related causes, such as lack of access to vaccines, medications or rehydration packs (WHO 2004).

This is the context in which developing country activists, such as Shiva, have raised their concerns about the misappropriation of resources. When the above shocking figures about human suffering are combined with the knowledge that economically rich but biodiversity-poor countries are profiting from resources found in economically poor but biodiversity-rich ones, the question indeed arises: should biological resources be regarded as the common heritage of humankind?

Let me compare this situation with medical research that leads to patents and new treatments. The author of this chapter is a well-off academic in a permanent post. If I were asked whether a blood sample of mine could be used for research purposes, I would probably say 'yes' and forget all about it. I would not ask for special benefit sharing, even though I have contributed something that might lead to benefits for others. In other words, I would have given something but would not expect anything in exchange. This apparent altruism or common-spiritedness relies on the fact that any direct benefits of the research in the form of potentially therapeutic treatments and accessible new health care products and services would be available to me in the future. Whether my health insurance, my salary or the state paid for it, I would have access to it when needed. In a wider perspective, my fellow citizens and I also receive indirect benefits in the form of jobs and affluence generated by a high-tech industry.[5] It is easy to show some altruism in this context. But what if one showed the same altruism and the results of the research never became available in one's country, or one could not afford them?

---

[5]The issue of alleged excessive profits is a different matter outside the scope of this chapter.

The main issue that has thrown doubt on the fairness of the altruism model in medical research is the potential exploitation of research subjects in developing countries (Schroeder and Lasen-Diaz 2006). In poor countries, one cannot take the above-mentioned benefits for granted. On the contrary, reasonable availability of newly developed products cannot be guaranteed, and neither can a match to the population's health needs nor the existence of secondary benefits (e.g. jobs). It is in this context that the demand for benefit sharing becomes obvious, and it is here that distributive justice issues link in.

When it comes to resources, be they blood samples or plants, the ideal scenario would be that both could be accessed and used for the benefit of humankind without any inherent exploitation. The idea of a common heritage of humankind would be appropriate to such ideal circumstances. Whether my blood or your plant knowledge leads to medical progress does not matter as long as we all have access to the benefits of their use. To impose highly bureaucratic barriers on the use of resources (other than for reasons of achieving sustainability) and require benefit sharing through, for instance, royalty payments, would be counterproductive if the international economic order and today's context resembled the island of affluent citizens described earlier. Imposing the CBD on such an island would reduce social utility. Distributive justice issues would not exist, with everybody having access to the fruits of some people's ingenuity, and it would not make sense to impose access restrictions on resources that are better seen as the common heritage of humankind.[6] The latter perspective would improve overall social utility.

However, the CBD has rightly favoured national sovereignty over the common heritage of humankind principle with regard to biological resources. Its guidelines have created a justice in exchange issue for items that were not previously considered restricted property, such as germplasm and traditional knowledge. This was a contextual decision made at the end of the twentieth century, when biodiversity was being rapidly depleted and developing countries were justifiably concerned about the exploitation of their resources. While the common heritage of humankind principle would be preferable to that of fencing in resources with bureaucratic procedures when they could be used for the benefit of humankind, this ideal scenario cannot apply in the context of an international economic order that is unjust and leads to significant human suffering.[7] This understanding of benefit sharing, according to the CBD, resolves two issues that are brought forth as criticism.

First, the CBD imposes conceptions of property that are alien outside market capitalism. Traditional knowledge is not generally fenced in, most certainly not by its holders (see Saskia Vermeylen, Chapter 10). In fact, the generosity that holders of traditional knowledge used to show by sharing their knowledge widely with Northern

---

[6]I am assuming here that products derived from biological resources would not be priced out of the range of some islanders through a system giving monopoly powers to the inventors for a considerable period.

[7]It is beyond the scope of this chapter to outline why today's international economic order is unjust. For a detailed justification of this claim (see Pogge 2008).

botanists points to the general acceptability of the common heritage of humankind idea. However, this idea becomes dubious when there is not enough altruism and community spirit on the part of the recipients to ensure that the benefits reach those who contributed to the advancement of knowledge and resulting products – especially when these contributors are seriously impoverished and marginalized communities. In such a context, imposing a property conception on traditional knowledge in order to secure some benefit to the impoverished contributors, as the CBD did, is a lesser evil than leaving the area unregulated.

Second, it has been argued that benefit sharing is a charade because nobody rewards, for instance, Bavarian cheese makers for their traditional knowledge[8] (Schuklenk 2003; Schuklenk and Kleinsmidt 2006). If one treats like cases alike, one of the foremost premises of justice (Aristotle 1934), one cannot favour the San community over the Bavarians if one aspires to achieve justice. Yet, with the position outlined in this chapter, one can distinguish between Bavarians and the San, and assign benefit sharing rules in favour of the latter but not the former. If the international economic order were as favourable to the San as it is to the Bavarians, the move from the common heritage of humankind principle to national sovereignty and justice in exchange would not be necessary or even justifiable.

2.4 Conclusion

Benefit sharing as envisaged by the CBD is a relatively new idea in international law. Within the context of non-human biological resources, it aims to guarantee the conservation and sustainable use of biodiversity by ensuring that its custodians are rewarded and hence encouraged to promote conservation. Prior to the adoption of the CBD, biological resources were frequently regarded as the common heritage of humankind. Bioprospectors were able to take resources out of their natural habitat and develop commercial products without sharing benefits with states or local communities. Since 1992, states have been required to obtain prior informed consent, as well as satisfactory benefit-sharing agreements on mutually agreed terms, in order to be granted access to biological resources.

This chapter has asked how benefit sharing fits into debates on justice. It argues that the CBD is a set of social rules designed to increase social utility, not an example of natural law with natural rights that apply universally and in every context. It also argues that while the common heritage of humankind principle would ideally

---

[8] It has to be noted that the comparison is not ideal, because one tends to think of trademarks or geographic indications when dealing with cheese, whilst one thinks more of patents on pharmaceuticals when dealing with traditional knowledge. However, the criticism remains pressing when distributive justice is excluded from the debate, as a patent for a quark remedy against fever based on German traditional medical knowledge would also not be treated as a CBD benefit-sharing case, in my view.

be preferable to assigning bureaucratic property rights to biological resources, this claim has to take account of context, and in particular of the justice of the current international economic order. As long as this economic order is characterized by serious distributive injustices, reflected in the enormous and avoidable death toll in developing countries from food insecurity and lack of access to essential medicines, any ethical attempt to redress the balance in favour of the disadvantaged has to be welcomed. By legislating for a justice in exchange system, covering non-human biological resources and traditional knowledge, in preference to the tacit common heritage of humankind principle, the CBD has provided a small step forward in redressing the balance. However, it must be seen as a means to an end, and a fairly humble means at that, if the end is a just economic order in a world where basic needs are fulfilled for all.

References

Aristotle (1934). *Nicomachean ethics* (trans: Rackham, H.). Cambridge, MA: Harvard University Press.
Beveridge, W. (1942). *Social insurance and allied services*. London: H. M. Stationery Office.
Brown, B. F. (1960). *The natural law reader*. New York: Oceana Publications.
Castle, D., & Gold, E. R. (2006). Traditional knowledge and benefit sharing: from compensation to transaction. In P. W. B. Phillips & C. B. Onwuekwe (Eds.), *Accessing and sharing the benefits of the genomics revolution*. Dordrecht: Springer.
CBD (1992). Article 1: objectives. In Convention on Biological Diversity. www.cbd.int/convention/convention.shtml. Accessed 14 April 2008.
Cole, P. (2001). *Philosophies of exclusion: liberal political theory and immigration*. Edinburgh: Edinburgh University Press.
COP (2000). Article 8(j) and related provisions, Decision V/16, Decisions Adopted by the Conference of the Parties to the Convention on Biological Diversity at Its Fifth Meeting, Nairobi, 15–26 May, UNEP/CBD/COP/5/23. www.cbd.int/doc/decisions/COP-05-dec-en.pdf. Accessed 11 July 2008.
De Jonge, B., & Korthals, M. (2006). Vicissitudes of benefit sharing of crop genetic resources: downstream and upstream. *Developing World Bioethics, 6*(3), 144–157.
D'Entrèves, A. P. (1970). *Natural law (2nd ed)*. London: Hutchinson University Library.
FIC (2003). Strategic plan: fiscal years 2000–2003: executive summary: reducing disparities in global health. Bethesda, MD: John E. Fogarty International Center for Advanced Studies in the Health Sciences. www.fic.nih.gov/about/plan/exec_summary.htm. Accessed 1 May 2008.
Finnis, J. (1980). *Natural law and natural rights*. Oxford: Clarendon Press.
Frankfurt, H. (1987). Equality as a moral ideal. *Ethics, 98*(1), 21–42.
HUGO (2000). Statement on benefit sharing. Human Genome Organisation Ethics Committee. www.hugo-international.org/img/benefit_sharing_2000.pdf. Accessed 13 July 2008.
Pogge, T. (2006). Justice. In D. M. Borchert (Ed.), *Encyclopedia of philosophy (2nd ed)* (Vol. 4). Detroit, MI: Macmillan Reference.
Pogge, T. (2007). Montréal statement on the human right to essential medicines. *Cambridge Quarterly of Healthcare Ethics, 16*(1), 97–108.
Pogge, T. (2008). *World poverty and human rights (2nd ed)*. Cambridge: Polity Press.
Raz, J. (1986). *The morality of freedom*. Oxford: Clarendon Press.
Rolston, H., III. (1995). Environmental protection and an equitable international order: ethics after the Earth Summit. *Business Ethics Quarterly, 5*(4), 735–752.
Schroeder, D. (2007). Benefit sharing: it's time for a definition. *Journal of Medical Ethics, 33*, 205–209.

Schroeder, D., & Lasen-Diaz, C. (2006). Sharing the benefit of genetic resources: from biodiversity to human genetics. *Developing World Bioethics, 6*(3), 135–143.

Schuklenk, U. (2003) Presentation on benefit sharing, 6th Global Forum for Bioethics in Research, April, Paris.

Schuklenk, U., & Kleinsmidt, A. (2006). North-South benefit sharing arrangements in bio-prospecting and genetic research. *Developing World Bioethics, 6*(3), 122–134.

Sheridan, C. (2005). EPO neem patent revocation revives biopiracy debate. *Nature Biotechnology, 23*(5), 511.

Shiva, V. (1991). *The violence of the green revolution.* London: Zed Books.

Srinivas, K. R. (2008). Traditional knowledge and intellectual property rights: a note on some issues, some solutions and some suggestions. *Asian Journal of WTO and International Health Law and Policy, 3*(1), 81–120.

UIS (n.d.) UNESCO Institute for Statistics. www.uis.unesco.org. Accessed 1 May 2008.

UN (1948). Universal Declaration of Human Rights, United Nations. www.un.org/Overview/rights.html. Accessed 15 April 2008.

UN (1979). Agreement governing the activities of states on the moon and other celestial bodies, United Nations. www.unoosa.org/oosa/SpaceLaw/moon.html. Accessed 15 April 2008.

UN (1982). Convention on the Law of the Sea, United Nations. www.un.org/depts/los/convention_agreements/texts/unclos/closindx.htm. Accessed 15 April 2008.

UNDP (2007). *Human development report 2007–2008.* Published for United Nations Development *Programme by Palgrave Macmillan, Basingstoke.* hdr.undp.org/en/reports/global/hdr2007-2008. Accessed 22 July 2008.

WHO (2004). *The world health report 2004.* Geneva: World Health Organization. www.who.int/whr/2004. Accessed 1 May 2008.

Chapter 3
Informed Consent: From Medical Research to Traditional Knowledge

Doris Schroeder

Abstract Obtaining informed consent has become an essential part of modern medical practice. Today, patients and research subjects are actively involved in medical decision-making and are no longer expected to defer responsibility to paternalistic, benevolent doctors. Since the early 1990s, the concept of informed consent has also been employed systematically in connection with indigenous peoples' rights of self-determination. The Convention on Biological Diversity (CBD), for instance, requires that prior informed consent be obtained from indigenous communities before accessing their traditional knowledge, innovations and practices.

This chapter outlines the four steps necessary to conclude a consent process ethically and successfully, namely legitimization to consent, full disclosure, adequate comprehension and voluntary agreement. It concludes that the similarities between obtaining informed consent in the medical context and obtaining prior informed consent in terms of CBD requirements are strong enough to warrant mutual learning. Such learning is particularly appropriate when dealing with the inherent power imbalances between medical staff and research subjects on the one hand, and bioprospectors and indigenous communities on the other hand. Importantly, in both fields, the autonomy of research subjects and the right to self-determination of indigenous peoples need to be upheld and strengthened with clear, enforceable legislation at national and international level.

Keywords benefit sharing • Convention on Biological Diversity • informed consent • prior informed consent • research ethics

---

D. Schroeder
UCLAN, Centre for Professional Ethics, Brook 317, Preston PR1 2HE, United Kingdom
e-mail: dschroeder@uclan.ac.uk

3.1 Introduction

Informed consent procedures seem to be on the increase. Recently, I had to sign an informed consent form to receive a yellow fever vaccination. Apparently I had, as a result, a 1:200,000 chance of dying from severe organ failure and a 1:130,000 chance of dying from a severe allergic reaction. I was also meant to look out for fever, dizziness and a variety of other complications for 30 days after the vaccination. Finally, the doctor disclosed that the last person who had died from the vaccine lived in Barcelona and was 28 years old. Not the most cheering of news prior to an injection, but I signed and so do millions of others every day.

Informed consent processes have become an essential part of modern medical practice. As the Nuremberg Code (1947) stated categorically in its first sentence: 'The voluntary consent of the human subject is absolutely essential.' Today, this requirement for consent applies to all interventions on human subjects that are experimental in nature, i.e. that involve research. In addition, it applies to standard medical procedures that involve some level of risk. So, for instance, one might be asked to sign an informed consent form before a chiropractor manipulates one's neck, but one will not be asked to sign a form before a blood sample is taken for routine testing.

Although the concept of informed consent developed from the relationship between doctors and patients,[1] it has since gained significance outside the medical field. Since the late 1980s, it has been employed between states to control the movement of hazardous materials across borders. Today, it is no longer permissible to ship hazardous chemicals or waste from one country to another without the consent of the receiving country.[2] And since the early 1990s, the concept has been employed more systematically in connection with indigenous peoples' rights of self-determination, in particular in the context of logging, mining, dam building, resettlement and access to genetic resources and traditional knowledge (Perrault 2004).

In 1992, the Convention on Biological Diversity (CBD 1992a) was adopted at the Earth Summit in Rio de Janeiro, Brazil. The CBD was the first international treaty to recognize that the destruction of biodiversity can only be stopped if the custodians of this natural wealth are allowed to benefit from its sustainable use and conservation. The custodians of biodiversity are often indigenous or local communities in developing countries whose rights were strengthened by CBD

---

[1] For ease of reading, I shall use the word 'patient' in the medical context exclusively, even though healthy volunteers can enrol in research studies and are, of course, protected through informed consent procedures.

[2] Such shipments are regulated through the 1989 Basel Convention on the Control of Transboundary Movements of Hazardous Wastes and their Disposal (www.basel.int/), which entered into force on 5 May 1992, and the 1998 Rotterdam Convention on the Prior Informed Consent Procedure for Certain Hazardous Chemicals and Pesticides in International Trade (www.pic.int/), which entered into force on 24 February 2004.

parties in 2000 when it was decided, according to decision V/16 (CBD 2000), that access to traditional knowledge should be subject to formal prior informed consent requirements.

> Access to the traditional knowledge, innovations and practices of indigenous and local communities should be subject to prior informed consent or prior informed approval from the holders of such knowledge, innovations and practices.

This chapter will compare the concept of informed consent in the medical field with that in the context of traditional knowledge. In the latter, we find two main expressions of the consent principle, namely 'prior informed consent' as used in the CBD and 'free, prior and informed consent', as used in the United Nations Declaration on the Rights of Indigenous Peoples (UN 2007).[3] A brief definition of the different expressions will be followed by a step-by-step analysis, looking at the four separate elements (free, prior, informed and consent) in reverse order.

3.1.1 Informed Consent in the Medical Context

Informed consent is the 'voluntary, uncoerced decision, made by a sufficiently competent or autonomous person on the basis of adequate information and deliberation, to accept rather than reject some proposed course of action that will affect him or her' (Gillon 1985:113). As this standard definition shows, informed consent consists of four basic elements:

- The capacity to consent.
- The full disclosure of all the relevant information.
- The adequate comprehension of the disclosed information by the patient.
- Their voluntary decision to agree to suggested treatment or research (Andanda 2005).

I look at these elements in detail below. But first, the following diagram illustrates informed consent procedures by looking at the roles different parties have with the example of experimental heart surgery (Fig. 3.1).

In this ideal example, the proposed course of action that will affect the patient is experimental heart surgery. The patient is a sufficiently competent or autonomous person able to make decisions about the surgery. Ideally, 'informed consent should be obtained by a well-informed physician who is not engaged in the investigation and who is completely independent of this [doctor–patient] relationship' (WMA 2002). This third party, the information provider, will ensure that any conflicts of interests for the surgeon do not bias the disclosure procedure. Conflicts of interest can arise because researchers (e.g. surgeons running a study on experimental heart surgery)

---

[3] The difference between the two expressions within the context of traditional knowledge will be explained in the section 'Free prior informed consent' below.

Fig. 3.1 Informed Consent Roles: Ideal Scenario[4]

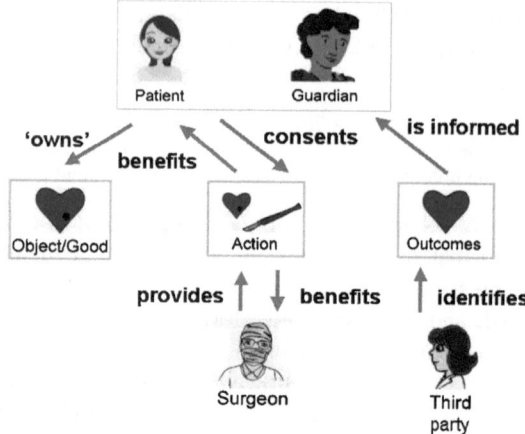

Table 3.1 International Guidelines Requiring Informed Consent

1947	Nuremberg Code, www.hhs.gov/ohrp/references/nurcode.htm
1964/2008	Declaration of Helsinki, www.wma.net/e/policy/b3.htm
1993/2002	International Ethical Guidelines for Biomedical Research Involving Human Subjects, www.cioms.ch/frame_guidelines_nov_2002.htm

have obligations to sponsors, institutions and medical progress that can compete with the obligation of caring for a patient (NBAC 2001). The third party will provide adequate information to form the basis for the patient's decision and will give her enough time for deliberation. At the end of the deliberation period, the patient voluntarily accepts or rejects the proposed intervention without having been unduly influenced (e.g. no cash incentives were offered for research participation other than expenses). To conclude the process, she signs a consent form. If the patient is unable to provide consent herself, proxy arrangements have to be made. This is most often the case for under-age patients or those without the required decision-making capacity, such as patients with a serious mental illness. A guardian – for instance the parents, in the case of a child – is then appointed to make the necessary decisions. Once the consent form has been signed by the patient or guardian, the surgeon uses the appropriate skills to undertake the operation. Given that we are dealing with research, the surgeon can benefit in terms of lessons learned and techniques developed in experimental heart surgery, publications and so on. We would hope that the surgery is successful and that the patient becomes a research beneficiary too.

This ideal informed consent process is required by various international guidelines, the most important of which are listed in the table above (Table 3.1).

---

[4] The role of the state – which is not depicted in this diagram – is to provide the necessary legal guidelines and enforcement mechanisms to enable the system of informed consent.

3.1.2 Prior Informed Consent in Accessing Traditional Knowledge

No internationally agreed definition of prior informed consent yet exists (Haira 2006). However, the above definition for the health care context can be adapted as follows (adaptation in italics).

> *Prior* informed consent is the voluntary, uncoerced decision *made by a subgroup[5] that legitimately represents an indigenous community*, on the basis of adequate information and deliberation, to accept rather than reject some proposed course of action that will affect *the community*.

As in the health care context, prior informed consent consists of four basic elements:

- The legitimate authorization to consent.
- The full disclosure of all the relevant information.
- The adequate comprehension of the disclosed information by the representatives.
- Their voluntary decision to agree to the proposed course of action.

Let me take the example of the *Hoodia* plant to draw a diagram analogous to the surgeon-patient diagram (Fig. 3.2).

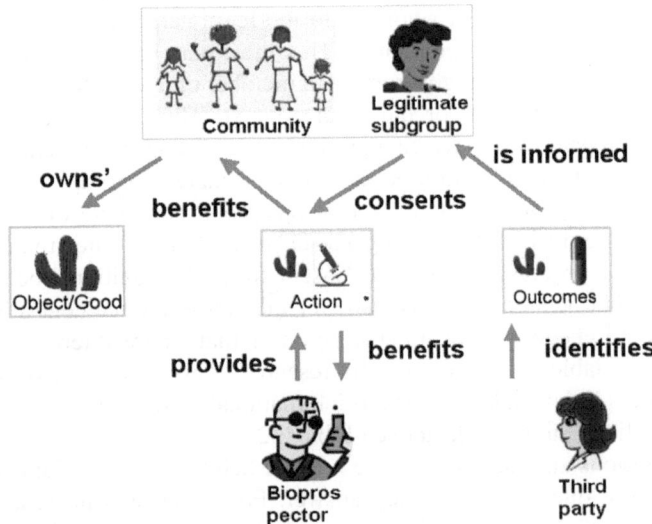

Fig. 3.2 Prior Informed Consent Roles: Simplified Scenario[6]

---

[5] I shall not discuss the highly unusual, if ideal, case where it is possible to obtain consent from every single member of a community. In most cases, some form of legitimate subgroup will have to consent on behalf of a group.

[6] This diagram, like the first one, does not explicitly depict the role of the state. As will be seen later, the CBD gives sovereignty over biological resources and traditional knowledge to the nation state rather than indigenous populations. The diagram presumes the ideal situation, in which the nation state discharges its duty to enable the obtaining of consent from its subpopulations, such as indigenous communities.

Table 3.2 International Guidelines Requiring (Free) Prior Informed Consent from Indigenous Communities

1989	Convention 169 on Indigenous and Tribal Peoples (International Labour Organization), entered into force 5 September 1991, www.ilo.org/ilolex/cgi-lex/convde.pl?C169
1992	Convention on Biological Diversity (CBD), entered into force 29 December 1993, www.cbd.int/doc/legal/cbd-un-en.pdf
2002	Bonn Guidelines on Access to Genetic Resources and Fair and Equitable Sharing of the Benefits Arising out of their Utilization (Bonn Guidelines, non-binding), www.cbd.int/doc/publications/cbd-bonn-gdls-en.pdf
2007	United Nations Declaration on the Rights of Indigenous People, www.un.org/esa/socdev/unpfii/documents/DRIPS_en.pdf

In this simplified example, the proposed course of action that will affect the community is the use of traditional knowledge by a bioprospecting company with the aim of developing a commercial product. A specified subgroup legitimately represents the community, which is the holder of the traditional knowledge, and can make competent decisions on its behalf. Usually, a third party that has no conflict of interest will provide adequate information to form the basis for this decision either to the community as a whole or to the legitimate representatives, who, in turn, will consult with the community. This third party could, for instance, be a representative of the national government, a member of a local NGO or an intermediary with good knowledge of the local context. The third party and the bioprospecting company will leave enough time for community deliberation.

At the end of the deliberation period, the legitimate subgroup, on behalf of the community, voluntarily accepts or rejects the proposed course of action without having been unduly influenced (e.g. no cash incentives other than expenses). To conclude the process, they sign a consent form. An additional benefit-sharing agreement might be signed at this stage, or the consent agreement might include a commitment to negotiate benefit sharing on mutually agreed terms[7] if something commercially viable emerges from the research. This simplified prior informed consent process is demanded by various international guidelines, the most important of which are listed in the table above (Table 3.2).

The definitions and diagrams above may be helpful for an initial overview, but they raise more questions than they answer. For instance, who is a sufficiently competent person? Who is a legitimate subgroup? How much and which information

---

[7] One might ask how benefit-sharing arrangements fit with the prerequisite not to unduly influence decision-makers. This is an unresolved issue, both in medical research when involving severely disadvantaged research subjects, for whom even an initial health assessment (as required for many research projects) could be regarded as undue inducement, and for marginalized communities for whom the commercialisation of traditional knowledge could be the only way, for instance, to obtain funds for their children's health care. In medical ethics, early answers are currently being developed. See, for instance, Ezekiel et al. (2005). In the context of access to traditional knowledge, this is an important gap in the research.

is adequate? How much time is required for deliberation? This chapter cannot fully answer these questions, but it can shed light on the components of consent processes. It will do so by discussing the four elements (free, prior, informed, consent) in reverse order.

3.2 Consent

In everyday language, 'consent' stands for the acceptance of, or agreement to, a particular course of action. For instance, the expression 'age of consent' refers to the minimum age at which a person can agree to engage in sexual acts. In the United Kingdom, until recently, the age of consent was 18 between men, but 16 between women and for heterosexual acts. This rule changed with the adoption of the Sexual Offences (Amendment) Act in 2000, when the ages were harmonized (16 for all groups). This meant that before 2000, 17-year-old men could not agree to have sex with other men without committing a criminal offence. They were not deemed mature enough to make such a decision.

This example highlights the main underlying prerequisite for successful consent. In the medical context, the process requires a sufficiently competent and autonomous person (Gillon 1985). In the context of traditional knowledge, it requires legitimate representatives who can make decisions on behalf of a group. What does this mean exactly? Let us look at the medical context first.

3.2.1 Consent and Health Care

In the twentieth century, *autonomia*, Greek for 'self-rule', entered Western medicine as one of its cornerstones. Patients have a right to be actively involved in medical decision-making and are no longer expected to defer that responsibility to paternalistically benevolent doctors. In particular, they have the final say in decisions concerning their own health care. A doctor can educate and inform her patients, but she cannot make decisions for them. This means that, for instance, a 19-year old patient in Norway with treatable leukaemia can refuse chemotherapy and neither his parents nor his doctors can overrule his decision. If he dies as a result of his refusal, it was his autonomous choice. 'Competent patients have the moral and legal claim to make their own decisions, and these decisions take precedence over those of the doctor or the family' (Pellegrino and Thomasma 1993:129).

Patient autonomy did not instantly materialize fully formed, ready to be applied by doctors in their informed consent procedures. The principle developed slowly, with the strongest impetus coming from horrendous medical experiments on humans (see Box 3.1) and an increasing awareness of cultural differences (see Box 3.2). The consent requirement for medical procedures recognizes that the

Box 3.1 Human Guinea Pigs[8]

There has been research involving human subjects since the end of the nineteenth century. However, the momentum for introducing formal international guidelines for its regulation only built up after the most horrendous experimentation on humans. Such experiments were exposed in many countries worldwide, but the cases most often cited are the Nazi experiments and the Tuskegee study.

Nazi experiments

During the Second World War, German doctors carried out at least 26 different types of experiments on prisoners in concentration camps, including the forced ingestion of seawater, exposure to extremely low water temperatures, the transplantation of limbs, injection with infectious bacteria and deliberate infection with typhoid (Loue 2000). At the same time, doctors carried out forced sterilizations (a practice that was legal in most European countries and the United States at the time) in order to maintain the 'superiority of the Aryan race' (Loue 2000:18). The so-called 'Doctors Trial' in Nuremberg, which preceded the adoption of the Nuremberg Code (see Table 3.1), ruled on 23 defendants, of whom 7 were acquitted, 7 received death sentences and the remainder were sentenced to imprisonment ranging from 10 years to life.

The Tuskegee study

This is the name given to what was, in the view of many commentators, the most infamous medical experiment carried out in United States history (Loue 2000). The study, conducted between 1932 and 1972, aimed to examine the natural history of untreated syphilis in black males. Initially, in the early 1930s, the treatment available for syphilis was highly toxic, and non-treatment was held to be a potentially useful alternative. The purpose of the study was to test this claim. By 1945, however, it was clear that syphilis could be treated effectively with penicillin. Yet the study continued for almost 30 years, depriving several 100 men, without their consent, of effective treatment for syphilis (Loue 2000). Since then, the Tuskegee experiment has become for many blacks, a symbol of racism, which seriously damaged their belief in medical authorities (Jones 1993:241).

---

[8] The title *Human Guinea Pigs* was chosen by M.H. Pappworth for a highly influential book published in 1967. Pappworth argued that medical experimentation without relevant consent was still conducted widely in the UK almost two decades after the Nuremberg trials.

> **Box 3.2 Consent and Cultural Differences**
>
> Humans cannot always understand or even be aware of their fellow humans' values and world views, even those most sacredly held. The best illustration of this is in an often cited example from the fifth century BC recorded by Herodotus (1995).
>
> When Darius was king, he summoned the Greeks who were with him and asked them for what price they would eat their fathers' dead bodies. They answered that they wouldn't do it for any amount of money. Then Darius summoned those Indians who are called Callatiae, who eat their parents, and asked them (the Greeks being present and understanding through interpreters what was said) what would make them willing to burn their fathers at death. The Indians cried aloud, that he should not speak of so horrible an act.
>
> What one tribe considered a respectful and moral way to treat their dead was regarded as horrendous by the other. Today, the most well-known case of value differences in health care occurs when a doctor's commitment to help and cure patients stands in direct conflict with a Jehovah's Witness's refusal to receive certain blood products. Such a refusal by a competent adult is normally respected, even if it leads to an early death (Beauchamp 2003). What is worth noting here is that it might never occur to a doctor who has not heard of Jehovah's Witnesses that somebody could reject a life-saving blood transfusion. This is where informed consent procedures are essential in diverse societies, to take account of values that would be diametrically opposed to certain treatments.

participant is the doctor's equal, that her set of values must be respected and that she must not be used to further somebody else's interests. For instance, even if a medical experiment conducted on Mr Miller would not be life-threatening to him and might provide a non-invasive cure for leukaemia, it cannot be undertaken without his consent. No individual must be used to further society's ends with regard to medical knowledge without agreeing to the procedure. Or, in the words of Campbell et al. (2001:224):

> The consent process recognizes that the participant is a person in his or her own right, with a set of personal values that must be respected. Research without consent constitutes an invasion of personal integrity of the research participants, even if no physical harm is caused to them.

Self-rule, or the ability to make independent, self-determined decisions, is not available to everybody. For instance, veterinary surgeons do not ask for consent from their patients. We do not expect dogs, pigs and hamsters to agree to any of their treatments. How then does a doctor determine whether the human patient he is talking to is a sufficiently competent and autonomous person to be able to consent to a proposed intervention? To answer this question, one has to distinguish between legal competency and capacity to consent.

In most Western countries, patients are legally competent to provide informed consent at the age of 18. However, this does not mean that they necessarily have the decision-making capacity to consent. Severely afflicted Alzheimer's patients who are well over 18 do not, for instance. Conversely, a 16-year-old may well have the maturity, intelligence, education and rational capacity to make autonomous decisions, but is not legally competent to provide informed consent. Hence a doctor first needs to establish whether a patient is legally competent to consent, and then has to determine whether that patient has the appropriate level of decision-making capacity. Alas, the latter is not without its difficulties. It would go beyond the scope of this chapter to describe mechanisms in detail. Suffice it to say that capacity is usually presumed and non-capacity determined on an individual basis (Loue 2000). For instance, doctors are entitled to presume that individuals have the appropriate level of decision-making capacity if they change their behaviour to adapt to new situations, talk comprehensibly, remember what they were told, respond in a meaningful way to questions, and have not been declared legally incompetent (Wear 1998). In borderline cases, formal assessment procedures are used, ranging from general intelligence tests such as the Wechsler Adult Intelligence Scale[9] to more specialized neuropsychological tests that assess abilities such as attention span and memory (Sullivan 2003). A doctor who has established that a patient does not have the decision-making capacity to be involved in an informed consent procedure needs to establish who the person's legal guardians are. Since proxy arrangements, as they are called, differ from country to country, they will not be outlined here.

To summarize: the ethical basis for consent in the medical field is the doctor's respect for patients' personal values and their related right of self-determination. It is the doctor's task to determine whether patients are able to make self-determined decisions. If they are not, they might either be legally incompetent to do so (e.g. a minor) or their decision-making powers might be impaired (e.g. an Alzheimer's patient).

3.3 Consent and Traditional Knowledge

Traditional knowledge has been defined as

> the knowledge, innovations and practices of indigenous and local communities around the world. Developed from experience gained over the centuries and adapted to the local culture and environment, traditional knowledge is transmitted orally from generation to generation. It tends to be collectively owned and takes the form of stories, songs, folklore, proverbs, cultural values, beliefs, rituals, community laws, local language, and agricultural practices, including the development of plant species and animal breeds. (CBD 1992b)

More recently, traditional knowledge has been described as

---

[9] The Wechsler Adult Intelligence Scale is a general test of intelligence published in 1955 and developed by David Wechsler from tests used to assess military personnel.

encompassing traditional and tradition-based literary, artistic or scientific works; performances; inventions; scientific discoveries; designs; marks, names and symbols; undisclosed information; and all other traditional and tradition-based innovations and creations resulting from intellectual activity in the industrial, scientific, literary or artistic fields. (WIPO 2001)

In the twenty-first century, anybody who wants to use traditional knowledge and promote its wider application, for instance through the development of pharmaceutical products, needs to ensure that knowledge holders have approved access and use. Article 8(j) of the CBD (1992a) prescribes the requirements as follows.

> Each Contracting Party shall ... [s]ubject to its national legislation, respect, preserve and maintain knowledge, innovations and practices of indigenous and local communities embodying traditional lifestyles relevant for the conservation and sustainable use of biological diversity and promote their wider application with the *approval and involvement* of the holders of such knowledge, innovations and practices and encourage the equitable sharing of the benefits arising from the utilization of such knowledge, innovations and practices .... (my emphasis)

Traditionally, the ethical basis for obtaining consent from indigenous communities before engaging in activities such as mining, logging or dam building was identical to that in the medical field. Indigenous communities have a right of self-determination as outlined, for instance, in the International Covenant on Economic, Social and Cultural Rights (ICESCR 1966). Article 1 of the covenant determines: 'All peoples have the right of self-determination. By virtue of that right they freely determine their political status and freely pursue their economic, social and cultural development.' In addition to the right of self-determination, the CBD adds another ethical foundation for obtaining consent before accessing and using traditional knowledge, namely the issue of fairness. The CBD's three main objectives are the conservation of biological diversity, the sustainable use of its components and the fair and equitable sharing of benefits from the use of genetic resources (Article 1). Complying with the latter objective involves mutual agreements between parties involved, i.e. individually negotiated contracts between individuals and/or groups. To obtain consent to access traditional knowledge is the first step in arriving at fair benefit-sharing agreements. A consent procedure acknowledges that the ownership of traditional knowledge rests with its traditional guardians and that fairness requires both permission from owners for access and subsequent compensation.

From an ethical perspective, therefore, there are two reasons why consent is essential when accessing traditional knowledge. The first is respect for indigenous communities' right to self-determination in the sense of freely pursuing their own economic and cultural development. Entering into benefit-sharing agreements as required by the CBD, for instance, can imply a commitment to the money economy, which is one way of pursuing economic development. By contrast, blocking access to traditional knowledge could be a protection mechanism for cultural development, if it is impossible to establish relationships of trust with bioprospecting companies. Or it could be a community reaction against the commercial use of knowledge, which may, for instance, be regarded as sacred to its holders. The second reason is respect for the commitment to fairness outlined in the CBD, which requires permission from owners and subsequent benefit-sharing arrangements (Page 2004).

Having established the ethical foundation for obtaining consent from indigenous communities, we must now ask from whom consent is obtained. In the case of experimental heart surgery, this is very easy to decide. The 'owner' of the research resource, to use analogous phrasing, is the patient. If this 'owner' cannot make legally valid decisions, the law prescribes very clearly who can do so on her behalf (e.g. her parents, partner or children). Hence, in all standard cases, the consentee (person giving consent) is straightforwardly identified. Not so in the case of traditional knowledge holders. Two main difficulties are usually raised. First, it can be difficult to distinguish traditional knowledge, as falling under CBD requirements, from other commonly held knowledge. For instance, it has been known for generations in southern Germany that wasp stings can be treated with fresh onions, cut in half and applied on the sting. This has not been developed into a pharmaceutical product. But if it were, I doubt that anybody would ask southern Germans for prior informed consent. Their knowledge would be regarded as in the public domain. Yet is this type of knowledge easily differentiated from the *Hoodia* knowledge? In other words, it is unclear when consent is required to access knowledge or when one might be able to use knowledge freely, removing the need to identify a consentee. Second, if it has been established that traditional knowledge is to be accessed, it can be difficult to find a legitimate organization or group able to provide consent.

Let me give another example for the first difficulty. If you have ever eaten olives straight from an olive tree, you will know (a) that this is not a good idea and (b) that olives as sold in shops have gone through some sort of processing. Consider now that I retire from my academic post to Tuscany, acquire a small estate and start my own olive production. I can talk to local growers, read about olive-growing and generally make myself familiar with the knowledge required to make olives palatable. And in the medium term I might even be able to apply for a trademark with my highly successful olive business. The knowledge will have originated from a group in the Mediterranean and will have made its way to other groups and finally into books, as it did in the case of the *Hoodia*.[10] This seems to fit the CBD definition of traditional knowledge: agricultural practices of local communities gained over the centuries and adapted to the local culture. But would I have to ask for consent from local Mediterranean communities before I can use olive-growing knowledge? Probably not. How traditional knowledge is marked out exactly as relevant to the CBD is thus unclear, as critics of the concept have pointed out (Schuklenk and Kleinsmidt 2006).

The second difficulty arises when a potential research group or company knows or assumes that it is dealing with traditional knowledge as falling under CBD requirements but does not know whom to approach for consent. The *Hoodia* case is famous for unawareness on a grand scale. The media reported that a beneficiary to the patent (for details on the *Hoodia* case, see Wynberg and Chennells, Chapter 6) had declared the San to be extinct (IRIN 2003). Hence there was allegedly no group to obtain consent from or negotiate a benefit-sharing deal with. Such erroneous public pronouncements are rare, but difficulties in establishing who the correct consentee is and how they could be approached are not (Perrault 2004).

---

[10] The earliest written record of *Hoodia* use was by botanist Francis Masson in 1796. Its appetite suppressant qualities were first recorded in 1936 (Wynberg 2004).

In an effort to operationalize CBD requirements, the Bonn Guidelines (CBD 2002) provide nations that are party to the CBD[11] with options on how to enable successful consent processes. The guidelines suggest that a national focal point be established to provide information on relevant laws and legislation as well as procedures for obtaining consent from indigenous communities (Perrault 2004). Communities should articulate their decision-making processes, presumably in writing, including how consent can be obtained (Perrault 2004). Information exchange between nations would be facilitated through a clearing-house mechanism.[12]

To summarize: the ethical basis for consent in the area of traditional knowledge is twofold, comprising, first, the communities' right of self-determination and, second, their right to fairness when their resources are being accessed and used. Those who seek access to traditional knowledge need to establish whom to contact – according to the Bonn Guidelines, with the help of national governments and through a national focal point.

At this point, our doctor and our bioprospecting company have found their respective consentees. This means we have dealt with the first element of the informed consent process as outlined in the introduction (capacity to consent). Now they need to transmit information to them, which leads us to the next section and to the second and third elements of the informed consent process (full disclosure and adequate comprehension).

3.4 *Informed* Consent

In order to facilitate a consent procedure, information must be transmitted. Only on the basis of adequate information can a decision be made on whether to accept or reject a proposed course of action. The process has two components. First, an adequate amount of information has to be disclosed. Second, the disclosed information must be comprehended by the recipient. How does this work in the health care sector and in the context of traditional knowledge?

3.4.1 *Informed Consent and Health Care*

Most of us try our best to keep out of surgeries and hospitals. This is not surprising. One of the strongest predictors for subjective human well-being is general health (Diener 2000), while the use of medical facilities is correlated negatively with well-being

---

[11] The only states that are not party to the CBD are Andorra, Somalia, the United States of America and the Vatican (Holy See).

[12] Traditionally, a clearing house was an institution that provided clearing services for financial transactions by helping member banks to add up reciprocal debits and credits and only settle net balances in cash. Today, a clearing house is mostly understood to be an institution that collects and distributes information. The latter applies to the CBD's clearing-house mechanism.

(Arrindell et al. 1999). In addition, some of us might approach the medical profession with doubts based on past misconduct (e.g. see the reaction to the Tuskegee case in Box 3.1). This sets the scene for a potentially very unequal relationship between doctor and patient. The patient is often physically weakened or mentally fragile and the doctor is seen as the main person whose beneficence can restore the patient's well-being. Let us bear the resulting power imbalance in mind as the background to the information disclosure process.

In most legislative jurisdictions, a doctor needs to disclose the following information to a patient prior to any intervention: the diagnosis, the recommended intervention and its alternatives, the intervention's risks and benefits, and the prognosis if the intervention is refused (Wear 1998; Loue 2000). In addition, several assurances must be given, the most important of which are, first, that refusal to participate will not alter the health care a person is receiving (Campbell et al. 2001) – patients have a right to say 'no' – and, second, that one can withdraw at any time without reprisal (Campbell et al. 2001).

The second element in our discussion of the items relevant to the informed consent process was full disclosure. (Just to remind you, the first was capacity to consent.) We have listed the main items that need to be disclosed to a patient before he can consent to a proposed course of action. The third element in the informed consent process is adequate comprehension. It is more difficult to achieve adequate comprehension in a situation of inbuilt power imbalances than in an equal relationship. For instance, patients might have an acquiescent attitude to a doctor (Andanda 2005; Chokshi et al. 2007) because of the hopes they invest in the doctor or their respect for the medical profession. (One commonly used German phrase for medical doctors is 'gods in white'.) As a result, patients might be reluctant to question details or ask for explanations of technical phrases used by the doctor. It is *essential* that information be disclosed in a language the patient is comfortable with and without technical jargon. In addition, sufficient opportunity for clarification and questions must be provided (Andanda 2005) to enable adequate comprehension.

Bearing this all in mind, the disclosure part of the informed consent procedure is usually concluded with the consentee's signature on a written consent form. But it needs to be remembered that the final signature should 'merely be the formal acknowledgement of a process whereby the person giving consent has come to a full understanding of what is entailed' (Campbell et al. 2001:223).

3.4.2 Informed Consent and Traditional Knowledge

Before going into the disclosure process in connection with accessing traditional knowledge, let us have a brief look at power balances. Indigenous populations are acutely affected by the legacy of colonialism. Although blatant imperialist expropriation of lands and resources may have come to an end, large-scale economic development projects continue to disregard indigenous values, interests and rights to participate.

In many instances, this has led to the perception among indigenous communities that development is a very negative concept (Daes 2001).

Given that the sustainable *use* of biodiversity as envisaged by the CBD has connotations of development, two potential sources for conflict or inequity become apparent: first, the possibility that attempts at gaining access to traditional knowledge represent a continuation of colonialist exploitation, with its inherent imbalances of power; and second, the perception that the terms of engagement regarding access to traditional knowledge were defined by remote governments and tailored for corporations and their large legal teams (Nakagawa 2004; Page 2004). One only has to look as far as the first *Hoodia* case to see an illustration of power imbalances and the *potential* for exploitation.[13]

South Africa's Council for Scientific and Industrial Research (CSIR), which identified the appetite suppressant qualities of the *Hoodia*, is the largest research and development agency in Africa, with a track record spanning more than 50 years. The CSIR's first *Hoodia* licensee, Pfizer, is the world's largest research-based pharmaceutical company. The second, Unilever, is one of the world's largest consumer goods companies, with over 400 brands (e.g. Dove, Ben & Jerry, Flora, Knorr). Their would-be addressees in informed consent processes following CBD requirements for the *Hoodia*, the San peoples, are among the most marginalized and impoverished populations in the world. Serious poverty among the San has led to endemic tuberculosis fuelled by malnutrition. Alcohol and cannabis (*dagga*) abuse is widespread, and education levels are extremely low (e.g. 1.1% of South African San receive post-school training or tertiary education) (Kollapen 2004). By comparison, the power imbalances between an average doctor and a seriously ill patient seem minuscule, so this is something that has to be borne in mind in relation to any disclosure procedures between bioprospectors and traditional communities.

What are the exact disclosure requirements for adding 'informed' to the consent procedure in the context of accessing traditional knowledge? The CBD's Ad Hoc Inter-Sessional Working Group on Article 8(j) did provide a list of information items that must be disclosed as part of an informed consent process. However, the list is mainly relevant to development activities such as mining and logging. It includes 'personnel likely to be involved in ... construction', 'the locality of areas that will be affected' and 'the duration of the ... construction phase' (CBD 2001).

Although the Bonn Guidelines' list of disclosure requirements focuses on access to genetic resources rather than traditional knowledge, it is helpful. It includes (CBD 2002):

- Legal entity and affiliation of the applicant and/or collector and contact person when the applicant is an institution
- Quantity of genetic resource to which access is sought
- Starting date and duration of the activity

---

[13] 'Exploitation' can be defined as the act of taking unfair advantage of another party to serve one's own interests, usually facilitated through power imbalances (see Wertheimer 1996; Macklin 2004).

- Geographical prospecting area
- Accurate information regarding intended use (e.g. taxonomy, collection, research, commercialization)
- Identification of where the research and development will take place
- Information on how the research and development is to be carried out
- Identification of local bodies for collaboration in research and development
- Possible third party involvement
- Purpose of the collection, research and expected results
- Kinds/types of benefits that could come from obtaining access to the resource
- Indication of benefit-sharing arrangements
- Budget
- Treatment of confidential information

Representatives of indigenous communities are not fully satisfied with the Bonn Guidelines because they emphasize cooperation with national authorities rather than indigenous groups. For instance, the Second International Indigenous Forum on Biodiversity (SAIIC 1997) identified as its main concern regarding the implementation of Article 8(j) of the CBD the 'lack of recognition of indigenous peoples as peoples with inalienable a priori rights and therefore as parties to the Convention and its implementation'. However, the guidelines emphasize that governments should obtain 'the consent of relevant stakeholders, such as indigenous and local communities, as appropriate to the circumstances and subject to domestic law' (CBD 2002). This phrasing has been criticized by indigenous representatives because the term 'stakeholder' fails to acknowledge that indigenous peoples are 'rights holders' in international law. And likewise the phrasing ignores the fact that their relevant rights are derived from international law and should therefore not be dependent on the national recognition of such rights (Perrault 2004).

A different list of disclosure requirements was compiled by the International Society of Ethnobiology in its code of ethics (ISE 2006):

- The full range of potential benefits (tangible and intangible) to the communities, researchers and any other parties involved
- The extent of reasonably foreseeable harms (tangible and intangible) to such communities
- All relevant affiliations of the individual(s) or organization(s) seeking to undertake the activities, including where appropriate the contact information of institutional research ethics boards and copies of ethics board approvals for research
- All sponsors of the individual(s) or organization(s) involved in the undertaking of the activities
- Any intent to commercialize outcomes of the activities, or foreseeable commercial potential that may be of interest to the parties involved in the project, and/or to third parties who may access project outcomes directly (e.g. by contacting researchers or communities) or indirectly (e.g. through the published literature)

And indigenous communities themselves have listed the following disclosure elements (summarized by Perrault 2004):

- Proposed objectives of the suggested activity
- Foreseeable consequences on the local community, including impact statements, for instance, impact on the environment
- The potential of the suggested activity for commercial applications
- Relevant legal and financial information, including affiliations and where funds will be sourced from
- Benefits to be shared with the community
- Previous or related activities undertaken by the same individual/research group/company

These requirements satisfy the second element in informed consent processes, namely full disclosure. But full disclosure does not mean that the relevant community can come to an informed decision. The information must be understood by the community and their representatives, our third procedural element. In this context, the seventh meeting of the Conference of the Parties to the CBD adopted the Akwé: Kon Voluntary Guidelines (CBD 2004). These prescribe that sufficient time needs to be allocated for communication with indigenous communities and that information exchange should take place using appropriate language and processes.

Given that consent processes in the medical field are older than those in relation to traditional knowledge, it is interesting to see what medical researchers do when they have to obtain community permission for research. The following outlines a process developed for a malaria vaccine study site by the Faculty of Medicine at the University of Bamako in Mali in collaboration with the School of Medicine of the University of Maryland in the United States. The process comprises six steps (Diallo et al. 2005) (Table 3.3).

1. A study of the community
2. An introductory meeting with leaders
3. A formal meeting with leaders
4. Personal visits to leaders
5. Meetings with traditional healers
6. Recognition that obtaining permission is a dynamic process

As in the medical field, informed consent processes for accessing traditional knowledge are concluded with signatures on formally agreed and written statements.

3.5 *Prior* Informed Consent

So far, I have outlined what *consent* implies and what information needs to be disclosed in an understandable manner to a patient or an indigenous community in an *informed* consent process. The term 'prior' was added to the phrase in the context of traditional knowledge only. A doctor or researcher simply does not turn around *after* conducting an experiment on a person and say: 'By the way, I used a new combination of antibiotics on you to reduce the overall duration of your tuberculosis

Table 3.3 Process for Obtaining Community Permission to Develop a Vaccine Study Site

Steps	Procedures	Resources involved
Study of the community	Elucidate community's socio-cultural structure	Nine months, principal investigator, medical anthropologist, two local guides, car with driver
Introductory meeting with leaders	Introduce research team, solicit best process for community permission	Meeting (2–3 h), principal investigator, director of district hospital, two local guides, car
Formal meeting with leaders	Explain research project in detail, take and respond to questions	Meeting (2–3 h), principal investigator, director of district hospital, two local guides, car, cost of broadcasting message on local radio
Personal visits to leaders	Visit leaders in their homes for opportunity to answer further questions	Meeting (30 min to 1 h), principal investigator, chief of centre for traditional medicine, one local guide, car with driver, gift per meeting
Meetings with traditional healers	Develop formal agreement with traditional healers for collaboration	meeting (30 min to 1 h), principal investigator, one local guide, car with driver, gift per meeting
Obtaining permission as a dynamic process	Conduct a modified consultation process at each modification in the protocol	Meeting (2–3 h), principal investigator, director of district hospital, two local guides, car, cost of broadcasting message on local radio

Source: Table 1, Diallo et al. (2005) (shortened)

treatment. Was that OK?' The term 'informed consent' is all one needs to express the requirement that consent must come before any intervention. The 'prior' is taken for granted in the medical field.

By contrast, in the context of accessing traditional knowledge, post-access attempts at obtaining approval are not infrequent. In the *Hoodia* case, international news coverage and interventions by local and international NGOs created pressure on the CSIR to acknowledge both that traditional knowledge of the San peoples had been used and that benefit-sharing negotiations would be entered into (Wynberg 2004). At this point, San representatives could have contested the patent that had already been filed by the CSIR and thereby expressed their refusal to share their traditional knowledge with the research institute. This would have been almost the opposite of prior informed consent, namely post-informed refusal (see Wynberg and Chennells, Chapter 6). But they decided to opt for the negotiation of a benefit-sharing agreement. At which point, then, should consent be obtained? What does 'prior' mean exactly?

According to MacKay (2004), consent must be sought *before* the commencement of activities by bioprospecting parties: that much is clear. But the question is: before what activities? In the *Hoodia* case, the CSIR first included the plant in a study in 1963. It was 32 years before the research institute filed for a patent (Wynberg 2004). The Bonn Guidelines (CBD 2002) include a two-sentence item on timing and deadlines.

3 Informed Consent: From Medical Research to Traditional Knowledge

Prior informed consent is to be sought adequately in advance to be meaningful both for those seeking and for those granting access. Decisions on applications for access to genetic resources should also be taken within a reasonable period of time.

Although this is not specific, it clearly means that consent must be sought before the first use of traditional knowledge. Hence, in the *Hoodia* case, the Bonn Guidelines would have required consent to be sought before the first experiment in 1963 – if the CBD had existed then, of course. However, how long before cannot be regulated inflexibly. The appropriate period depends on factors such as the number of people or communities concerned, the complexity of the suggested activity and the amount of information that needs to be disclosed. Important in all cases is to set clearly understood deadlines for the process (MacKay 2004). Complicating this recommendation about clear deadlines, an influential report by the World Bank (2003) focusing on extractive industries[14] noted that prior informed consent

should not be understood as a one-off, yes-no vote or as a veto power for a single person or group. Rather, it is a process by which indigenous peoples, local communities, government, and companies may come to mutual agreements in a forum that gives affected communities enough leverage to negotiate conditions under which they may proceed and an outcome leaving the community clearly better off. Companies have to make the offer attractive enough for host communities to prefer that the project happens and negotiate agreements on how the project can take place and therefore give the company a 'social license' to operate.

Similarly, a discussion paper by the World Conservation Union (2004) argued that the dialogue with indigenous communities 'should occur prior to, *and continue throughout*, the time that the activity is conducted' (my emphasis). Hence, these recommendations talk about obtaining consent as a *continuous* process rather than a single action with a clear end.

The 'prior' in the medical context is easily settled because of the immediacy of the relationship between doctor and patient and the long-standing and high degree of regulation. Setting unusual cases aside (e.g. a patient in a comatose state arriving in an emergency unit without a guardian), the doctor will always know whom to ask for consent and when to do it, as this procedure will have had to be outlined for local ethics committee approval. In the case of traditional knowledge, regulations are only just emerging (e.g. South Africa's Biodiversity Act entered into force in 2004) and non-compliance with the CBD and the Bonn Guidelines might occur through ignorance rather than intent. It is essential that the legal clarity prevailing in the medical context be replicated in the context of traditional knowledge, preferably at the international as well as national level.

We have now looked at three of four elements of prior informed consent procedure: first, the legitimate authority to consent; second, the full disclosure of relevant information; and third, the adequate comprehension of the disclosed information by the representatives. The last element is the *voluntariness* of any decisions.

---

[14] In contrast to extractive industries, such as mining and logging, bioprospecting usually has minimal environmental impact at the point of extraction. But since the issue of the timing of informed consent is very similar in both cases, it is worth looking at recommendations from this sector.

3.6 *Free* Prior Informed Consent

For many philosophers, 'free consent' is a pleonasm, like 'dead corpse' or 'unmarried bachelor'. 'Consent' means agreement and acceptance, thereby excluding coercion by definition. This explains why the term 'free' is not added to the phrase in the context of health care. In fact, any coercion or inducement to participate against one's better judgement would invalidate consent (Campbell et al. 2001).

Of course, coercion is not unknown in the medical field. On the contrary, undue inducement is a much debated concept (Wertheimer 1996; Ezekiel et al. 2005; Harris 2005). The fear of persuading patients to risk their health against their better judgement in order to obtain other benefits has led to an almost blanket ban on any payment in money or in kind for research participation, except for expenses (CIOMS 2002, Guideline 7).[15] The second area where guidelines try to avoid the potential for coercion is situational pressure, which can create a sense of obligation to take part in research or submit to a risky medical procedure. Such pressure usually arises when dependency relationships exist between doctors and patients. To avoid such situational pressure, a treating physician should not in person seek consent from a patient to enrol in a research study (Loue 2000). You may remember that the first diagram separated the information provider from the heart surgeon. That way our patient, who needs to decide whether she wants to allow experimental heart surgery, does not feel pressure to give her consent because she is grateful to her treating physician.

To summarize, coercion is avoided in the medical context through strict guidelines, and the term 'free' is not added to the phrase 'informed consent', as coerced consent would simply not be called consent.

Given the legacy of colonialism and the potentially serious power imbalances between bioprospecting companies and indigenous communities, an emphasis on the voluntariness of consent might be appropriate in relation to traditional knowledge, though. And adding 'free' to 'prior informed consent' might achieve that emphasis. *Free* prior informed consent is then understood as an agreement concluded without coercion or undue pressure. Yet it is important to note that neither the CBD nor the Bonn Guidelines talk about *free* prior informed consent. Of the main international documents on the rights of indigenous peoples, it is notably the recently adopted United Nations Declaration on the Rights of Indigenous Peoples (UN 2007) that adds this additional term. Article 10 uses it in relation to relocation.

> Indigenous peoples shall not be forcibly removed from their lands or territories. No relocation shall take place without the *free, prior and informed consent* of the indigenous peoples concerned and after agreement on just and fair compensation and, where possible, with the option of return.

---

[15] The gifts mentioned in Table 3.3 were given to community leaders who were not asked to enrol in research themselves.

Article 11(2) refers directly to consent for access to traditional knowledge.

> States shall provide redress through effective mechanisms, which may include restitution, developed in conjunction with indigenous peoples, with respect to their cultural, intellectual, religious and spiritual property taken without their *free, prior and informed consent* or in violation of their laws, traditions and customs.

In spirit, however, there is no difference between the terms 'free, prior informed consent' and 'prior informed consent'. Evidently neither allows coercion to influence the outcome. The former is mostly used in human rights law, the latter in biodiversity legislation.

Gifts, hospitality, bribery and coercion are not unfamiliar, in practice, in the exercise of obtaining consent from indigenous communities. For instance, in an attempt to 'ease' negotiations between indigenous communities and mining companies in the Philippines, all of the following have occurred: adding relatives of community elders to mining company payrolls; lavish hospitality in Manila nightclubs for decision-makers; offering money to secure support; and beatings of protestors by company security forces (Cariño 2005). In the terminology of the United Nations Declaration on the Rights of Indigenous Peoples, consent given after the above efforts would not have been given freely. In the terminology of the medical context, consent would not have been given at all; it would have been invalidated. The same applies when dependency relationships exist, and the adding of relatives to company payrolls is an example of both undue inducement and the creation of situational pressure. First, it provides a benefit that is likely to induce a decision-maker to accept a proposal that might otherwise be rejected (job and salary for a family member, probably in a situation of high unemployment and serious poverty). Second, by linking the family to the company, future dependency issues are created: for instance, if the company does not get a particular mining contract, jobs are at risk, so a negative decision by a community elder would have an immediate impact on the family.

Let me conclude this brief tour through consent in the medical field and how that relates to traditional knowledge with a few words on enforcement. Although the guidelines introduced for obtaining consent are international in outlook, enforcement mechanisms are not, either in the medical field or in the field of traditional knowledge. Unless a country has promulgated national laws with enforcement mechanisms, there is no legal recourse for local communities when their consent has not been obtained prior to their knowledge being accessed. In the San case, for instance, it was media pressure that led to benefit-sharing negotiations. However, there have been efforts at resolving this matter at the international level. In brief, these include the following proposals: a mandatory disclosure of origin of genetic resources and traditional knowledge in patent applications; a global treaty (Global Bio-Collection Society) to develop an international enforcement pyramid; and a sui generis system for traditional knowledge only, which separates it from the international intellectual property rights system. (For an excellent summary on current proposals, see Srinivas 2008).

3.7 Conclusion

Fifty years ago, the idea of shifting decision-making power in medicine from doctors to patients was a radical one. It seemed unthinkable that doctors could not be allowed to override patients' wishes if they assumed their own decisions would serve patients' interests better. Today it is presumed that one's own body is a sphere of autonomous decision-making allowing no interference from external parties, including family and doctors. In Western countries at least, this is fairly uncontroversial. Even though long debates can be had about how much information and choice a scared, seriously ill, potentially dying patient can cope with, doctors, on the whole, subscribe fully to the principle of patient autonomy. Experiments such as those carried out by the Nazis or in the Tuskegee study are unimaginable in Germany and the United States today. Patient autonomy and informed consent have won.

The same cannot be said for (free) prior informed consent in connection with accessing traditional knowledge. The illegitimate use of biodiversity and its related traditional knowledge is still frequent.[16] However, as this chapter has shown, the parallels are numerous.

First, the ethical basis for obtaining consent in the fields of medicine and traditional knowledge is the internationally accepted and legally binding right of patients and indigenous communities to self-determination. Strikingly, the ethical basis for requiring consent before accessing traditional knowledge is even broader than the equivalent basis in the medical field. In addition to self-determination rights, indigenous communities have a property-based right of fairness when their knowledge is being used.

Second, the inherent power imbalances between doctors and patients are mirrored and magnified when bioprospecting companies meet highly marginalized indigenous communities. It is essential that 'survival of the fittest' strategies (whoever undertakes an experimental procedure on this patient first, or whoever uses this knowledge to register a patent first, wins) be countered through clear legislation.

Third, the clear legislation and procedures in the medical field were not developed overnight. For instance, tests to establish adequate levels of decision-making competencies in patients are still being developed. And it took decades to establish the proxy-consent requirements in European countries that we have today. It is therefore not surprising that procedures for establishing who is a legitimate partner to negotiate with and how such negotiations can take place are not necessarily available in all countries.

It is encouraging that the belief in the ethical foundations of patient autonomy led to reliable and legally binding legislation on informed consent over time. The belief in the autonomy and right of self-determination of indigenous communities will, one hopes, achieve the same for (free) prior, informed consent in the accessing of traditional knowledge.

---

[16] See Third World Network reports on biopiracy at www.twnside.org.sg/access_7.htm.

References

Andanda, P. (2005). Module two: informed consent. *Developing World Bioethics, 5*(1), 14–29.

Arrindell, W. A., Heesink, J., & Feij, J. A. (1999). The satisfaction with life scale (SWLS): appraisal with 1700 healthy young adults in the Netherlands. *Personality and Individual Differences, 26*(5), 815–826.

Basel Convention (1992). Basel Convention on the Control of Transboundary Movements of Hazardous Wastes and Their Disposal. www.basel.int/. Accessed 15 August 2008.

Beauchamp, T. L. (2003). Methods and principles in biomedical ethics. *Journal of Medical Ethics, 29*, 269–274.

Campbell, A., Gillett, G., & Jones, G. (2001). *Medical ethics (3rd ed)*. Oxford: Oxford University Press.

Cariño, J. (2005). Indigenous peoples' right to free, prior, informed consent: reflections on concept and practice. *Arizona Journal of International and Comparative Law, 22*(1), 19–39.

Chokshi, D. A., Thera, M. A., Parker, M., Diakite, M., Makani, J., Kwiatkowski, D. P., et al. (2007). Valid consent for genomic epidemilogy in developing countries. *Public Library of Science Medicine, 4*(4), e95.

CBD (1992a). Convention on Biological Diversity. www.cbd.int/doc/legal/cbd-un-en.pdf. Accessed 15 August 2008.

CBD (1992b). *Traditional Knowledge and the Convention on Biological Diversity*, p. 1. www.cbd.int/doc/publications/8j-brochure-en.pdf. Accessed 15 August 2008.

CBD (2000). Article 8(j) and Related Provisions, Decision V/16, Decisions Adopted by the Conference of the Parties to the Convention on Biological Diversity at Its Fifth Meeting, Nairobi, 15–26 May, UNEP/CBD/COP/5/23. www.cbd.int/doc/decisions/COP-05-dec-en.pdf. Accessed 14 April 2008.

CBD (2001). Report of the Second Meeting of the Ad Hoc, Open-Ended, Inter-Sessional Working Group on Article 8(j) and Related Provisions of the Convention on Biological Diversity, UNEP/CBD/WG8J/2/6 ADD1. www.cbd.int/doc/meetings/tk/wg8j-02/official/wg8j-02-06-add1-en.doc. Accessed 14 August 2008.

CBD (2002). *Bonn Guidelines on Access to Genetic Resources and Fair and Equitable Sharing of the Benefits Arising out of Their Utilization*, Decision VI/24, 2002, Secretariat of the Convention on Biological Diversity, Quebec. www.cbd.int/doc/publications/cbd-bonn-gdls-en.pdf. Accessed 22 March 2008.

CBD (2004). *Akwé: Kon Voluntary Guidelines for the Conduct of Cultural, Environmental and Social Impact Assessments Regarding Developments Proposed to Take Place on, or Which are Likely to Impact on, Sacred Sites and on Lands and Waters Traditionally Occupied or Used By Indigenous and Local Communities*, Secretariat of the Convention on Biological Diversity, Quebec. www.cbd.int/doc/publications/akwe-brochure-en.pdf. Accessed 15 August 2008.

CIOMS (2002). International ethical guidelines for biomedical research involving human subjects, Council for International Organizations of Medical Sciences, Geneva. www.cioms.ch/frame_guidelines_nov_2002.htm. Accessed 15 August 2008.

Daes, E.I.A. (2001). *Prevention of discrimination and protection of indigenous peoples and minorities: indigenous people and their relationship to land*, E/CN.4/Sub.2/2001/21, United Nations Economic and Social Council. www.unhchr.ch/Huridocda/Huridoca.nsf/(Symbol)/E.CN.4.Sub.2.2001.21.En?Opendocument. Accessed 13 August 2008.

Diallo, D. A., Doumbo, O. K., Plowe, C. V., Wellems, T. E., Emanuel, E. J., & Hurst, S. A. (2005). Community permission for medical research in developing countries. *Clinical Infectious Diseases, 41*, 255–259.

Diener, E. (2000). Subjective well-being: the science of happiness and a proposal for a national index. *American Psychologist, 55*(1), 34–43.

Ezekiel, J. E., Currie, X. E., & Herman, A. (2005). Undue inducement in clinical research in developing countries: is it a worry? *Lancet, 366*, 336–340.

Gillon, R. (1985). *Philosophical medical ethics*. Chichester: Wiley.

Haira, A. (2006). *Prior informed consent an introduction.* www.med.govt.nz/templates/MultipageDocumentTOC____19320.aspx#P112_14121. Accessed 15 August 2008.

Harris, J. (2005). Scientific research is a moral duty. *Journal of Medical Ethics, 31*, 242–248.

Herodotus. (1995). *The histories III 38, vol 2.* Cambridge, MA: Harvard University Press.

ICESCR (1966). International Covenant on Economic, Social and Cultural Rights. www.unhchr.ch/html/menu3/b/a_cescr.htm. Accessed 15 August 2008.

ILO (1989). Convention (No. 169) Concerning Indigenous and Tribal Peoples in Independent Countries, International Labour Organization, Geneva. www.ilo.org/ilolex/cgi-lex/convde.pl?C169. Accessed 30 July 2008.

IRIN (2003). Marginalised San win royalties from diet drug. *Science in Africa.* www.scienceinafrica.co.za/2003/may/san.htm. Accessed 15 August 2008.

ISE (2006). *Code of ethics,* International Society of Ethnobiology. http://ise.arts.ubc.ca/_common/docs/ISECodeofEthics2006_000.pdf. Accessed 29 April 2008.

IUCN (2004). *Facilitating prior informed consent in the context of genetic resources and traditional knowledge,* World Conservation Union. http://pdf.wri.org/ref/perrault_04_facilitating.pdf. Accessed 15 August 2008 (parts also published as Perrault, 2004 – see below).

Jones, J. H. (1993). *Bad blood: the Tuskegee experiment.* New York: Free Press.

Kollapen, J. (2004). *Report on the inquiry into human rights violations in the Khomani San community.* Johannesburg: South African Human Rights Commission.

Loue, S. (2000). *Textbook of research ethics: theory and practice.* New York: Kluwer.

MacKay, F. (2004). *Indigenous peoples' right to free, prior and informed consent and the World Bank's extractive industries review,* Forest Peoples Programme.

Macklin, R. (2004). *Double standards in medical research in developing countries.* Cambridge: Cambridge University Press.

Nakagawa, M. (2004). Overview of prior informed consent from an international perspective. *Sustainable Development Law and Policy, Special Issue: Prior Informed Consent, 4*(2), 27–28.

NBAC (2001). Ethical and policy issues in international research: clinical trials in developing countries, National Bioethics Advisory Commission, Bethesda, MD. www.bioethics.gov/reports/past_commissions/nbac_international.pdf. Accessed 15 August 2008.

Nuremberg Code (1947). www.hhs.gov/ohrp/references/nurcode.htm. Accessed 15 August 2008.

Page, A. (2004). Indigenous peoples' free prior and informed consent in the inter-American human rights system. *Sustainable Development Law and Policy, Special Issue: Prior Informed Consent, 4*(2), 66–74.

Pappworth, M. H. (1967). *Human guinea pigs experimentation on man.* London: Routledge & K Paul.

Pellegrino, E. D., & Thomasma, D. C. (1993). *The virtues in medical practice.* Oxford: Oxford University Press.

Perrault, A. (2004). Facilitating prior informed consent in the context of genetic resources and traditional knowledge. *Sustainable Development Law and Policy, Special Issue: Prior Informed Consent, 4*(2), 27–28.

RC (2004) Rotterdam Convention on the Prior Informed Consent Procedure for Certain Hazardous Chemicals and Pesticides in International Trade. www.pic.int/. Accessed 15 August 2008.

SAIIC (1997). *Working document on the implementation of article 8(j) and related articles,* Second International Indigenous Forum on Biodiversity, Madrid, 20–23 November, South and Meso American Indian Rights Center. http://saiic.nativeweb.org/biodiv2.html. Accessed 29 April 2008.

Schuklenk, U., & Kleinsmidt, A. (2006). North-South benefit sharing arrangements in bio-prospecting and genetic research. *Developing World Bioethics, 6*(3), 135–143.

Srinivas, K. R. (2008). Traditional knowledge and intellectual property rights: a note on some issues, some solutions and some suggestions. *Asian Journal of WTO and International Health Law and Policy, 3*(1), 81–120.

Sullivan, K. A. (2003). Neurological assessment of mental capacity. *Neuropsychology Review, 14*(3), 131–142.

UN (2007). *United Nations Declaration on the Rights of Indigenous Peoples*, Adopted by United Nations General Assembly Resolution 61/295 on 13 September. www.un.org/esa/socdev/unpfii/documents/DRIPS_en.pdf. Accessed 29 April 2008.

Wear, S. (1998). *Informed consent: patient autonomy and clinician beneficence within health care.* Washington, DC: Georgetown University Press.

Wertheimer, A. (1996). *Exploitation*. Princeton, NJ: Princeton University Press.

WIPO (2001). *Survey on existing forms of intellectual property protection for traditional knowledge*, World Intellectual Property Organization, WIPO/GRTKF/IC/2/5. www.wipo.int/edocs/mdocs/tk/en/wipo_grtkf_ic_2/wipo_grtkf_ic_2_5.doc. Accessed 29 April 2008.

WMA (2002). Declaration of Helsinki, World Medical Association, Article 23. www.wma.net/e/policy/b3.htm. Accessed 14 August 2008.

World Bank (2003). *Striking a better balance: the World Bank Group and extractive industries*, Washington, DC: World Bank. http://siteresources.worldbank.org/INTOGMC/Resources/finaleirmanagementresponse.pdf. Accessed 14 August 2008.

Wynberg, R. (2004). Rhetoric, realism and benefit sharing. *Journal of World Intellectual Property*, 7, 851–876.

Chapter 4
Protecting the Rights of Indigenous Peoples: Can Prior Informed Consent Help?

Graham Dutfield[*]

Abstract This chapter assesses the meaning, origins and uses of prior informed consent and the assumptions underlying its application to traditional knowledge and biological resource transactions. It also deals with the complexities that need to be overcome before it can become a workable policy tool.

Using a case study approach, the chapter shows why applying prior informed consent requirements in very diverse and extremely different cultural settings, and in very tense political contexts, can be immensely challenging. Even with the best intentions and the most carefully drawn up plans, things go wrong. It also shows that the concept may in many cases be inapplicable because a great deal of knowledge and resources is already in free circulation and can no longer be attributed to a single originator community or country. This should not, however, lead us to conclude that there can be no moral obligations even in the absence of legal ones.

As a consequence of the manifold and complicated linkages between drug discovery and marketing, obtaining prior informed consent may do little to resolve biopiracy in its broadest sense. However, this is not to suggest that it is a useless concept. Indigenous peoples have a right to expect bioprospectors to request their consent formally. Still, obtaining prior informed consent is not a substitute for respect of basic human rights. Prior informed consent should be seen as a necessary but not a sufficient requirement for the establishment of more equitable bioprospecting arrangements – but only if it is acquired according to procedures that are effective, culturally appropriate, transparent and flexible.

Keywords benefit sharing • Convention on Biological Diversity • indigenous peoples • prior informed consent • traditional knowledge

---

G. Dutfield
School of Law, University of Leeds, 20 Lyddon Terrace, Leeds LS2 9JT, United Kingdom
e-mail: g.m.dutfield@leeds.ac.uk

---

[*] I am grateful to the editors for their comments on an earlier verson of this chapter. Any remaining shortcomings are the author's responsibility alone.

4.1 Introduction

This chapter explores the concept of prior informed consent as applied to the knowledge and biological resources of indigenous peoples. The discussion is confined mainly to transactions involving, on the one side, people to whom the International Labour Organisation's Convention 169 (ILO 1989) applies, namely 'indigenous and tribal peoples in independent countries',[1] and on the other, commercial, governmental and scientific entities. Accordingly, prior informed consent is covered in the context of consent being sought from members of groups very different from the seekers in terms of culture, world-view, expertise and power.

It is not self-evident that a concept originating in modern health care situations should have any relevance to the search for fairer ways to trade in traditional knowledge and associated biological resources. Nonetheless, prior informed consent has been central to much of the discussion on access and benefit sharing since the Convention on Biological Diversity (CBD) came into force. What are the concerns that led to the promotion of prior informed consent as a means of ensuring fairness in transactions involving traditional knowledge and biological resources found in developing countries, including those known about, used, and in many cases managed and improved, by indigenous peoples? And how have such concerns translated into law and policy? As this chapter explains, these concerns all tend to be regarded as aspects of 'biopiracy', a popular catch-all term that has focused attention on perceived inequities in the ways that the benefits of biodiversity-based commerce are distributed.

As to the development of responsive law and policy, this chapter shows that prior informed consent tends, as a consequence of the problem being seen through the 'lens' of biopiracy, to be linked to proposals to reform patent law in ways that are intended to make the patent system more transparent and fair, namely those relating to disclosure of origin (Chouchena-Rojas et al. 2005). The benefits of this approach for indigenous peoples are far from clear. Having said that, this need not be a cause for concern if prior informed consent for indigenous peoples continues to be seen as a fundamental right to which they are entitled, and not an issue to be pursued exclusively in the context of patent reform.

---

[1] ILO Convention 169 defines them as follows in article 1:

This Convention applies to:

(a) Tribal peoples in independent countries whose social, cultural and economic conditions distinguish them from other sections of the national community, and whose status is regulated wholly or partially by their own customs or traditions or by special laws or regulations
(b) Peoples in independent countries who are regarded as indigenous on account of their descent from the populations which inhabited the country, or a geographical region to which the country belongs, at the time of conquest or colonisation or the establishment of present state boundaries and who, irrespective of their legal status, retain some or all of their own social, economic, cultural and political institutions (ILO 1989)

It is, of course, one thing to demand that prior informed consent be central to the achievement of equitable relationships between business, government and universities on the one side and indigenous peoples on the other; it is quite another to put such a demand into practice.

In order to assist the process of effectively implementing prior informed consent, this chapter assesses the meaning, origins and uses of prior informed consent, and the assumptions underlying its application to traditional knowledge and biological resource transactions. The key assumption is that communities, or groups of communities, are bounded political entities with systems of governance that allow for direct and definitive negotiating and deal-making between indigenous groups and bioprospectors. This chapter also deals with the complexities that need to be overcome before prior informed consent can become a workable policy tool in the present context. Using a case study approach, the chapter shows why applying prior informed consent requirements in very diverse and extremely different cultural settings, and in very tense political contexts, can be immensely challenging. Even with the best intentions and the most carefully drawn-up plans, things go wrong, and misunderstanding, confusion, inappropriate exclusion, disappointment, resentment and even internal conflict can ensue. It also shows that prior informed consent may in many cases be inapplicable because a great deal of knowledge and resources is already in free circulation and can no longer be attributed to a single originator community or country. This should not, however, lead us to conclude there can be no moral obligations – even in the absence of legal ones.

I contend that as a consequence of the manifold and complicated linkages between drug discovery and marketing, the prior informed consent concept may do little to resolve biopiracy in its broadest sense. This is not to suggest that prior informed consent is not a useful concept. In my view, indigenous peoples have a right to expect that bioprospectors will formally request their prior informed consent. However, prior informed consent is not a substitute for respect of their basic human rights as individuals and as peoples. Prior informed consent should be seen as a necessary, but not sufficient, requirement for the establishment of more equitable bioprospecting arrangements – but only if it is acquired according to procedures that are effective, culturally appropriate, transparent and flexible.

4.2 'Biopiracy'

The vast majority of countries formally recognize that the cross-border exchange of genetic resources and traditional knowledge must be carried out in compliance with the principles of the CBD. For a number of reasons, intellectual property rights – particularly patents, but also plant variety protection – have become central to discussions on this matter. The reasons for the centrality of intellectual property rights relate to the following:

1. The conviction – widely held among developing countries and non-governmental organizations (NGOs) – that biodiversity and associated traditional knowledge have tremendous economic potential.
2. The fact that patent claims in various countries may incorporate biological and genetic material including life forms within their scope.
3. The belief, also shared by developing countries and NGOs, that this feature of the patent system enables corporations to misappropriate genetic resources and associated traditional knowledge, or at least to unfairly free-ride on them.[2]
4. The ability of modern intellectual property law to protect innovations produced by industries based mainly in the developed world, and its *in*ability to protect adequately those innovations with which developing countries are relatively well endowed.
5. The perception that as a consequence of reasons 2–4, the unequal distributions and concentrations of patent ownership and the unequal share of benefits obtained from the industrial use of biogenetic resources are closely related.

'Biopiracy' has emerged as a term to describe the ways that corporations from the developed world free-ride on the genetic resources and traditional knowledge and technologies of developing countries. While corporations complain about 'intellectual piracy' in developing countries, developing nations counter that their biological, scientific and cultural assets are being 'pirated' by these same businesses. 'Intellectual piracy' is a political term, and as such is inaccurate – and deliberately so. The assumption behind it is that the copying and selling of pharmaceuticals, music compact discs and films anywhere in the world is intellectual piracy, irrespective of whether the works in question have patent or copyright protection under domestic laws. In truth, if drugs cannot be patented in a certain country, their copying by local companies for the domestic market, or for overseas markets where the drugs in question are also not patented, is hardly piracy in the *legal* sense of the word.

Similarly, biopiracy is an imprecise term, But such 'strategic vagueness' is not a helpful approach for those working on legal solutions in such forms as national laws and regulations or international conventions.

Let us start by elucidating, as far as we can, the actual meaning of the word (see also, Schroeder, Chapter 2; Wynberg and Laird, Chapter 5; and Chennells and Vaalbooi, Chapter 11). To start with the obvious, 'biopiracy' is a compound word consisting of 'bio', which is an abbreviation for 'biological', and 'piracy'. According to the *Concise Oxford Dictionary*,[3] 'piracy' means the following: (1) the practice

---

[2] The distinction I seek to draw between misappropriation and unfair free-riding is that with misappropriation, there must be victims as well as beneficiaries for the word to apply. However free-riding is not necessarily harmful to anybody, and there is likely to be considerable disagreement about where to draw the line between fair and unfair free-riding.

[3] 10th revised edition, 2001.

or an act of robbery of ships at sea; (2) a similar practice or act in other forms, especially hijacking; and (3) the infringement of copyright.

Apart from being useful for its rhetorical effect, the word 'piracy' does not seem to be applicable to the kinds of act referred to as biopiracy. But let us now turn to the verb 'to pirate'. The two definitions given in the same dictionary are: (1) appropriate or reproduce (the work or ideas etc. of another) without permission for one's own benefit; and (2) plunder.

These definitions seem to be more appropriate, since inherent to the biopiracy rhetoric are misappropriation and theft. In essence, 'biopirates' are individuals and companies accused of one or both of the following acts: (a) the misappropriation of genetic resources or traditional knowledge through the patent system; and (b) the unauthorized collection for commercial ends of genetic resources or traditional knowledge. But since biopiracy is not just a matter of law, but also one of morality and fairness, we need to acknowledge that the line between an act of biopiracy and a legitimate practice may not always be easy to draw. This difficulty is compounded by the vagueness with which the term is applied.

To illustrate this point, a wide range of acts that have been considered acts of biopiracy of traditional knowledge are listed below (Dutfield 2005).

Collection and Use

- The unauthorized use of common traditional knowledge
- The unauthorized use of traditional knowledge only found among one indigenous group
- The unauthorized use of traditional knowledge acquired by deception or failure to fully disclose the commercial motive behind the acquisition
- The unauthorized use of traditional knowledge acquired on the basis of a transaction deemed to be exploitative
- The unauthorized use of traditional knowledge acquired on the basis of a conviction that all such transactions are inherently exploitative ('all bioprospecting is biopiracy')
- The commercial use of traditional knowledge on the basis of a literature search

Patenting

- A patent claiming traditional knowledge in the form in which it was acquired
- A patent covering a refinement of the traditional knowledger
- A patent covering an invention based on traditional knowledge *and* other modern or traditional knowledge

How much biopiracy actually goes on? This is by no means clear. Apart from the lack of information, the answer depends on how one differentiates between legitimate and unfair exploitation – a distinction that is not always obvious. The answer also depends on whether resources are considered to be wild and unowned or domesticated and owned. A common view among critics of conventional business practice is that companies may have a moral obligation to compensate communities providing genetic material for their intellectual contribution, even when such material is assumed to be 'wild'. Often genetic resources considered 'gifts of nature' are in fact the results of many generations of selective crop breeding and landscape management. Essentially the argument is that failing to recognize and compensate for the past and present intellectual contributions of traditional communities is a form of intellectual piracy.

The likely response from industry is that this is not piracy, since the present generation may have done little to develop or conserve these resources. The argument might continue that this is, at worst, a policy failure, and that measures outside the intellectual property rights system could be put into place to ensure that traditional communities are rewarded.

As for the patent-related version of 'biopiracy', there is little doubt that companies are in an advantageous position in the sense that, while a useful characteristic of a plant or animal may be well known to a traditional community, the community cannot obtain a patent even if it could afford to do so or could describe the phenomenon in the language of chemistry or molecular biology.[4] While it is unlikely that a company could then obtain a patent simply by describing the mode of action or the active compound,[5] it could claim a synthetic version of the compound or even a purified extract. In the absence of a contract or specific regulation, the company would have no requirement to compensate the communities concerned.

It is important to understand that the emergence of the prior informed consent concept is tightly linked to the emergence of the biopiracy concept and the concerns regarding biopiracy. Without biopiracy concerns, we would not be talking about prior informed consent, whether of developing country governments or of indigenous peoples. But if biopiracy is such a vague and elastic notion, how can prior informed consent do anything about it? That is a question that has not yet been answered. Arguably this is because very few people have even thought to pose the question in the first place!

4.3 The Concept of Prior Informed Consent

As is well known (see Schroeder, Chapter 3), informed consent has its origins in medical practice. Anybody undergoing medical treatment in a hospital, especially surgery, is likely to be requested to sign a consent form. Since consent should be informed, doctors and carers have responsibilities to their patients and research

---

[4] It may be able to do so if it can describe a specific formulation, even in fairly non-technical terms.
[5] In some circumstances this may be allowable under the US patent system.

subjects that may be legally enforceable. In the now famous John Moore case,[6] the failure of a doctor to fully disclose the commercial motivations of his interest in bodily substances donated by a former patient resulted in litigation in which informed consent came up as a key issue.

According to Moore, a former sufferer of hairy-cell leukaemia, he made repeated trips to hospital at the request of his former physician, Dr. David Golde, to give body fluid samples. He was told that these outpatient visits and extractions were necessary for his health and well-being. In fact, Golde had been aware when Moore was his patient that the latter's white blood cells (T lymphocytes) produced unusually large quantities of proteins called lymphokines, involved in immunity, giving them and the cells which produced them potential commercial value. So when Moore consented to having his spleen removed on the grounds that it was necessary to save his life, Golde had arranged to acquire parts of the spleen for reasons, as later became known, that were unrelated to Moore's medical care. During the period in which Moore was making return trips to the hospital and giving samples, Golde not only filed a patent application on a cell line comprising Moore's T lymphocytes, but also negotiated a lucrative financial deal with the private sector; all of this without telling Moore. Although the court controversially rejected Moore's claim that Golde had interfered with his personal property (i.e. the extracted cells), it did find that the doctor had breached a fiduciary duty he had to Moore, namely to obtain genuinely *informed* consent.

Informed consent in its original context therefore concerns information exchange between *individuals*. Informed consent turned out to be a useful concept in other contexts and consequently found its way into international environmental law. Thus the principle was incorporated into the Basel Convention on the Control of Transboundary Movements of Hazardous Wastes and Their Disposal, and then the actual expression 'prior informed consent' was used in the CBD. Article 15(5) of the latter agreement states: 'Access to genetic resources shall be subject to prior informed consent of the Contracting Party providing such resources, unless otherwise determined by that Party.' Clearly, the intention is to apply the prior informed consent principle not to individuals or non-state groups, but only to competent government agencies. Nonetheless, article 8(j) requires contracting parties, inter alia, to promote the wider application of traditional knowledge 'with the approval and involvement of the holders of such knowledge, innovations and practices'.

'Approval and involvement' are clearly not the same as prior informed consent. While 'approval' may be synonymous with 'consent', there is no explicit requirement that such approval must be based on the full disclosure of relevant information beforehand.

However, in response to lobbying from indigenous peoples' organizations and their supporters, at the Fifth Ordinary Meeting of the Conference of the Parties to

---

[6] *John Moore v Regents of the University of California*, 51 Cal. 3d 120; 271 Cal. Rptr. 146; 793 P.2d 479.

the Convention on Biological Diversity in Nairobi in 2000, a Decision V/16 (COP 2000) was adopted. This stated the following as a general principle:

> Access to traditional knowledge, innovations and practices of indigenous and local communities should be subject to prior informed consent or prior informed approval from the holders of such knowledge, innovations and practices.

But how can one operationalize the concept? In a 1996 book, the esteemed ethnoecologist and campaigner for indigenous peoples' rights, Darrell Posey and I came up with a working definition in the hope that it would contribute to effective implementation:

> Prior informed consent is consent to an activity that is given after receiving full disclosure regarding the reasons for the activity, the specific procedures the activity would entail, the potential risks involved, and the full implications that can realistically be foreseen. Prior informed consent implies the right to stop the activity from proceeding, and for it to be halted if it is already underway. The following types of activity should be subject to the prior informed consent condition:
>
> - Medical or other research carried out on a human body, whether or not it involves extraction of material, such as organs, body fluids, etc., and whether or not it is for commercial purposes;
> - Medical treatment, especially where it entails risk;
> - The extraction of biogenetic material and minerals from local communities or the territories of traditional communities, whether or not the communities have legal title to these lands;
> - The acquisition of knowledge from a person or people;
> - All projects affecting local communities, such as construction works, colonization schemes, and protected areas.
>
> Requests for consent should be accompanied by full disclosure of the following, in writing **in the local language**:
>
> - The purpose of the activity;
> - The identity of those carrying out the activity and its sponsors, if different;
> - The benefits for the people or person whose consent is being requested and for the sponsors;
> - The costs and disadvantages for the people whose consent is being requested;
> - Possible alternative activities and procedures;
> - Any risks entailed by the activity;
> - Discoveries made in the course of the activity that might affect the willingness of the people to continue to cooperate;
> - The destination of knowledge or material that is to be acquired, its ownership status, and the rights of local people to it once it has left the community;
> - Any commercial interest that the performers and sponsors have in the activity and in the knowledge or material acquired; and
> - The legal options available to the community if it refuses to allow the activity (Posey and Dutfield 1996).

Few efforts have been made since then to further elaborate the concept as a clear set of guidelines for well-intentioned scientists to follow. One shining exception to this is the International Society of Ethnobiology, which over a decade developed and adopted a code of ethics that was mainly drafted by Maui Solomon, an indigenous lawyer from New Zealand (ISE 2006).

4.4 Prior Informed Consent in Practice: Is It Workable? And What Good Can It Do Anyway?

4.4.1 The Peru International Cooperative Biodiversity Groups Project

The first case study, which is meant to address the first question ('Is prior informed consent workable?'), is a project titled Peruvian Medicinal Plant Sources of New Pharmaceuticals. It ran from 1994 to 2000 and was funded by four US government agencies under a programme known as the International Cooperative Biodiversity Groups (ICBG).[7] The funding agencies were the National Institutes of Health, the National Science Foundation and the US Department of Agriculture. The ICBG programme was intended to support the principles of the CBD while promoting the industrial use of biodiversity. As anthropologist Shane Greene explains, 'ICBG grants are based on a collaborative funding, research, and mutual-benefits relationship between U.S. and developing-country institutions, commercial partners, and, in a few cases, specific indigenous/local communities' (Greene 2004).

One of the awardees was Washington University in St Louis, Missouri, whose consortium included the Universidad Peruana Cayetano Heredia, the Museo de Historia Natural de la Universidad San Marcos, G. D. Searle & Co. (a pharmaceutical firm then part of Monsanto) and the Aguaruna people. The Aguaruna are a large Amazonian population, over 45,000 strong, living in more than 180 communities, most of which are affiliated to at least 13 organizations run by the Aguaruna alone or jointly with neighbouring ethnic groups. The primary aim of the project was to collect and study medicinal plants used by the Aguaruna for both scientific and commercial ends.

It soon became apparent that prior informed consent would be a major challenge. Initially, Walter Lewis of Washington University, who was the project leader, had identified the Organización Central de Comunidades Aguarunas del Alto Marañón (OCCAAM) as a potential partner organization. Presumably, the consent of this organization would have been taken to mean the consent of the whole Aguaruna people. However, once the grant was awarded, Lewis was apparently advised to approach a bigger and better-known organization, the Consejo Aguaruna Huambisa (CAH), which he did. A rather basic written agreement was made between the ICBG team and the CAH promising annual payments for plant collections and royalties. Having done this deal, Washington University negotiated a more formal arrangement with Searle according to which the university would receive the payments due from Searle and then pass on a share to the CAH. The latter organization was unhappy that such a separate deal had been made without their direct involvement, and they began to object. Once the two agreements were

---

[7] This case study draws heavily on Greene's (2004) work.

made public, Washington University and Searle found themselves condemned as biopirates for offering the Aguaruna too small a share of the proceeds, for keeping them out of the substantial negotiations and for not being transparent with them. One issue apparently overlooked at the time was that the CAH was hardly sufficiently representative to negotiate on behalf of the entire population of 45,000 plus Aguaruna anyway.

The following year, 1995, the CAH withdrew from the project, leaving the consortium without any Aguaruna representation. Washington University decided to approach OCCAAM, which turned out to be much more receptive, and a detailed written agreement was signed enabling OCCAAM to join the ICBG consortium. The CAH publicly condemned this, but OCCAAM came out in defence of the ICBG's activities with the support of two other Aguaruna organizations, which jointly rejected the representativeness of the CAH's leader.

At this point, a national indigenous peoples' confederation, to which OCCAAM and the other two organizations were affiliated, became involved. This was the Confederación de Nacionalidades Amazónicas del Perú (CONAP). CONAP organized a meeting of community and organization leaders including representatives of OCCAAM and several other Aguaruna organizations and one from a neighbouring group, the Huambisa (but not the CAH), and representatives of the ICBG consortium, including Searle and other interested individuals. The outcome was the formation of a consortium comprising CONAP and several Aguaruna organizations, and an agreement that three individuals should go to Searle's headquarters in St Louis to negotiate a contract directly.

Putting to one side the question of the extent to which the parties represented all Aguaruna people – in fact, they represented fewer than half of the Aguarana – did these organizations have sufficient legitimacy to give the consent of members of communities that *were* represented? Greene has this to say:

> Acceptance by CONAP and Affiliates [a consortium of Aguaruna organizations] of course did not automatically mean acceptance by all Aguaruna communities formally affiliated with those organizations. In many instances, individual communities challenged CONAP and Affiliates' authority to accept the project on their behalf and refused to permit the ICBG researchers to work in their communal territory despite their affiliation with one of the participating organizations. While there is not enough space to document all this local dissent, it is important that it be mentioned, since even the apparent incorporation of these organizations provoked substantial internal debate, discussion, and disagreement among Aguaruna community leaders (Greene 2004).

In the event, a contract was agreed including a know-how licence agreement. Members of the CONAP-led consortium formed the parties on the Aguaruna side. The agreement provided for payments for plant samples, and for licence fees to be paid as long as Searle used plant extracts accompanied by the collective medicinal know-how of the Aguaruna people. Also promised were milestone payments and royalties dependent on the research and development progress of discovered therapeutic agents acquired from these plants. Can this be right? Notwithstanding the formality of the agreement, one can argue that it was presumptuous of CONAP and its partners to license the collective know-how of *all* the Aguaruna people.

4 Protecting the Rights of Indigenous Peoples: Can Prior Informed Consent Help?   63

There was a separate biological collection agreement with the ICBG researchers. However, to meet the objection that the CONAP group was unfairly monopolizing the benefits of the unrepresented Aguaruna people, the agreement expressly committed the parties 'to ensur[ing] the fair and equitable sharing of benefits among the Aguaruna People' (Greene 2004, quoting article 3.01 of the agreement). As Greene explains, with reference to article 7:

> The agreement thus remains open to other Aguaruna communities, provided that they apply for inclusion by affiliating themselves with an existing Aguaruna organization, and to other Aguaruna organizations, provided that they are approved by CONAP and Affiliates in a traditional assembly and dialogue called the Ipaamamu that has become central to CONAP's strategy for dealing with the local constituency (Greene 2004).

This does not seem sufficient since '[t]he legal arrangement clearly moves in the direction of contractual and financial legitimation of CONAP and Affiliates as representatives of the "Aguaruna People"'(Greene 2004).

In the event, financial benefits were generated from the project. Most of these went to the two Peruvian universities involved. However, money did get channelled to the Aguaruna. Of the money from Searle, CONAP and its Aguaruna consortium used part of it to support themselves. The rest went 'to their affiliated communities in the form of small loans, scholarships for Aguaruna students, and individual reimbursement to field informants who worked with the ICBG researchers in identifying medicinal plants' (Greene 2004). Needless to say, the CAH was frozen out. However, it turned out that Searle, while keen to screen the plants, was not interested in the associated traditional knowledge. Its agreement with the Aguaruna was good public relations for the company, but contributed nothing scientifically to the company, which chose not to renew the agreement when it expired in 2000.

There are some very positive things to say about the novelty of the project. First, the direct negotiations between an indigenous group and a major pharmaceutical corporation were undoubtedly groundbreaking. Second, the know-how licence was unprecedented and had some significant implications in terms of knowledge ownership. As Brendan Tobin, legal counsel to the Aguaruna, explained to Greene, 'the know-how license is a truly novel step in contract law, for the first time giving a group of indigenous peoples control and full ownership of its traditional knowledge' (Greene 2004).

On the negative side, for all the efforts to put the prior informed consent concept into practice, the results were decidedly unsatisfactory levels of representation, sharp divisions among the Aguaruna, rather limited benefits and an abrupt end to the commercial relationship.

4.4.2   The Rosy Periwinkle

That leaves us with the second question posed above: what good can prior informed consent do anyway? In the 1950s the rosy periwinkle (*Catharanthus roseus*), a plant originally found in Madagascar, yielded two anti-cancer alkaloids, vincristine and vinblastine, which have generated huge profits for pharmaceutical giant Eli

Lilly since they came on the market around four decades ago. To some this is a classic case of biopiracy, with Madagascar and its people the unfortunate victims (Stone 1992). In fact, while the plant is thought to originate from Madagascar, it exists throughout the tropics and has grown in the Caribbean long enough to be considered a native plant there. It has been many years since the company relied on Madagascar for supplies of the plant, and most now come from plantations in Texas. The Eli Lilly researchers who discovered and patented vincristine[8] and its anti-cancer properties decided to study the plant when a literature search uncovered its use by rural populations in the Philippines. Those at the University of Western Ontario who discovered and patented vinblastine[9] received plant samples from Jamaica that were considered worth testing – again, because people used the plant for therapeutic purposes. In both countries the plant was used by rural communities to treat diabetes, not cancer.[10] Neither research team made any secret in their publications of the fact that they were inspired by traditional knowledge. On the other hand, only the University of Western Ontario team was reliant upon both overseas sources of plant material and unpublished ethnobotanical information when it began research on the periwinkle. Since then two further vinca compounds have come on the market: GlaxoSmithKline's vinorelbine and Eli Lilly's vindesine.

The rosy periwinkle case exemplifies the fact that portraying pharmaceutical development as a linear process taking place over a relatively short period[11] is a gross oversimplification with many if not most drugs. It also suggests that in many cases, a prior informed consent requirement is neither practicable nor, strictly speaking, applicable. It was only with vinblastine that ethnobiological information and plant samples were directly acquired from local people. In that case a prior informed consent requirement could have benefited the local healers and their communities, but not in the other cases.

4.5 Discussion

The Aguaruna case leaves one wondering how, and indeed whether, things could have been done better. Certainly, avoidable mistakes were made. But the case study casts serious doubt on whether the prior informed consent concept can always translate

---

[8] US Patent No 3,205,220 (issued 7 September 1965) ('Leurosidine and leurocristine and their production').

[9] US Patent No 3,097,137 (issued 9 July 1963) ('Vincaleukoblastine'). The patent was assigned by the inventors, Charles T. Beer, James H. Cutts and Robert L. Noble, to Canadian Patents and Development, Ltd., who made a deal with Eli Lilly allowing the latter company to commercially exploit the invention.

[10] As expressed by three medical researchers at the University of Western Ontario, 'the disease of cancer was certainly far from our thoughts when we learned of a tea made from the leaves of a West Indian shrub that was supposedly useful in the control of diabetes mellitus' (Noble et al. 1958).

[11] A common estimate of the average duration is 10–15 years from initial discovery to marketing.

successfully into very different social and cultural settings and produce the intended outcomes, even with the noblest intentions. This is especially the case where there is utter confusion about representation and indigenous governance structures.

To make matters even more difficult, the case study revealed that the company in question, G.D. Searle, was not interested in traditional knowledge in any direct sense, but preferred random screening of plants and found no compelling economic justification for maintaining a long-term relationship with the Aguaruna. Assuming that this is a common state of affairs, it is quite disappointing for those who back relationships between indigenous peoples and companies involving exchanges of knowledge and resources on the grounds that these can lead to the generation of substantial benefits for indigenous peoples. The potential may indeed be there, but it continues to be unproven.

In the rosy periwinkle case study, the question of moral obligations becomes salient. A tremendous sum of money has been made out of the vinca alkaloids over the past half century. Should we be concerned that Madagascar did not benefit from being the original habitat of the plant? Given the catastrophic loss of biodiversity on that biologically unique island and the abject poverty of most of its people, one can argue that the companies have some moral obligation to extend financial or other support for sustainable development there. But these are not, I believe, strong obligations, given the extensive dispersion of the rosy periwinkle over two centuries, and might fall into the realm of beneficence.

As for prior informed consent, we might draw a moral distinction between the case that could be made for the rural populations of the Philippines, who held relevant knowledge that became available through the scientific literature, and the one that could be made for the Jamaican rural dwellers, who had the same knowledge, which was acquired directly. But is such a distinction sustainable? And in attempting to make it, are we being unfair by applying our moral standards to those who operated at a time when nobody thought that what they were doing entailed any moral responsibilities?

In my view, such a distinction *is* sustainable. Academic scientists ought to reflect on the implications of publishing ethnobiological information. But once knowledge has been freely circulated, it falls out of anybody's control. We cannot then 'un-know' it any more than the proverbial genie can be put back in the lamp. However, an exception can be made when the knowledge is specifically attributable to a single group, community or locality. This is the case with the San and *Hoodia*. Thus, even though information about *Hoodia* had been published in the past, moral obligations for prior informed consent remained. As with the Jamaicans, in my view, the prior informed consent requirement was morally obligatory, as they had supplied plant samples to the developer in question.

As to whether it is fair to impose our moral standards on people of another time, I am inclined to the view that it is not. This is why I cannot condemn the behaviour of the University of Western Ontario researchers, even if I think that they ought to have done more.

Some will argue, justifiably, that I have cherry-picked my cases deliberately to put the prior informed consent 'solution' in a sceptical light. They are right; I have.

But I maintain that my reasons for doing so are sound. 'Methodological scepticism' is the only way to assess the true potential of prior informed consent. We need to get our doubts and objections out of the way first. In my view we still have not done that.

4.6 Conclusions

In the introduction to this chapter I stated its objective as being to assist the process of effectively implementing prior informed consent by assessing the meaning, origins and uses of prior informed consent, and the assumptions underlying its application to traditional knowledge and biological resource transactions.

Doing so comprehensively would require a whole book and not a mere chapter! But let me offer a few points to sum up. As I indicated, prior informed consent originates in health care. However, its application has been stretched quite radically, the apparent assumptions being, first, that it is a rather clear and obvious concept and, second, that what works for a patient and her doctor can also work for an indigenous group and their corporate visitors. My feeling is that this stretching has been done without sufficiently thinking through the practicalities, without much theoretical reflection and without the necessary consideration of political economy.

The case studies, in my view, support such concerns. Specifically they highlight, first, the point that we are still some way from developing workable prior informed consent procedures for equitable cross-cultural traditional knowledge transactions. Even the most sincere and painstakingly worked-out efforts to do prior informed consent right can lead to unfortunate unforeseen complications. As the poet Robert Burns reminds us, even '[t]he best laid schemes o' Mice an' Men/Gang aft agley'.[12] Second, emphasizing prior informed consent over other approaches may be unhelpful. Prior informed consent should be part of a broader regulatory framework; it should not be mistaken for the framework itself.

References

Chouchena-Rojas, M., Ruiz Muller, M., Vivas, D., & Winkler, S. (Eds.) (2005). *Disclosure requirements: ensuring mutual supportiveness between the WTO TRIPS agreement and the CBD.* IUCN, Gland, Switzerland, and Cambridge, UK; and ICTSD, Geneva, Switzerland.
COP (2000). Article 8(j) and Related Provisions, Decision V/16, Decisions Adopted by the Conference of the Parties to the Convention on Biological Diversity at Its Fifth Meeting, Nairobi, 15–26 May, UNEP/CBD/COP/5/23. www.cbd.int/doc/decisions/COP-05-dec-en.pdf. Accessed 11 July 2008.

---

[12] The best-laid schemes of mice and men often go awry.

Dutfield, G. (2005). What is biopiracy? In M. Bellot-Rojas & S. Bernier (Eds.), *International Expert Workshop on Access to Genetic Resources and Benefit Sharing: Record of Discussion*, Cuernavaca, Mexico, October 24–27, 2004, CONABIO and Environment Canada.

Greene, S. (2004). Indigenous people incorporated? Culture as politics, culture as property in pharmaceutical bioprospecting. *Current Anthropology, 45*(2), 211–238.

ILO (1989). Convention (No. 169) Concerning Indigenous and Tribal Peoples in Independent Countries, International Labour Organisation, Geneva. www.unhchr.ch/html/menu3/b/62.htm. Accessed 10 May 2008.

ISE (2006). Code of ethics, International Society of Ethnobiology, 8 November. http://ise.arts.ubc.ca/_common/docs/ISECodeofEthics2006_000.pdf. Accessed 29 November 2007.

Noble, R. L., Beer, C. T., & Cutts, J. H. (1958). Role of chance observation in chemotherapy: Vinca rosea. *Annals of the New York Academy of Sciences, 76*(3), 882–894.

Posey, D. A., & Dutfield, G. (1996). *Beyond Intellectual Property: Toward Traditional Resource Rights for Indigenous Peoples and Local Communities*. Ottawa: International Development Research Centre.

Stone, R. (1992). The Biodiversity Treaty: Pandora's box or fair deal? *Science, 256*(5064), 1142.

Chapter 5
Bioprospecting, Access and Benefit Sharing: Revisiting the 'Grand Bargain'

Rachel Wynberg and Sarah Laird

Abstract This chapter sets out the wider international context of bioprospecting, access and benefit sharing, and describes the fraught policy process that has evolved since the adoption of the Convention on Biological Diversity (CBD) in 1992. Notwithstanding the abundance of new policies and laws to control access to genetic resources and ensure fair benefit sharing, their effectiveness has been questionable. The complexity and diversity of bioprospecting activities and commercial players are often poorly recognized, and policy has lagged behind the practice of biprospecting. Moreover, the vast range of issues involved – from trade to conservation, intellectual property, biotechnology and traditional knowledge – has resulted in the policy process becoming a forum for much wider concerns dealing with globalization, corporate behaviour and the disparities between rich and poor.

Some of the key issues that remain unresolved in the run-up to finalizing an international regime on access and benefit sharing revolve around compliance, and whether or not patent holders should be obliged to disclose the origin of biological resources and knowledge in patent applications, the scope of the agreement, and whether or not it should go beyond the CBD to address biochemicals and derivatives. Expectations of what bioprospecting can deliver are unrealistic and overly optimistic and no 'grand bargain' has actually been possible.

Keywords access and benefit sharing • commercial use of biodiversity • Convention on Biological Diversity • policy research and development • technology transfer • traditional knowledge • A version of this article appeared in *Environment*, vol. 49, no. 10 (2007). Permission to reproduce parts of this article is gratefully acknowledged.

R. Wynberg (✉)
Environmental Evaluation Unit, University of Cape Town, Private Bag X3, Rondebosch 7701, Cape Town, South Africa
e-mail: rachel@iafrica.com

S. Laird
People and Plants International Incorporated, PO Box 73, Essex Junction, VT 05452, USA
e-mail: sarahlaird@aol.com

5.1 Introduction

Bioprospecting – the exploration of biological material for commercially valuable genetic and biochemical properties (Reid et al. 1993) – has sparked the public and policy imagination in recent decades. Located at the interface of leading genetic and information technologies, it promises a lot: new drugs to cure diseases; innovative cosmetic, food, plant and health care products; technology for developing countries; incentives to conserve biodiversity[1] in poor countries; and potentially rich rewards for those providing the biological material and knowledge.

In 1992, at the Earth Summit in Rio de Janeiro, countries negotiated an agreement, the Convention on Biological Diversity (CBD), which they anticipated would bring these benefits. But today both providers and users of genetic resources find themselves caught up in an environment characterized by misunderstanding, mistrust and regulatory confusion. Cries of 'biopiracy' abound from those concerned about the misappropriation of genetic resources and knowledge without the consent of traditional knowledge holders or countries of origin. Industry and scientists, on the other hand, vent frustration about the bureaucracies created by new regulations and perceived hurdles to research placed in their way by biodiversity-rich countries. Now scientists, industry, policymakers and traditional communities are negotiating anew in an attempt to develop an international regime for 'access and benefit sharing' – the term used to explain the way in which genetic resources are accessed and used – that many hope will resolve some of these intractable issues.

In today's hyperconnected world, this debate is especially significant. In 2007, Indonesia, which has had more human cases of avian flu than any other country, stopped sending samples of the H5N1 virus to the World Health Organization (WHO) on the grounds that it wanted a more equitable system of access to vaccines for developing countries (McNeil 2007). Although this decision was reversed after the WHO agreed to develop a new global mechanism for vaccine-sharing that would be fairer to poorer nations (WHO 2007), the case catapulted access to genetic resources and benefit sharing onto the global agenda. How can we make sure that the biological riches of the earth remain accessible for scientific exploration and research while ensuring that their commercial development yields benefits that are distributed fairly and equitably (in other words, achieve access and benefit sharing)? How can a balance be struck between conducting and regulating ethical science? How does the increasing privatization of biodiversity affect food and health security? And who has the right to own innovations in biological resources: the countries from which those resources originate, traditional knowledge holders and/or the companies that develop these resources and this knowledge into products? An evolving policy process has been seeking to address these questions since the Earth Summit, but it is still grappling with many difficult issues today.

---

[1] 'Biodiversity' here refers to the number and variety of living organisms on earth.

5.2 Bioprospecting through History

Bioprospecting is typically associated with the contemporary exploration and development of biodiversity using sophisticated technologies in research-intensive industries. However, the practice of collecting, analyzing and commercializing biological material is as old as human civilization. One of the earliest recorded plant collection expeditions took place in 1495 BC, when Queen Hatshepsut of Egypt sent a team to the land of Punt (in the vicinity of modern-day Ethiopia, Eritrea and north-eastern Sudan) to obtain species of *Boswellia*, a plant whose fragrant resins produced frankincense (Juma 1989). Similar expeditions occurred elsewhere in the world, involving the collection of trees, figs, vines, roses and citrus fruits. These expeditions were based on an explicit understanding of the economic value of the plants encountered and their close links with scientific enquiry and economic growth. Indeed, plant transfers were central to the economic and scientific development of Europe and North America, helped drive the expansion of colonial empires and continue to contribute to economic disparities between countries today (Crosby, 1972, 1986; Juma 1989).

The arrival of Christopher Columbus in the New World in 1492, the connection of the Old and New Worlds, and the transfer of biological material between these worlds profoundly changed the scale, nature and political significance of exchanges – and also the course of human development – by initiating an expansive trade in vast numbers of new plant species (Crosby 1986; Headrick 1990). In eighteenth-century Europe, plant transfers between countries became a carefully planned and strategic activity subsidized by governments through botanical gardens, with the intention to investigate plants of potential use (Juma 1989; Headrick 1990). Professional scientists were an integral part of the new trading companies (Grove 1996), and botanical gardens would not only collect and classify plants, but also develop them for agricultural purposes.

By the early twentieth century virtually all the world's primary arable lands were under cultivation, and new discoveries of commercially useful wild plants for agriculture had almost ceased. Research and development focused on enhancing the productivity of a few, familiar species and on developing new inputs and technologies to increase agricultural production (Tuxill 1999). In the field of medicine, however, interest in products derived from plants, microorganisms and other natural sources waxed and waned throughout the twentieth century in response to scientific and technological developments (Balick and Cox 1997).

In recent decades, wild species and the genetic resources they contain have become of increasing interest as leads towards new types of foods, medicines, ornamental plants and other useful products. For some companies – including many in the food, botanical medicine, and personal care and cosmetics sectors – this interest has been driven by consumer demand for all things 'natural'. For others – including pharmaceutical, biotechnology and seed companies – developments in science and technology make it possible to study and use genetic resources in ways previously unimagined (Koehn and Carter 2005; Rubenstein et al. 2005; Smolders 2005).

Today, 'genome mining' of even well-known species is an important new approach to natural-product drug discovery (McAlpine et al. 2005). Although government research institutes undertake important research on biodiversity, the wide range of private companies involved in bioprospecting have largely taken over the role historically occupied by the state in pursuit of new economic opportunities from biodiversity.

5.3 New Regulatory Frameworks for Bioprospecting

By the late 1980s, it was clear that with scientific and technological advances, genetic resources were a valuable starting point for research and development in extremely profitable industries. At the same time, the rights of companies to claim ownership over innovations related to biodiversity were expanding, alongside global intellectual property rights systems for agriculture, food and health care introduced through the TRIPS (Trade-Related Aspects of Intellectual Property Rights) Agreement of the World Trade Organization (WTO). In particular, Article 27.3(b) of TRIPS required intellectual property protection for microorganisms, non-biological and microbiological processes, and plant varieties (Dutfield 2002) .The growing integration of the global economy, the steady increase in size and importance of a few multinational corporations positioning themselves as 'life science' giants and the rapidly expanding biotechnology sector led to fundamental changes in the way in which biodiversity was used and developed (ten Kate and Laird 1999).

This period also witnessed an escalation in global concern about the loss of biodiversity. Public attention focused on the threats to biodiversity and the vanishing 'medicinal riches' of the rainforest, and international negotiations commenced to set in place a treaty to conserve biodiversity. Using their leverage as the main repositories of biodiversity, the biologically rich countries of the developing world argued that in order to allow companies access to their biodiversity – and indeed to justify the conservation of economically important biological resources in developing countries – the technologically rich developed world should transfer technology and share benefits from biodiversity commercialization (Sanchez and Juma 1994; Macilwain 1998). This was considered especially crucial given the historical accrual by colonial powers and companies from the North of benefits derived from the commercialization of resources from the South.

In what has been described as the 'grand bargain' (Gollin 1993), the 1992 Convention on Biological Diversity laid down a new way of treating trade in genetic resources and regulating bioprospecting: to gain access to genetic resources, users needed to provide fair and equitable benefits to the provider country, including technology transfer; and to receive such benefits, a provider country needed to facilitate access to genetic resources (hence 'access and benefit sharing') (ten Kate and Laird 1999; CBD 2002); Svarstad and Dhillion 2004). What this meant in practice was that companies and signatory countries now had an obligation to get permission before collecting resources and knowledge (prior informed consent), agree on the terms for exchange (mutually agreed terms) and share benefits fairly with local providers and countries (fair and equitable benefit sharing).

The CBD thus represented a fundamental change in the way in which genetic resources were exchanged and viewed: no longer were they seen as the 'common heritage' of humankind, but instead countries now increasingly asserted sovereign rights over their biological resources and control over their access.

In a series of deliberations among parties to the CBD, these concepts were further elaborated and refined through adoption of the voluntary *Bonn Guidelines on Access to Genetic Resources and Fair and Equitable Sharing of the Benefits Arising Out of their Utilization* in 2002 (CBD 2002). The primary intention of the guidelines was to assist governments to develop an access and benefit-sharing strategy, as well as the necessary legal, administrative or policy measures.

Accompanied by a capacity-building programme and a suite of donor-funded projects, the access and benefit-sharing provisions of the CBD led to a variety of initiatives to develop national legislation and appropriate standards: at least 58 countries are in the process of developing, or have already adopted, access and benefit-sharing measures, and a number of regions have set out approaches for access and benefit sharing (CBD 2007a).

Despite this apparent progress, most countries have failed to implement their obligations under the CBD and the Bonn Guidelines, and opinions are mixed as to the efficacy of those regulatory measures that have been adopted (Laird and Wynberg 2006). On the one hand, the procedures being put in place are perceived to be too restrictive, but on the other there is a belief that such measures are insufficient to curb the misappropriation of resources and knowledge. At the World Summit on Sustainable Development in 2002, developing country governments – and particularly the so-called Like-Minded Megadiverse Countries[2] – pushed jointly for a legally binding international regime on access and benefit sharing in relation to biological resources and traditional knowledge. The Johannesburg Plan of Implementation adopted by the World Summit on Sustainable Development required action to negotiate 'an international regime to promote and safeguard the fair and equitable sharing of benefits arising out of the utilization of genetic resources' (WSSD 2002). Such negotiations have been going on since 2003 under the auspices of the CBD's Ad-Hoc Open-Ended Working Group on Access and Benefit Sharing, but little progress has been made on either the scope or objectives of the new agreement.

---

[2] The group of Like-Minded Megadiverse Countries comprises Bolivia, Brazil, China, Colombia, Costa Rica, Ecuador, the Philippines, India, Indonesia, Kenya, Malaysia, Mexico, Peru, South Africa and Venezuela, representing 70% of the earth's biodiversity. The group was formally constituted through the Cancun Declaration of 18 February 2002 as a 'consultation and cooperation mechanism' to promote common interests and priorities related to the conservation and sustainable use of biodiversity. The development of an international regime to promote and safeguard the fair and equitable sharing of benefits arising out of the utilization of genetic resources has been adopted by the group in its action plan as one of five areas of priority and action (see also http://lmmc.nic.in/).

5.4 The Commercial Use of Biodiversity

Some of the reasons for this intractability lie in an inherent conflict in views between the 'biodiversity-rich' developing country providers and 'technology-rich' developed country users of biodiversity. Developing countries are resentful of centuries of colonialism and uncompensated export of genetic material and traditional knowledge, and want to address these injustices and prevent further misappropriation (Sanchez and Juma 1994). The world view of developed countries, by contrast, is to seek unimpeded access to genetic resources within a softer legal framework of corporate social responsibility and contractual agreements for benefit sharing.

Confusion also results from the complexity and diversity of the activities and players involved in the commercial use of biodiversity, the divergent objectives of each and the fact that many participating in the policy process do not fully understand the sectors they seek to regulate (Laird and Wynberg 2008).

For example, the pharmaceutical, biotechnology, seed, crop protection, horticulture, cosmetics and personal care, fragrance and flavour, botanicals, and food and beverage industries all undertake research and develop commercial products from genetic resources. Each of these sectors has unique markets, undertakes research and development in distinct ways, and uses genetic resources and demands access to these resources very differently. Drug discovery and development typically take more than 10 years, for example. Only very rarely will an individual compound result in a commercial product, and the cost could be in excess of US$800 million (PhRMA 2007). At the same time, blockbuster drugs can generate over a billion dollars in sales a year for large multinational companies.

The cyclical nature of industry interest in natural products is also significant. The recent surge of interest in natural products, for instance, is driven both by failures in alternative approaches like combinatorial chemistry, which involves the rapid synthesis or computer simulation of a large number of different but structurally related molecules, and scientific and technological developments that allow researchers to better study natural products already in their collections (Cragg et al. 2005; Koehn and Carter 2005; Handelsman 2005). Similarly, advances in DNA extraction technology have made available 99% of the microbial diversity previously inaccessible through traditional cultures and have led to a heightened interest in the economic potential of microorganisms (Handelsman 2005; McAlpine et al. 2005).

In contrast, botanical medicine companies, which produce natural medicines directly from whole plant material, work intensively on a handful of carefully selected species and might take just a few years to develop a product, the annual sales of which will likely not exceed a few million dollars. The industry as a whole is also much smaller than the pharmaceutical industry, with the annual US market for all botanical products not much bigger than the sale of a few blockbuster pharmaceuticals (*Nutrition Business Journal*, 2003). As with the personal care and cosmetics, food and beverage, and horticulture industries, botanicals are less research-intensive than the pharmaceutical and biotechnology sectors. They also tend to generate a far larger number of commercial products with significantly smaller markets than the pharmaceutical and biotechnology industries, which produce smaller numbers of high-value products.

The US$54.6 billion (Ernst and Young 2006) biotechnology industry is in itself a study in diversity. It is made up of industrial, agricultural and health care biotechnology companies that range in size and scope from those that are small, dedicated and research-intensive to large, diversified ones that have greater in-house resources. The ways in which biotechnology companies use genetic resources vary significantly. For example, some develop speciality enzymes, enhanced genes or small molecules for use in crop protection and drug development; others develop enzymes that act as biological catalysts in the production of polymers and speciality chemicals or for use in industrial processing; and others might insert genes that impart desirable traits to crops. Biotechnology companies have a particular interest in the astounding biochemical diversity found in genetic resources from diverse and extreme environments and ecological niches (for example, salt lakes, deserts, caves, hydrothermal vents and cold seeps in the deep seabed) as well as areas with microbial diversity associated with endemic flora and fauna (Lange 2004; Arico and Salpin 2005).

While the sectors and companies that demand access to genetic resources are clearly diverse, the nature of demand for access is also constantly changing in response to markets, laws, and scientific and technological advances. For example, in the seed industry, there has been reduced demand for wild genetic resources and greater reliance on *ex-situ* and private collections. However, demand for wild material continues to meet consumer pressures to reduce the use of chemicals and vulnerability to pests and diseases (Rubenstein et al. 2005; Laird and Wynberg 2008). Similarly, the ornamental horticulture industry has a low dependence on wild genetic resources, but some companies continue to hunt for wild material with a view to introducing novel ornamental species or providing new variations of colour and other character traits (Laird and Wynberg 2008). As described earlier, technological advances in the pharmaceutical and biotechnology industry have stimulated renewed interest in natural products, but have also made it possible to look anew at what is found in companies' 'backyards'.

Box 5.1 Regulating the Protection and Commercial Use of Traditional Knowledge

The commercial use of traditional knowledge raises a range of complex issues. For example, is all knowledge, including that which is widely known, subject to access and benefit-sharing regulations? Who should provide prior informed consent, enter into a benefit-sharing agreement and receive benefits? How are the owners of traditional knowledge identified? What if knowledge is shared by a number of communities? And, as Saskia Vermeylen asks (Chapter 10, this volume), how do concerns and conflicts about the commodification of traditional knowledge get addressed?

(continued)

Box 5.1 (continued)

Within a suite of global instruments and institutions, negotiated texts and processes have evolved to address these concerns, primarily the Convention on Biological Diversity (CBD), the United Nations Permanent Forum on Indigenous Issues and the World Intellectual Property Organization (WIPO).

Through Article 8(j), the CBD requires member parties to 'respect, preserve and maintain' the biodiversity-related knowledge, innovations and practices of indigenous peoples and local communities'. It also establishes that the 'wider application' of this knowledge should be promoted with the 'approval and involvement of the holders of such knowledge'. The CBD also encourages the equitable sharing of benefits derived from the use of knowledge, innovations and practices related to the conservation or sustainable use of biodiversity (CBD 2000). These principles are taken further in the 2002 Bonn Guidelines, which aim 'to contribute to the development by Parties of mechanisms and access and benefit-sharing regimes that recognize the protection of traditional knowledge, innovations and practices of indigenous and local communities, in accordance with domestic laws and relevant international instruments' (CBD 2002). An Ad Hoc Open-ended Working Group on Article 8(j) and Related Provisions provides advice on the protection of traditional knowledge, by legal and other means, and is undertaking work to identify priority elements of *sui generis* (unique) systems for traditional knowledge protection, fair benefit sharing and prior informed consent. The recently adopted United Nations Declaration on the Rights of Indigenous Peoples is another important instrument in support of indigenous peoples' rights over their biodiversity-related traditional knowledge, stating that:

> Indigenous peoples have the right to maintain, control, protect and develop their … traditional knowledge and … the manifestations of their sciences, technologies and cultures, including genetic resources, seeds, medicines … [and] knowledge of the properties of fauna and flora. … They also have the right to maintain, control, protect and develop their intellectual property over such cultural heritage, traditional knowledge, and traditional cultural expressions. (UN 2007, Article 31.1)

Traditional knowledge is also a matter increasingly under consideration in relation to the Agreement on Trade Related Aspects of Intellectual Property Rights (TRIPS) of the World Trade Organization. A proposed amendment to TRIPS would bring it in line with obligations under the CBD, adding a requirement for disclosure of origin in patent applications and possibly requiring benefit sharing with communities to deter biopiracy.

Intellectual property rights issues in genetic resources also figure predominantly in the mandate of WIPO, which has set up an Intergovernmental Committee on Traditional Knowledge, Genetic Resources and Folklore (IGC). The IGC gives countries guidance, based on research and the work of fact-finding missions, on strategies for the protection of traditional knowledge and genetic resources.

(continued)

Box 5.1 (continued)

Some of the measures being adopted include the development of biodiversity registers or databases that record biodiversity use and knowledge in particular regions. These defensive methods of protection of traditional knowledge may be complemented by the legal recognition of collective ownership of resources and knowledge, co-ownership of patents and products, and certificates of prior informed consent, benefit sharing and/or origin of the resource or knowledge in patent applications.

In practice, however, many of these tools and approaches are still in their early stages and present significant challenges. Many companies have therefore adopted a hands-off approach to the use of traditional knowledge, while others have little awareness of the need to enter into access and benefit-sharing arrangements when using traditional knowledge. The diverse ways in which companies use and interpret traditional knowledge adds a further layer of complexity. In cases where traditional knowledge is used, companies typically rely heavily on intermediary institutions such as research institutions, NGOs or governments to resolve difficult issues. The intractable nature of many of these issues means that projects involving traditional knowledge are often inherently controversial.

5.5 Perceptions of Access and Benefit Sharing

Complexity, change and diversity dominate the field of bioprospecting. Understanding these distinct challenges and acquiring knowledge of the market, legal, scientific and technical realities of bioprospecting are vital for effective regulation. In part because policymakers have been unable or unwilling to acquire this expertise, policy has typically lagged behind both the science and the practice of bioprospecting. As a result, regulatory frameworks seldom reflect the reality of bioprospecting, and measures have been poorly formulated and implemented. In fact, the objectives that law and policy on access and benefit sharing are intended to serve – equitable benefit sharing, biodiversity conservation, the promotion of domestic biodiversity research and technology transfer – are rarely achieved by these measures.

Industry and researcher perceptions of the CBD, and of access and benefit sharing in particular, have become increasingly negative in the past decade, with companies often loath to access genetic resources or undertake partnerships in more than a handful of what they consider 'safe' countries with strong institutions and relatively clear approaches to access and benefit sharing that provide 'legal certainty' (ten Kate and Laird 1999; Laird and Wynberg 2006). Academic researchers have also expressed serious concerns about the impact of access and benefit-sharing requirements on basic science and traditions of trust and collaboration (ten Kate and Laird 1999; Laird and Wynberg 2006).

As scientific and technological developments have dramatically improved our ability to understand and use genetic and biochemical resources, the availability of organisms for research has diminished, sometimes in countries with extremely threatened ecosystems where the future of these organisms is uncertain. In Brazil, for example, the arrest and imprisonment of Marc van Roosmalen, a renowned Dutch primatologist accused of collecting samples in the Amazon without permission, led to outrage among the scientific community and a concern that unjustly severe limitations were being introduced at the expense of basic biological research (Rohter 2007).

At the other end of the spectrum are increased concerns about 'biopirates', a term that has evolved to describe the ways in which corporations claim ownership of, or misappropriate, the genetic resources and traditional knowledge and technologies of developing countries without consent or compensation (Dutfield 2004; Zedan 2005; Dutfield, Chapter 4). Although the term has a multitude of interpretations, it also represents the view held by many developing countries, civil society organizations and indigenous peoples that the intellectual property rights system is inimical to traditional knowledge protection and just reparation (CIPR 2002; Mgbeoji 2007). A number of high-profile cases have reinforced this perception. That of the succulent plant *Hoodia*, the focus of this book, involved patents stemming from the use of traditional knowledge of the indigenous San peoples of southern Africa about the plant's appetite-suppressing properties. A patent on a product derived from seeds of the neem tree (*Azadirachta indica*), whose fungicidal properties have been long known in India, led to a legal challenge and a decision to revoke the patent (Sheridan 2005). And the South American vine *Banisteriopsis caapi*, used widely in traditional religious and healing ceremonies there (Dobkin de Rios 1992; Metzner 1999; Shah 2001) and the subject of a US patent, has also been under legal challenge (Wiser 2002).

Notwithstanding these cases, traditional knowledge is not currently a major research tool in bioprospecting, and partners in access and benefit-sharing agreements are unlikely to be indigenous peoples or local communities, except as local stewards of biodiversity. Some exceptions exist – such as the Kani in India, who receive a proportion of the licence fee and royalty from the commercialization of Jeevani, a herbal drug which rejuvenates and builds strength (Anuradha 1998; Chaturvedi, Chapter 13), and the San in southern Africa, who receive a proportion of royalties from *Hoodia* product sales (Wynberg and Chennells, Chapter 6) – but for the vast majority of cases, community involvement and benefits are negligible.

To a large degree, the expectations of what bioprospecting can bring are both unrealistic and misdirected. Most bioprospecting activities do not yield commercial products, many do not use traditional knowledge and in most cases – particularly those involving partnerships with companies in research-intensive industries – benefits are most significant in the research or discovery phase (ten Kate and Laird 1999; Rosenthal and Katz 2004; Laird and Wynberg 2008; Laird et al. 2008). Bioprospecting is therefore far more likely to help build scientific and technological capacity in biodiversity-rich countries than it is to alleviate rural poverty or improve biodiversity conservation and its contributions to the latter tend to be through the generation of critical scientific information rather than large sums of money (Laird et al. 2008). But despite this reality, there remain high expectations and a deeply embedded belief that bioprospecting represents the proverbial goose that will deliver the golden egg.

5.6 Coming Back to Earth? Today's Key Policy Issues

Since its inception in 1992, the CBD has brought together a complex mix of scientific, conservation, trade and legal elements that fit uneasily into a regulatory whole (Hodges and Daniel 2005). Access and benefit-sharing regulations exist at the juncture of many interlacing bodies of law that criss-cross the same biological material, including international agreements on trade, environment, biodiversity, agriculture and intellectual property (Thornstrom 2005). Moreover, the access and benefit-sharing policy process has provided a forum for a wide range of concerns about the ethical, legal, and political implications of new biotechnologies, the commercialization and ownership of life forms, the patenting of gene sequences, the Human Genome Project and broader concerns about globalization and corporate behaviour (Laird 2002; Parry 2004; Rosenthal and Katz 2004; Laird and Wynberg 2006).

While these are critical issues to debate and resolve as part of international and national policy processes, the effect of combining so many different issues into a single policy process has been divisive and has drained the access and benefit-sharing policy process of the goodwill necessary to come to agreement. Rather than having come together over the past 16 years to create simple, workable legal and regulatory frameworks for access and benefit sharing, the providers and users of genetic resources are increasingly estranged.

Bridging this divide represents a major challenge in CBD negotiations for the international access and benefit-sharing regime. A central theme in negotiating sessions to date has been the extent to which user countries comply with the terms and conditions of access and benefit sharing, and how this compliance can be effectively monitored. Two proposals have been put forward to address these concerns. The first is that intellectual property laws be modified, possibly through Article 29 of TRIPS, to include a strong disclosure mechanism, which would require all applicants for intellectual property rights to disclose the country of origin of genetic resources, the source of relevant traditional knowledge and positive proof of benefit sharing and prior informed consent. The second, which could occur concurrently with the disclosure mechanism, would require an international certificate demonstrating origin, the source or legal provenance of genetic resources and possibly also proof of prior informed consent and benefit sharing.

Both have met with strict opposition from industry and some user countries, but the so-called disclosure proposal is especially contentious given its clear implications for TRIPS and envisaged impediments to innovation. Proponents of this proposal – for example, a number of developing countries led by Brazil and India and supported by indigenous peoples' organizations – have argued that no protection of genetic resources and traditional knowledge will be effective unless and until international mechanisms are found and established within the framework of the TRIPS agreement to require patent applicants to disclose the origin of genetic material and traditional knowledge.[3] They suggest that other means, such as access contracts and databases for patent examinations, can only be supplementary to such international mechanisms, which must contain an obligation on members collectively and individually to prohibit, and to take measures to prevent, the misappropriation of genetic resources and traditional knowledge. This, it is argued, would increase transparency and assist in the enforcement of access and benefit-sharing agreements.

Arguments against the disclosure proposal are varied. At one extreme is concern that acceptance of the proposal could implicitly condone the practice of patenting life forms and natural products. At the other are strong concerns from industry about the uncertainties a mandatory requirement for disclosure could create among researchers and those developing commercial products, and the complications and costs of trying to identify what should and should not be disclosed (Rosenberg 2007). Sceptics also question whether disclosure will bring any practical benefit to national economies or populations (Dutfield 2005).

Despite these concerns, a number of provider and user countries, including India, Costa Rica, South Africa, Denmark and Norway, are already introducing disclosure of origin requirements in domestic legislation and many others are considering such measures (Chouchena-Rojas et al. 2005). However, the territorial nature of patents means that any requirements will apply only in respect of patents issued in those countries, justifying a more international solution to the issue (CIPR 2002).

The accompanying and/or alternative compliance mechanism of an international certificate is similarly fraught with unanswered questions, including its scope and purpose.[4] One proposal is that the certificate or 'passport' could accompany the genetic material along its life cycle, with verification at various points of that cycle, including the application for intellectual property rights. The certificate could therefore increase transparency and traceability, in particular throughout the research process, ensure legal certainty for users of genetic resources, and give providers the assurance that their resources are being used in compliance with legal obligations (CBD 2007b). But whether such a scheme would address the underlying

---

[3] See, for example, submission to TRIPS Council by Bolivia, Brazil, Colombia, Cuba, Dominican Republic, Eduador, India, Peru and Thailand, document IP/C/W/447 (2005); submission by Peru regarding the relationship between TRIPS and the CBD (2004); and submission to TRIPS Council by Brazil, India, Pakistan, Peru, Thailand and Venezuela supported by Cuba and Ecuador, Document IP/C/W/429 (2004).

[4] See, for example, International Chamber of Commerce (ICC), 'Issues for Consideration by the Group of Technical Experts Concerning a Certificate Relating to Genetic Resources', submission of ICC to the CBD Secretariat pursuant to Decision VIII/4 paragraph 1 Regarding the Form, Intent and Functioning of an Internationally Recognized Certificate, Including its Practicality, Feasibility and Costs, Document No 450/1020, 15 September 2006.

equity concerns of developing countries, or indeed be effective or practical given the complexities described, remains an unanswered question.

The scope of the international regime remains largely unresolved. While the CBD focused narrowly on genetic resources – defined as 'genetic material of actual or potential value' – bioprospecting entails the commercial use not only of genetic material, but also of chemical compounds found within the organism, as well as derivatives and products from the genetic material. Excluding derivatives, biochemicals or metabolic extracts from international and national laws therefore significantly curtails benefit-sharing opportunities and is, as one negotiator has put it, 'akin to a rose without its fragrance' (International Institute for Sustainable Development 2005). Biodiversity-rich countries are therefore increasingly drafting access and benefit-sharing laws to go beyond the CBD to address biochemicals and derivatives.[5] However, poorly defining what constitutes derivatives, biochemicals or metabolic extracts can lead to legal confusion and has created concern on the part of industries that use these resources in research and development.

Unclear definitions and a lack of legal understanding of these definitions combine to create even murkier waters when determining the ownership of genetic resources. Indeed, no country has yet found or developed a workable legal framework that clarifies who owns genetic resources (Chishakwe and Young 2003; CBD 2007a), a situation often compounded by the difficulties of finding claimants to work with – especially for resources with long-established traditional use. Bioprospecting in areas such as Antarctica, with unclear ownership and jurisdictional issues (Lohan and Johnston 2003), for pathogens such as the H5N1 virus that mutate continuously, or for resources with wide distribution ranges, makes sovereignty claims even more difficult to determine.

A potentially more significant question is the way in which access and benefit-sharing approaches relate to the large, significant trade in biological resources ('biotrade'). This broader category includes genetic resources, but also organisms or parts thereof, populations or any other biotic components of ecosystems with actual or potential use or value for humanity. These might include, for example, non-timber forest products harvested from the wild; medicinal, food, cosmetic and other plants grown on farms and sold as commodities in international trade; and even raw materials grown in bulk to supply the manufacture of pharmaceuticals. Historically, many of these biodiversity-based products have entered commodity markets similar to those for agricultural products, but it is becoming more difficult to distinguish between the categories of genetic and biological resources, more especially when holders of traditional knowledge are involved.

The case of *Hoodia* (Wynberg and Chennells, Chapter 6 and Wynberg, Chapter 7) illustrates well the overlapping and sometimes artificial boundaries between trade

---

[5] See, for example, South Africa's National Environmental Management: Biodiversity Act (10 of 2004), the Philippines' Executive Order 247 on Access to Genetic Resources and the Costa Rica Biodiversity Law 7788 (1998).

in genetic resources and in biological organisms, and the difficulties of prescribing legislation under such circumstances. A recent benefit-sharing agreement (Institute of Biodiversity Conservation et al. 2004) to develop the cereal crop *Eragrostis tef*, or tef, the staple diet of Ethiopia, as a gluten-free food reveals a similar 'grey area' between what constitutes a genetic resource and what a food product (Laird and Wynberg 2008).[6] Broadening the CBD concepts of access and benefit sharing to these categories of products may well be where the real economic benefits of biodiversity lie, although the regulation of such varied activities and products could present major challenges.

5.7 Conclusion

The access and benefit-sharing policy process for bioprospecting has sprung up from a largely unrealistic and overly optimistic foundation. As a result, no 'grand bargain' has actually been possible, and the billion-dollar drug cures from biodiversity have been few and far apart. Despite the early rationale that bioprospecting would enable biodiversity conservation to 'pay its way', the reality is that the high-technology industries engaging in this field are not interested in supporting biodiversity conservation as a way of protecting their research interests. For many, natural products and genetic resources are only one part of a complex research strategy that must compete with approaches that require fewer resources and are less legally ambiguous. For others, numerous *ex situ* sources of material exist, for example in private collections and seed banks and, increasingly, in a company's backyard. Some of the more carefully crafted bioprospecting partnerships have included payments to conservation funds or parks, and have supported research on biodiversity, but there have never been incentives for these industries to invest in conservation as part of their business model. Nor, unfortunately, do bioprospecting law and policy affect those companies that cause rampant biodiversity loss – such as logging, mining and industrial agriculture.

However, when done right, bioprospecting can yield valuable benefits for developing countries – mostly through building scientific and technological capacity. This requires innovative partnerships between companies, developing country research institutions and governments, indigenous communities and others – and a great deal of work to bring these groups together.

---

[6]Tef, unlike wheat, has a low gluten content and other attributes of interest to the food industry. An agreement that the Ethiopian Institute of Biodiversity Conservation and the Ethiopian Agricultural Research Organization signed with the Dutch-based company Health and Performance Food International sets out a framework for accessing tef varieties and sharing benefits derived from their commercial development. However, although tef products such as bread and sports bars are already being marketed and sold, disagreements between the contracting parties have prevented the distribution of benefits.

An intensely political and conflict-ridden process is envisaged as countries prepare for the final stretch of negotiations for the international regime on access and benefit sharing, due to conclude in 2010. At the time of writing there is still no common vision regarding the nature and scope of the regime or even its necessity, and virtually all proposals remain hotly contested, most especially those with implications for the intellectual property rights system. 'We are on two roads,' commented an Australian delegate to negotiations (ENB, 2007), reflecting the deadlock that had once again stalled progress: one of facilitating access to genetic resources and another of preventing biopiracy.

More and more, however, developing countries are speaking with one voice, and industry likewise is becoming more organized. The coalescing of positions could make bargaining easier, and the bridging role increasingly played by the European Union may expedite the adoption of pragmatic proposals. The recent and historic adoption by the United Nations General Assembly of the United Nations Declaration on the Rights of Indigenous Peoples has also given a major boost to the demands of indigenous people in the access and benefit-sharing discussions.

Bridging already polarized views presents a major challenge, particularly in the absence of informal processes to facilitate informed dialogue, consensus and understanding between stakeholders. But finding a solution is essential, given that all countries are potentially both providers and users of genetic resources, and that in today's globalized world, each is integrally dependent on the other.

References

Anuradha, R. V. (1998). Sharing with the Kanis: A Case Study from Kerala, India. Submitted to the Conference of the Parties to the Convention on Biological Diversity, Fourth Meeting, Bratislava, Slovakia, 4–15 May.

Arico, S., & Salpin, C. (2005). *Bioprospecting of genetic resources in the deep seabed: scientific, legal and policy aspects.* Yokohama, Japan: United Nations University Institute of Advanced Studies.

Balick, M. J., & Cox, P. A. (1997). *Plants, people and culture: the science of ethnobotany.* New York: Scientific American Library.

CBD (2000). *Annex III: Decisions Adopted by the Conference of the Parties to the Convention on Biological Diversity at Its Fifth Meeting*, Nairobi, Kenya, 15–26 May. www.cbd.int/doc/decisions/COP-05-dec-en.pdf. Accessed 31 October 2008.

CBD (2002). *Bonn Guidelines on Access to Genetic Resources and Fair and Equitable Sharing of the Benefits Arising out of Their Utilization*, Secretariat of the Convention on Biological Diversity, Montreal, QC. www.cbd.int/doc/publications/cbd-bonn-gdls-en.pdf. Accessed 31 October 2008.

CBD (2007a). *Analysis of Gaps in Existing National, Regional, and International Legal and Other Instruments Relating to Access and Benefit Sharing.* Secretariat of the Convention on Biological Diversity, draft for peer review prepared for the Ad-Hoc Open-Ended Working Group on Access and Benefit Sharing, Fifth Meeting, Montreal, QC, 8–12 October.

CBD (2007b). *Report of the Meeting of the Group of Technical Experts on an Internationally Recognized Certificate of Origin/Source/Legal Provenance.* Ad-Hoc Open-Ended Working Group on Access and Benefit Sharing of the Convention on Biological Diversity, 20 February. www.cbd.int/doc/meetings/abs/abswg-05/official/abswg-05-07-en.pdf.

Chishakwe, N., & Young, T. R. (2003). *Access to genetic resources, and sharing the benefits of their use: international and sub-regional issues, IUCN–The World Conservation Union.* Switzerland: Gland.

Chouchena-Rojas, M., Ruiz Muller, M., Vivas, D., & Winkler, S. (Eds.) (2005). Disclosure requirements: ensuring mutual supportiveness between the WTO TRIPS agreement and the CBD. IUCN, Gland, Switzerland and Cambridge; ICTSD, Geneva; CIEL, Washington, DC and Geneva; IDDRI, Paris; QUNO, Geneva.

CIPR (2002). Chapter 4, Traditional knowledge and geographical indications. In *Integrating intellectual property rights and development policy: traditional knowledge and geographical indications.* Report of the Commission on Intellectual Property Rights, London. www.iprcommission.org/papers/pdfs/final_report/Ch4final.pdf. Accessed 30 October 2008.

Cragg, G. M., Kingston, D. G. I., & Newman, D. J. (Eds.) (2005). *Anticancer agents from natural products.* Boca Raton, FL: CRC Press, Taylor & Francis.

Crosby, A. W., Jr. (1972). *The Columbian exchange: biological and cultural consequences of 1492.* Westport, CT: Greenwood Press.

Crosby, A. W., Jr. (1986). *Ecological imperialism: the biological expansion of Europe, 900–1900.* Cambridge: Cambridge University Press.

Dobkin de Rios, M. (1992). *Amazon healer.* Bangalore, India: Prism Books.

Dutfield, G. (2002). *Intellectual property rights, trade and biodiversity: seeds and plant varieties.* London: Earthscan.

Dutfield, G. (2004) What is biopiracy? Paper presented at the International Expert Workshop on Access to Genetic Resources and Benefit Sharing, Cuernavaca, Mexico, 24–27 October. www.canmexworkshop.com/documents/papers/I.3.pdf. Accessed 31 October 2008.

Dutfield, G. (2005). Disclosure of origin: time for a reality check? In M. Chouchena-Rojas, M. Ruiz Muller, D. Vivas, & S. Winkler (Eds.), *Disclosure requirements: ensuring mutual supportiveness between the WTO TRIPS agreement and the CBD.* IUCN, Gland, Switzerland and Cambridge; ICTSD, Geneva; CIEL, Washington, DC and Geneva; IDDRI, Paris; QUNO, Geneva.

Earth Negotiations Bulletin (2007). Fifth meeting of the ad-hoc open-ended working group on access and benefit sharing, 8–12 October, 2007, Montreal, Canada. www.iisd.ca/biodiv/abs5.

Ernst & Young (2006). Beyond borders: Global Biotechnology Report 2006. London: Ernst & Young Global Limited.

Gollin, M. A. (1993). An intellectual property rights framework for biodiversity prospecting. In W. V. Reid, S. A. Laird, C. A. Meyer, R. Gomez, A. Sittenfeld, D. H Janzen, M. A. Gollin, & C Juma (Eds.), *Biodiversity prospecting: using genetic resources for sustainable development.* World Resources Institute, Washington, DC; Instituto Nacional de Biodiversidada, Santo Domingo de Heredia, Costa Rica, FL; Rainforest Alliance, New York; African Centre for Technology Studies, Nairobi, Kenya.

Grove, R. H. (1996). *Green imperialism: colonial expansion Tropical Island Edens and the origins of environmentalism.* Cambridge: Cambridge University Press.

Handelsman, J. (2005). How to find new antibiotics. *The Scientist, 19*(19), 20.

Headrick, D. R. (1990). Technological change. In B. L. Turner, W. C. Clark, R. W. Kates, J. F. Richards, J. T. Mathews & W. B. Meyer (Eds.), *The earth as transformed by human action.* Cambridge: Cambridge University Press.

Hodges, T. J., & Daniel, A. (2005). Promises and pitfalls: first steps on the road to the International ABS Regime. *Review of European Community and International Environmental Law, 14*(2), 148–160.

Institute of Biodiversity Conservation, Ethiopian Agricultural Research Organization, Health and Performance Food International (2004). Agreement on access to, and benefit sharing from, Teff Genetic Resources. Addis Ababa, Ethiopia, December.

International Institute for Sustainable Development (2005). Third meeting of the ad-hoc open-ended working group on access and benefit sharing of the convention on biological diversity. *Earth Negotiations Bulletin, 9*(310). www.iisd.ca/download/pdf/enb09310e.pdf. Accessed 31 October 2008.

Juma, C. (1989). *The gene hunters: biotechnology and the scramble for seeds, ACTS Research Series No. 1*. Nairobi, Kenya: African Centre for Technology Studies.

Koehn, F. E., & Carter, G. E. (2005). The evolving role of natural products in drug discovery. *Nature Reviews Drug Discovery, 4*(3). www.nature.com/ nrd/journal/v4/n3/pdf/nrd1657.pdf. Accessed 31 October 2008.

Laird, S. A. (2002). *Biodiversity and traditional knowledge: equitable partnerships in practice*. London: Earthscan.

Laird, S., Monagle, C., & Johnston, S. (2008). *Queensland biodiscovery collaboration: the Griffith University AstraZeneca Partnership for natural product discovery: an access and benefit sharing case study*. Yokohama, Japan: United Nations University Institute of Advanced Studies.

Laird, S. A., & Wynberg, R. (2006). The commercial use of biodiversity: an update on current trends in demand for access to genetic resources and benefit sharing, and industry perspectives on ABS policy and implementation. Prepared for the Ad-Hoc Open-Ended Working Group on Access and Benefit Sharing, Fourth Meeting, Granada, Spain, 30 January to 3 February.

Laird, S. A., & Wynberg, R. (2008). Access and Benefit Sharing in Practice: Trends in Partnerships Across Sectors. CBD Technical Series No. 38, Secretariat of the Convention on Biological Diversity, Montreal, QC. www.cbd.int/doc/publications/cbd-ts-38-en.pdf. Accessed 31 October 2008.

Lange, L. (2004). Tropical biodiversity: an industrial perspective. *Luna, 2004*(30488) 01.

Lohan, D., & Johnston, S. (2003). *The international regime for bioprospecting. Existing policies and emerging issues for Antarctica*. Yokohama, Japan: United Nations University, Institute of Advanced Studies.

Macilwain, C. (1998). When rhetoric hits reality in debate on bioprospecting. *Nature, 392*(6676), 535–540.

McAlpine, J. B., Bachmann, B. O., Piraee, M., Tremblay, S., Alarco, A., Zazopoulos, E., et al. (2005). Microbial genomics as a guide to drug discovery and structural elucidation: ECO-02301, a novel antifungal agent, as an example. *Journal of Natural Products, 68*(4), 493–496.

McNeil, D.G. Jr (2007). Indonesia may sell, not give, bird flu virus to scientists. *New York Times*, 7 February.

Metzner, R. (Ed.) (1999). *Ayahuasca: human consciousness and the spirits of nature*. New York: Thunder's Mouth Press.

Mgbeoji, I. (2007). Lost in translation? The rhetoric of protecting indigenous peoples' knowledge in international law and the omnipresent reality of biopiracy. In P. Phillips & C. Onwuekwe (Eds.), *Accessing and sharing the benefits of the genomics revolution*. Dordrecht/Germany: Springer.

Nutrition Business Journal (2003). NBJ's Annual Industry Overview VIII. May/June.

Parry, B. (2004). *Trading the genome: investigating the commodification of bio-information*. New York: Columbia University Press.

PhRMA (2007). Pharmaceutical Industry Profile 2007. Pharmaceutical Research and Manufacturers of America, Washington, DC. www.phrma.org/files/Profile%202007.pdf. Accessed 31 October 2008.

Reid, W. V., Laird, S. A., Meyer, C. A., Gomez, R., Sittenfeld, A., Janzen, D. H., Gollin, M. A., & Juma, C. (Eds.) (1993). *Biodiversity prospecting: using genetic resources for sustainable development*. World Resources Institute, Washington, DC; Instituto Nacional de Biodiversidad, Santo Domingo de Heredia, Costa Rica; Rainforest Alliance, New York; African Centre for Technology Studies, Nairobi, Kenya.

Rohter, L. (2007) As Brazil defends its bounty, rules ensnare scientists. *New York Times*, 28 August.

Rosenberg, D. (2007). *Some business perspectives on the international regime*. United Kingdom: GlaxoSmithKline.

Rosenthal, J. P., & Katz, F. N. (2004). Natural products research partnerships with multiple objectives in global biodiversity hotspots: nine years of the International Cooperative

Biodiversity Groups Program. In A. T. Bull (Ed.), *Microbial diversity and bioprospecting*. Washington, DC: ASM Press.

Rubenstein, K. D., Heisey, P., Shoemaker, R., Sullivan, J., & Frisvold, G. (2005). Crop genetic resources: an economic appraisal. Economic Research Service Economic Information Bulletin Number 2. United States Department of Agriculture.

Sanchez, V., & Juma, C. (1994). *Biodiplomacy: genetic resources and international relations*. Nairobi, Kenya: African Centre for Technology Studies.

Shah, T. (2001). *Trail of feathers: in search of the Birdmen of Peru*. London: Phoenix.

Sheridan, C. (2005). EPO neem patent revocation revives biopiracy debate. *Nature Biotechnology, 23*, 511.

Smolders, W. (2005). Commercial practice in the use of plant genetic resources for food and agriculture. Background Study Paper No. 27, prepared for the Commission on Genetic Resources for Food and Agriculture, Food and Agriculture Organization of the United Nations, Rome.

Svarstad, H., & Dhillion, S. (2004). *Responding to bioprospecting: from biodiversity in the south to medicines in the north*. Oslo, Norway: Spartacus Forlag.

Ten Kate, K., & Laird, S. A. (1999). *The commercial use of biodiversity: access to genetic resources and benefit sharing*. London: Earthscan.

Thornstrom, C. G. (2005). The green blindness: microbial sampling in the Galapagos – the case of Craig Venter vs. the Darwin Institute and the lessons for the trip to China by 'S/V Gotheberg', unpublished manuscript.

Tuxill, J. (1999). Appreciating the benefits of plant biodiversity. In L. R. Brown & C. Flavin (Eds.), *State of the World 1999: special millennium*. London: Earthscan.

UN (2007). United Nations declaration on the rights of indigenous peoples. Adopted by United Nations General Assembly Resolution 61/295 on 13 September. www.un.org/esa/socdev/unpfii/documents/DRIPS_en.pdf. Accessed 29 April 2008.

WHO (2007). Pandemic influenza preparedness: sharing of influenza viruses and access to vaccines and other benefits. Sixtieth World Health Assembly, Agenda Item 12.1, WHA60.28, 23 May, World Health Organization.

Wiser, G. (2002). The *Ayahuasca* patent case: indigenous people's stand against misappropriation. In S. A. Laird (Ed.), *Biodiversity and traditional knowledge: equitable partnerships in practice*. London: Earthscan.

WSSD (2002). Johannesburg plan of implementation. World Summit on Sustainable Development, Johannesburg, South Africa. www.un.org/esa/sustdev/documents/WSSD_POI_PD/English/POIToc.htm. Accessed 30 October 2008.

Zedan, H. (2005). Patents and biopiracy: the search for appropriate policy and legal responses, *Brown Journal of World Affairs, 12*(1). www.bjwa.org/index.php?issue=12.1. Accessed 1 November 2008.

Part II
Learning from the San

We are thankful that the traditional knowledge of our forefathers is acknowledged by this important agreement, and that we are making it known to the world. As San leaders we are determined to protect all aspects of our heritage

(Petrus Vaalbooi, Chairperson of South African San Council, Press release at signing of Hoodia benefit-sharing agreement, Molopo, South Africa, 24 March 2003)

Chapter 6
Green Diamonds of the South: An Overview of the San-*Hoodia* Case

Rachel Wynberg and Roger Chennells

Abstract One of the most famous benefit-sharing initiatives to date is the San-*Hoodia* case. The San peoples are the oldest human inhabitants of southern Africa, but after centuries of genocide and marginalization by colonialists, they now number only about 100,000 people in Botswana, Namibia, South Africa and Angola. Their current lives are characterized by abject poverty, yet they still possess traditional knowledge about local biodiversity.

This chapter describes how San knowledge about the appetite-suppressant properties of *Hoodia* – a succulent plant used as a substitute for food and water during hunting expeditions – has led to agreements to share benefits arising from the use of this knowledge, and analyses the challenges in developing and implementing these agreements. It distils and synthesizes existing research, presents a review of new initiatives and, through the eyes of the San legal representative involved in negotiations and those of an activist and researcher monitoring developments, provides a critical analysis of the case study.

The chapter concludes that the challenges of implementation are substantial, in particular the distribution of benefits to impoverished communities in three different countries. Regional differences in benefit-sharing policies exacerbate these challenges, heightened by highly unstable *Hoodia* markets, more especially in light of the main licence holder's decision to terminate its involvement.

A crucial lesson to emerge from this case study is the need to obtain the prior informed consent of communities holding knowledge about biodiversity from the outset of a project and to engage communities as early as possible as active partners.

R. Wynberg (✉)
Environmental Evaluation Unit, University of Cape Town, Private Bag X3, Rondebosch 7701, Cape Town, South Africa
e-mail: rachel@iafrica.com

R. Chennells
Chennells Albertyn: Attorneys, Notaries and Conveyancers, 44 Alexander Street, Stellenbosch, South Africa
e-mail: scarlin@iafrica.com

Also emphasized is the importance of relationship building and of having in place a policy climate conducive to fair deliberation. The case has resulted in heightened interest about the importance of protecting traditional knowledge and ensuring that holders of such knowledge receive fair compensation.

Keywords benefit sharing • biopiracy • Convention on Biological Diversity • *Hoodia* trade • San indigenous communities • traditional knowledge protection

6.1 Introduction

The story of *Hoodia* is one that has been told many times (Geingos and Ngakaeaja 2002; Chennells 2003; Stephenson 2003; Wynberg 2004; Vermeylen 2007). Indeed, over the past 7 years no fewer than ten documentaries have been made about the case, more than a dozen PhDs and Master's dissertations registered to investigate it further, and hundreds of news items written. The involvement of the San, the oldest human inhabitants of Africa, and the intrigue of a plant that may simultaneously tackle the Western affliction of obesity and the developmental challenges of the San have triggered the public's imagination at a time when disparities between rich and poor have never been greater. For some, the case illustrates the possibilities of bioprospecting – the search for biological material with commercially valuable genetic and biochemical properties – and final, albeit tenuous, delivery on the long-standing promises of equitable benefit sharing in the Convention on Biological Diversity (CBD). For others, it typifies the problems of biopiracy, where traditional knowledge has been appropriated without the consent of holders of that knowledge.

This chapter presents an overview of the story to date. It distils and synthesizes existing research, presents a review of new initiatives and, through the eyes of the San legal representative involved in negotiations and an activist and researcher monitoring developments, provides a critical analysis of the case study. It begins by introducing the San, with their history of devastation and current developmental context. Then follows a review of the traditional use and knowledge of *Hoodia* by indigenous peoples in southern Africa and an overview of the commercial development of the plant. The next section describes how the benefit-sharing agreement was negotiated between the San and the Council for Scientific and Industrial Research (CSIR), and the key issues of these deliberations, and is followed by a review of current trends in *Hoodia* markets and the development of a second benefit-sharing agreement. The last part analyses current implementation challenges.

6.2 The San

The San peoples of southern Africa, also known as the 'Bushmen',[1] are generally regarded as having lived longer continuously in one location than any other population in history (Stephenson 2003). They are considered to be the progenitors of the rest of humankind (Deacon and Deacon 1999; Soodyall 2006) and certainly the oldest human inhabitants of southern Africa, having lived in small nomadic groups of hunters and gatherers for thousands of years as sole occupants of the region (Boonzaier et al. 1996; Lee et al. 2002). Unequivocal remains of their ancestors excavated just outside Cape Town date back approximately 120,000 years (Lee et al. 2002).

Humankind's fascination with our origins as hunter-gatherers and with the exotic or 'primitive' has made the San an icon of popular culture, a fixture in anthropological textbooks and films, and, more recently, a subject of anthropological and political controversy. To some they represent pristine hunter-gatherers, to others apartheid's[2] most oppressed and marginalized victims, but neither of these polarities captures the present realities (Hitchcock et al. 2006).

When settlers landed at the Cape in 1652, the San occupied an area stretching from the Congo-Zambezi watershed in Central Africa to the Cape in South Africa and numbered about 300,000 people (Lee 1976). Today the San comprise approximately 100,000 people, 55,000 of whom live in Botswana, 35,000 in Namibia, 8,500 in South Africa and 4,500 in Angola, with scattered populations in Zimbabwe and Zambia (SASI 2007). After centuries of genocide and marginalization, leading to loss of land and consequently loss of culture and identity, they occupy an unchallenged niche as the poorest of the poor in these countries (Suzman 2001), living in conditions of relative powerlessness.

The so-called 'Kalahari debate' articulates two positions on understanding the current vulnerable status of the San. The first is held by the 'traditionalists', who essentially see the San as primitive hunter-gatherers, relics of our forebears who have been isolated and have lived in harmony with nature, with a relatively resilient and static culture, until recent times (Wilmsen 1989). The 'revisionists', on the other hand, declare the San peoples to be an impoverished underclass, victims of an unrelenting class war against a host of more dominant peoples (Barnard 1996).

---

[1] The word 'San' was first used by the Harvard Kalahari Research Group as a replacement for the term 'Bushmen' in 1961 (Lee 1976). Whilst other terms are used in various contexts, for example 'Basarwa' in Botswana and 'Bushmen' by many including the San themselves, San leaders have agreed that the word 'San' is the only known overarching term that describes their peoples (Hitchcock et al. 2006).

[2] Meaning 'separateness' in Afrikaans, apartheid was a system of racial segegation in South Africa from 1948, and was dismantled in a series of negotiations from 1990 to 1993. These negotiations culminated in democratic elections in 1994.

Today, whilst a minority of San live in villages on their own land,[3] most reside in conditions of abject poverty on land to which they have no rights or traditional claim. Living in small rural villages in regions dominated by more powerful African cultures, in sterile government resettlement villages, or as labourers working on commercial ranches, they occupy an uneasy twilight zone between their former traditional ways and the modern world. A regional assessment of the status of the San concluded that despite decades of development assistance, they remain by far the most marginalized and dispossessed of all southern African communities (Suzman 2001).

Their former egalitarian and consensus-based hunter-gather lifestyles have had to adapt to rapid sedentarization, with predictable consequences. In common with other First Nations[4] elsewhere in the world, the San have to a large extent succumbed to societal breakdown and culture loss exacerbated by alcohol abuse and hopelessness (Silvain 2006). Representational leadership gives rise to the formation of new elites, with the concomitant jealousies and power struggles associated with modern political and social life. Some authors have suggested that it is the hunter-gather legacy that leaves societies such as the San with comparatively low capacity for bettering themselves materially (Diamond 1998). Others regard the consensual nature of decision-making in nomadic non-hierarchical societies as being central to their continued powerlessness (Colchester 2003).

The burden of the relatively recent genocidal predations on the San deserves mention. The collective trauma inflicted upon indigenous populations by colonial invasions has been remarkably similar, from the Americas to Australasia to Africa. Superior weaponry devastated entire populations, and the convenient *terra nullius*[5] doctrine gave comfort to governments responsible for atrocities committed in their name. Genocide of San peoples was rationalized as rightful retaliation against their theft of cattle, as imposing law and order on a 'lawless land' and clearing farming land of 'vagrant and treacherous savages'. *The Times* of London described the San as 'in appearance ... little above the monkey tribe, and scarcely better than the mere brutes of the field' (*History of the Bosjesmans, or Bush People*, 1847).

Penn's (1996) description of the systematic destruction of the Cape San by the authorities is breathtaking in its horror. The Cape colonial government was driven by a conviction that the San, being incompatible with the creation of a 'civilized society', needed to be eradicated. During the eighteenth century thousands of San were systematically exterminated by hunting parties, and their women and children taken into servitude. The following extract from Theal (1892–1919) is a fitting summary of this sad and recent history.

---

[3] Some 4,000 !Kung of the N=a Jaqna conservancy (formerly West Bushmanland) in Namibia, 5,000 Jun/uasi of the Nyae Nyae (formerly East Bushmanland) in Namibia and 800 ≠Khomani San of the Northern Cape, South Africa, have secured rights to live on their traditional land.

[4] 'First Nations' and 'First Peoples' are terms colloquially given to certain peoples, such as the Aboriginals of Australasia, the Inuits of Canada and the San of southern Africa, who inhabited their continents many millennia before the advent of subsequent colonizers.

[5] This doctrine of colonial empires held that land occupied by indigenous or local peoples, who did not maintain a recognized system of 'ownership' of the land, was in fact empty land and thus open to occupation by the civilizing invaders.

They [the San] could not adapt themselves to their new environment, they tried to live as their predecessors had lived, and therefore they were fated to perish. The wave of European colonisation was not to be stayed from rolling on by a group of savages who stood in its course.

The exhibition *Miscast* at the South African National Gallery (Skotnes 1996) shocked the world with photographs of dead San men and women hanging from trees after hunting parties, trophy heads and San body parts preserved for scientific research. The exhibition provided a shocking visual reminder of the sustained, merciless and unspeakable carnage wreaked on generations of San in the name of 'civilization'. San visitors to the museum, despite being aware of their history of subjugation, were equally horrified at the starkness of the visual record and reminder of their desolate past.

The San population today bears the scars of this devastating history. A number of dedicated non-governmental organizations (NGOs), collectively known as the Kuru Family of Organisations, that have evolved over the past 2 decades are grappling with the challenge of bringing appropriate development for the San (KFO 2006) (see also, Chennells et al. Chapter 9). In 1996, taking a leaf from the book of the Sami indigenous peoples of the Scandinavian north, the San formed their own advocacy organization, the Working Group of Indigenous Minorities in Southern Africa (WIMSA), charged with uniting and representing San communities from Botswana, Namibia and South Africa. San leaders in WIMSA ensured that their cultural and linguistic diversity was celebrated under a collective San cultural umbrella, which proved decisive in their aim to achieve San unity across national boundaries. As these organizations have developed, the capacity of the associated San employees and leaders to determine their own future has steadily risen. Chennells et al. (Chapter 9) describe the role played by these San organizations in San development and Vermeylen (Chapter 8) examines the degree to which San have achieved rights, both to their intellectual property and to their land.

6.3 Traditional Use and Knowledge of *Hoodia* Species

Use of *Hoodia* by the San probably dates back centuries, but the first recorded use of the plant was in all likelihood by the botanist Francis Masson (1741–1805), who visited the Cape from 1772 to 1774 and 1786 to 1795. He recorded finding 'Stapelia gordoni' (now called *H. gordonii*) (Masson 1796) and wrote that the stems of *Trichocaulon piliferum* were eaten by the 'Hottentots'. 'This is the real ghaap[6] of the natives,' wrote the South African naturalist Rudolf Marloth (1855–1931) of *T. piliferum*, 'who use it as a substitute for food and water. The sweet sap reminds one of licorice and, when on one occasion thirst compelled me to follow the example of my Hottentot guide, it saved further suffering and removed the pangs of hunger so efficiently that I could not eat anything for a day after having reached the camp' (Marloth 1932).

---

[6] A vernacular name for *Hoodia* and *Trichocaulon* species.

Who are the 'Hottentots' referred to by Masson, and how do they relate to the San earlier described? And what claim do they now have to knowledge about the properties of the plant? Strictly speaking, the 'Hottentots', or Khoe peoples, were herders who were related to the San, but this distinction is not recognized in the colonial botanical accounts, which cluster all groups as 'Hottentots', including the San.

The groups presumably used *Hoodia* for millennia, although the ways in which they did so are open to interpretation. A popular but perhaps simplistic account has the San using the plant for hunting purposes to give 'strength', and anecdotal accounts even suggest that hunters may have been given *Hoodia* to prevent their eating the kill. But San informants suggest that this would have been insulting to the hunter, whose skills and integrity negated the need for any external appetite suppressants.

What is undisputed, however, is use by the San of *Hoodia* and related species as a food and, especially, as a drink substitute and appetite suppressant, as well as for other purposes recounted variously as to improve virility; to cure or treat hangovers, haemorrhoids, high blood pressure, pulmonary tuberculosis, stomach pains, flu, asthma and eye pain; and, ironically, to stimulate the appetite (Watt and Breyer-Brandwijk 1962; Khoisis 1983; Dicks et al. as quoted in Van Wyk and Gericke 2000; Hargreaves and Turner 2002). Typically, such treatments would be prepared by scraping the spines off the succulent stems with a stone or stick and then eating the stem raw like a cucumber. It could also be cooked, to reduce the bitterness, or ground into a powder for treating certain ailments. In Botswana, Hargreaves and Turner (2002) note the use of *H. currorii* (known locally as *sekopane*) for purification after death and as part of a ritual to find the cause of death. *Hoodia* species are also mixed with various bulbs to wash the body to remove bad luck. A similar recipe promotes fertility in cattle. A variety of *Hoodia* species are also used in Botswana to increase crop yields, to prevent the sun from burning seedlings and to treat venereal diseases (Hargreaves and Turner 2002).

Some of these uses can undoubtedly be attributed exclusively and originally to the San, but the wide distribution of certain *Hoodia* species suggests extensive use by many other indigenous peoples in the region, including minority groups known as the Nama, Damara, and Topnaar in Namibia, both as a medicinal remedy and as a substitute for food and water. These Khoi-speaking peoples emerged in southern Africa many millennia after the San, occupied similar geographical regions and no doubt acquired San knowledge of plants and their uses, in addition to evolving their own knowledge. Steyn and du Pisani (1985) report use of *Hoodia* species by the Damara as a source of water. Van den Eynden et al. (1992) similarly indicate use of *H. currorii* as a thirst-quencher and medicinal remedy by the Topnaar of the Kuiseb Valley in Namibia. Among the Namibian Damara, reports Von Koenen (2001), *H. currorii* is known as a diabetes remedy, with a 'piece the length of a pencil cut off every day and one third eaten morning, noon and night', knowledge that has subsequently led to the filing of an international patent for the prevention and treatment of diabetes based on *Hoodia* species (EP1166792).

6.4 Research and Development of *Hoodia* for Commercial Application

The documented use of *Hoodia* species as a food and water substitute in colonial botanical accounts (Marloth 1932; White and Sloane 1937) is significant because it led directly to the CSIR, a South African research institution, including the plant for investigation in a 1963 project on edible wild plants of the region. A 1962 publication on medicinal and poisonous plants of southern Africa (Watt and Breyer-Brandwijk 1962) had inspired the CSIR project, which aimed to inform the South African Defence Force about the toxic and nutritional properties of wild foods and so ascertain their suitability for the army. Existing literature, combined with laboratory tests on mice which had been fed *Hoodia* species, led scientists to identify the potential of *Hoodia* species as a non-toxic appetite suppressant, although insufficient evidence existed to file for a patent. The lack of technology to isolate and identify active ingredients halted progress on the research, which commenced again in the early 1980s.

In 1986, the CSIR acquired high-field nuclear magnetic resonance spectroscopy equipment that made it possible to elucidate relevant molecular structures of *Hoodia* species (CSIR 2001), and in 1995, following 9 years of confidential development, a patent application was filed in South Africa by the CSIR for the use of the active components of the plant which were responsible for suppressing appetite (South African Patent No 983170).

In 1998, the CSIR signed a licensing agreement for the further development and commercialization of the product with Phytopharm, a small British company specializing in the development of phytomedicines (Phytopharm 1997), and this was followed in the same year by the granting of international patents in some countries (GB2338235 and WO9846243A2). The agreement granted Phytopharm an exclusive worldwide licence to manufacture and market *Hoodia*-related products and to exploit any other part of the CSIR's intellectual property rights (IPRs) relating to *Hoodia* species. Through a programme dubbed 'P57', Phytopharm developed this drug lead to a more advanced stage, leading to a licence and royalty agreement in August 1998 with Pfizer, the US-based pharmaceutical giant, for further development and commercialization.

In December 2001, Phase IIa/third-stage proof-of-principle clinical trials were reported to have been successfully completed in a double-blind, placebo-controlled clinical study, taking the drug one step closer to being commercially available (Phytopharm 2001). According to Phytopharm, the trials, which involved 18 overweight but healthy males, provided strong statistical evidence that the plant extract reduced daily calorie intake by an average of 1,000 cal.

In July 2002, Phytopharm announced a future development programme for P57, in which Pfizer would take responsibility for developing a botanical prescription pharmaceutical for the treatment of obesity and metabolic disorder, and Phytopharm would develop semi-synthetic versions of the active molecules and be free to seek other partners to commercialize these products (Phytopharm 2002).

During July 2003, Pfizer merged with Pharmacia and closed its Natureceuticals group, which had been responsible for the development of P57. This, combined with a variety of complex but poorly understood factors, led Pfizer to announce it was discontinuing clinical development of the drug and was returning the licensing rights to Phytopharm, leaving Phytopharm free to license P57 to other parties (Phytopharm 2003). Following the closure of the Natureceuticals group, Pfizer decided that the successful development and commercialization of P57 might 'be best achieved by another organisation'. Pfizer also stated that the positive clinical trial data of P57 encouraged further study of *Hoodia* as a therapy for obesity. Some critics saw the withdrawal of Pfizer from the development of *Hoodia* as the death knell for its commercialization, but Phytopharm and the CSIR remained confident of the possibility of finding other partners to take the project forward.

In December 2004, this optimism was borne out through the granting by Phytopharm of an exclusive global licence to consumer giant Unilever plc for *Hoodia gordonii* extracts, with their likely incorporation into existing food brands as a functional weight-loss product for the mass market (Phytopharm 2004). In terms of the agreement, Unilever would buy exclusive rights to the product for an initial £6.5 million, rising to £21 million once it had achieved certain milestones. Phytopharm would also receive an undisclosed royalty on sales of all products containing the extract. Through what was described by Phytopharm's then chief executive, Richard Dixey, as an 'aggressive programme', Unilever and Phytopharm would collaborate on a five-stage research and development programme of safety and efficacy studies, and Unilever would also take responsibility for the scaling up of agronomic capacity, through an expansion of cultivation efforts in both South Africa and Namibia (Dixey 2004). Unilever would lead the marketing of products, expected to be the factor that would 'win the day' (Dixey 2004). Consideration would also continue to be given to the possibility of developing an over-the-counter pharmaceutical product (Dixey 2004).

Many of these pronouncements were realized between 2004 and 2008 and developments reached an advanced stage, including clinical safety trials, manufacturing and the cultivation of some 300 ha of *Hoodia gordonii* in South Africa and Namibia (K. Povey, October 2007, Unilever, personal communication). Agreement was also reached between Unilever and the chemical company Cognis to develop a R750 million (US$94 million) extraction facility for *Hoodia* in the Western Cape province, South Africa (Department of Trade and Industry 2008). Unilever had plans to develop a *Hoodia*-based product for its line of Slim Fast® beverages, and submission to the US Food and Drug Administration for generally recognized as safe (GRAS) status was predicted for late 2009 for the use of *Hoodia* preparations as an additive in foods and beverages (Stafford 2009). This situation changed significantly in November, 2008, with the announcement by Unilever that it was to abandon plans to develop *Hoodia* as a functional food, because of safety and efficacy concerns (Douglas 2008; Phytopharm 2008). In further communication to South African government departments, Unilever announced that it would cease all 'drying, transport, trials and any other activity associated with *Hoodia* in South Africa' as from 31 March 2009, and that Phytopharm plc would take over a proportion of existing cultivation in South Africa and, to a limited extent, Namibia (Phytopharm 2009; Unilever 2009). Phytopharm in turn announced that it would now seek other partners to further develop *Hoodia* and bring products to market (Phytopharm 2008) and that it 'remained positive about opportunities for future commercialisation' (Phytopharm 2009).

Much is at stake if a successful product is developed: the global value of functional foods, defined as 'any modified food or food ingredient that may provide a health benefit beyond the traditional nutrients it contains' (Bloch and Thomson 1995) is estimated at US$65 billion (Phytopharm 2007), with the market value for the dietary control of obesity at over US$3 billion per annum in the United States alone (Phytopharm 2003). The growth potential of functional foods is predicted to be 50% from 2005 to 2010, with an accelerating trend towards new products.

Figure 6.1 graphically depicts the license agreements developed between the CSIR, Phytopharm and Unilever, and the benefit-sharing agreement between the CSIR and the San, discussed below in Section 6.5. A chronology of the use and commercial development of *Hoodia* follows in Table 6.1.

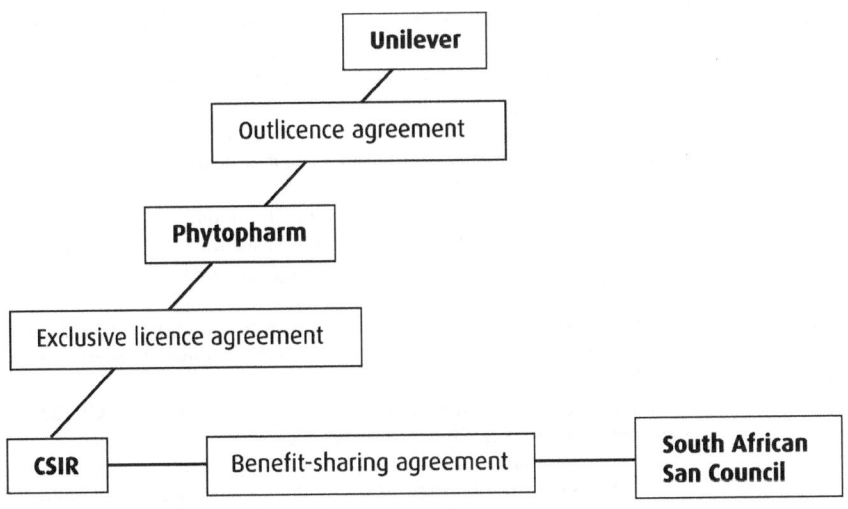

Fig. 6.1 Licence and Benefit-Sharing Agreements Developed Between the San, CSIR, Phytopharm and Unilever

Table 6.1 Chronology of the Commercial Development of *Hoodia*

C 25 000 BC to seventeeth century	The San use wild plants, including *Hoodia*, in a hunting and gathering economy
2000 BC	The earliest evidence of migration into southern Africa of pastoralists, regarded as ancestors of contemporary Khoi people (e.g. Nama, Griqua, Damara, Koranna), is from this period
AD 200 to AD 1200	Bantu-speaking (African) peoples, ancestors of southern Africa's majority populations (e.g. Zulu, Xhosa, Tswana, Herero, Ovambo), migrate south of the Zambezi River
AD 1200 to present	Extensive cultural and trade interaction, and some intermarriage, takes place between Bantu, Khoi and San peoples
1652–1900	Dutch settlers land at the Cape in 1652. The process of colonial settlement and subjugation of local tribes commences. Legalized hunting and extermination of San and Khoi peoples takes place as Afrikaner boers (farmers) drive their stock northwards and 'tame' the hinterland
1796	Use of *Hoodia* species by the 'Hottentots' is first recorded by the botanist Francis Masson
1910	The Union of South Africa is formed as a self-governing colony within the British Commonwealth
1937	The first publication of San traditional knowledge relating to the use of *Hoodia* for suppressing appetite, based on work by the German-born ethnobotanist Rudolf Marloth, appears
1945	The CSIR is established as South Africa's premier scientific research and development institute
1949	The Afrikaner-based National Party wins the election in South Africa and begins to enforce apartheid policies. San are forced to assimilate with the so-called coloureds, or people of mixed race
1955	The Population Registration Act is promulgated, forcing all indigenous people of colour to register either as Bantu or Coloured, thereby eliminating recognition of the San by government
1963	The CSIR includes *Hoodia* species in a project on edible wild plants, based on the ethnobotany of the San
1968	The death of a leading scientist on the *Hoodia* project and technical problems lead to the mothballing of the project
1983–1986	The acquisition of high-field nuclear magnetic resonance spectroscopy equipment allows for the relevant molecular structures of *Hoodia* species to be elucidated by the CSIR
1986–1995	The CSIR continues confidential work on the development of *Hoodia* species
1995	The CSIR files a patent application in South Africa for active components of *Hoodia* species responsible for suppressing appetite (South African Patent No 983170)
August 1998	CSIR and Phytopharm sign a licence agreement for the further development and commercialization of *Hoodia*, which they code-name Programme 57 (P57)
1998	International patents are granted to the CSIR in some territories (GB2338235 and WO9846243A2). Phytopharm sublicenses Pfizer to complete clinical development, obtain regulatory approval and commercialize the drug. The CSIR publishes its Bioprospecting Policy, declaring its commitment to sharing benefits with holders of traditional knowledge. However, in practice, this commitment is not implemented in the P57 project

(continued)

6 Green Diamonds of the South: An Overview of the San-*Hoodia* Case

Table 6.1 (continued)

2001	Phase IIa/third-stage proof-of-principle clinical trials for P57 are reported to be successfully completed. WIMSA passes a resolution at its annual general meeting that heritage is indivisible and that all benefits received from the shared San heritage are to be divided amongst all San in the region
June 2001	Through lobbying work by Biowatch and Action Aid, the British *Observer* newspaper reports commercial development of *Hoodia* without the involvement of the San and quotes Phytopharm's chief executive as stating that the CSIR had led him to believe that the San were 'extinct'. The San establish that a patent has been registered based on *Hoodia* use, and that the CSIR has granted Phytopharm a licence to exploit the patent. The San inform the CSIR through their lawyer that they intend to demand their legal intellectual property rights
June 2001 to March 2002	The South African San Council is mandated by WIMSA to negotiate with the CSIR, and negotiations between the CSIR and the San commence
March 2002	A memorandum of understanding is signed between the CSIR and the South African San Council, recognizing the San as the originators of knowledge about *Hoodia* and including a commitment to benefit sharing
February 2002 to March 2003	Negotiations continue between the CSIR and the South African San Council. Workshops are held with San leaders to debate issues relating to *Hoodia* and intellectual property and to agree on principles of benefit sharing, including confirmation of the collective ownership of heritage by all San
March 2003	The CSIR (represented by the Minister of Arts, Culture, Science and Technology) and the South African San Council sign a benefit-sharing agreement. The San are to receive 6% of CSIR royalties and 8% of milestone payments
July 2003	Pfizer withdraws from commercial development of P57 and returns the licensing rights to Phytopharm
2001–2004	In parallel to the CSIR-Phytopharm initiative, a growing market develops for *Hoodia* in herbal and dietary supplements, using knowledge of the San to promote products. Some products are later revealed to be fakes, with no *Hoodia* content
October 2003	The San meet in Upington to discuss benefit sharing and decide on allocations between San councils in each country and WIMSA
2004	Phytopharm announces its intention to develop P57 as a food supplement
May 2004	A proposal is tabled to list *Hoodia* species in Appendix II of the Convention on International Trade in Endangered Species of Wild Fauna and Flora (CITES), to allow for controlled commercial trade (CITES 2004)
June 2004	Namibia announces its intention to commercialize *Hoodia*
August 2004	The San apply for registration of the San *Hoodia* Benefit-Sharing Trust
September 2004	The National Environmental Management: Biodiversity Act 10 of 2004 (Biodiversity Act) is promulgated in South Africa, requiring a benefit-sharing agreement to be developed with holders of traditional knowledge where their knowledge is used for bioprospecting
October 2004	A proposal to list *Hoodia* species in CITES Appendix II is adopted by the 13th Conference of the Parties to CITES. The CSIR announces the initiation of a broader bioprospecting project with the San
December 2004	Phytopharm grants consumer giant Unilever an exclusive global licence to *Hoodia gordonii* extracts for incorporation into existing food brands
February 2005	The San-*Hoodia* Benefit-Sharing Trust is elected, formed and registered. First payments are made. Continued efforts are made to develop the capacity of the trust to manage anticipated payments to San councils

(continued)

Table 6.1 (continued)

December 2005	The *Hoodia* Growers Association of Namibia is launched
February 2006	The San, through WIMSA, enter into a benefit-sharing agreement with the South African *Hoodia* Growers (Pty) Limited (SAHG) which entitles the San to 6% of farmgate sales of raw *Hoodia*
March 2006–2007	Negotiations commence between the San, the Cape Ethno-botanical Growers Association (CEGA), the SAHG and environment departments of the Northern Cape and Western Cape provinces
January 2007	Unilever begins growing *Hoodia* in Namibia
January 2007	A memorandum of understanding is signed between WIMSA, CEGA and SAHG, with the involvement of the Western Cape and Northern Cape provincial governments
February 2007	Threatened or Protected Species Regulations are promulgated in South Africa under the Biodiversity Act. *Hoodia gordonii* and *H. currorii* are listed as protected species
March 2007	A benefit-sharing agreement is signed between WIMSA and the Southern African *Hoodia* Growers Association (SAHGA), with the approval of the South African government. The San are to receive R24 per dry kg of *Hoodia*
March 2007	Draft regulations on access and benefit sharing are tabled by the South African government in terms of the Biodiversity Act
July 2007	South Africa, Namibia and Botswana agree to prohibit the export of live *Hoodia* material from the region
September 2007	Phytopharm announces that stage 3 activities of the joint development agreement for *Hoodia* extract with Unilever have been initiated
October 2007	The US Federal Trade Commission initiates action against *Hoodia* e-mail spammers
2007	A Cabinet Directive establishes an Interim Bioprospecting Committee in Namibia
May 2008	Plans are uncovered for Cognis to build an R750 million extraction facility in southern Africa for *Hoodia*
April 2008	Bioprospecting, Access and Benefit-Sharing Regulations under the Biodiversity Act (10 of 2004) come into effect in South Africa requiring a benefit-sharing agreement in all cases where traditional knowledge is associated with an indigenous biological resource
14 November 2008	Unilever announces its withdrawal from the *Hoodia* project
31 March 2009	Unilever ceases all *Hoodia*-related operations and Phytopharm takes over a limited number of cultivation initiatives

6.5 Negotiating a Benefit-Sharing Agreement with the CSIR

6.5.1 Initiating Talks

What did these developments mean for the San, the original holders of knowledge about the properties of *Hoodia*? Up until 2001, agreements for the further development and commercialization of the *Hoodia* drug had proceeded apace without acknowledgement of the contribution of the San, let alone their prior informed consent. Indeed a newspaper report quotes Phytopharm's Richard Dixey as having been told by the CSIR that the 100,000 strong San 'no longer existed' (Barnett 2001). In defence of its position, the CSIR linked its initial reluctance to engage with the San to a concern that expectations would be raised with promises that could not be met and insisted that the organizational policy on bioprospecting was to eventually share benefits of research based on indigenous knowledge. But clearly, the realities of implementing this policy were complex and difficult. How, it was argued by the CSIR and Phytopharm, could the real owners of traditional knowledge be identified, and what if one group had historically stolen the knowledge from another group? The potential scenarios seemed endless and intricate.

While these concerns were undoubtedly valid and are common in such cases, they were also obfuscatory and to some extent provided a useful defence for the CSIR and Phytopharm. Such sentiments were also in flagrant disregard of the International Labour Organization's Convention 169, an international agreement for the protection of indigenous peoples' rights; the letter and spirit of the CBD; the African Union's Model Law for the Protection of the Rights of Local Communities, Farmers and Breeders and for the Regulation of Access to Biological Resources (Ekpere 2001); and the Bonn Guidelines on Access to Genetic Resources and Fair and Equitable Sharing of the Benefits Arising out of their Utilization, a voluntary guide to assist governments in developing an access and benefit-sharing strategy, as well as necessary legal, administrative or policy measures (CBD 2002). Although not stated in quite so many words by the San, who to a large degree remain on the fringes of international indigenous peoples' movements, they also ignored numerous indigenous peoples' declarations and statements that explicitly refer to the importance of obtaining prior informed consent from holders of traditional knowledge before commercialization of this knowledge and the need to ensure that benefits derived from commercialization are equitably shared with them (see Dutfield 2002 for a review of such statements).

In June 2001, the situation changed dramatically. Ongoing vigilance by a South African-based NGO, Biowatch South Africa, assisted by the international NGO Action Aid, alerted the foreign media to the potentially exploitative nature of the CSIR-Phytopharm agreement, and a British newspaper, *The Observer*, published a leading story about the case (Barnett 2001). This was not the first time that news about the patent had been made public (e.g. *Cape Times* 1997; CSIR 1999), but the international news coverage catalysed action on the case, heightened interest in links between patents, traditional knowledge and benefit sharing, and led to pressure for a rapid response on the part of both the San and the CSIR.

Ironically, the CSIR's failure to consult with the San prior to the patent application considerably strengthened the bargaining and political leverage of the San, who, having secured the moral high ground, now had a high-profile case being followed keenly throughout the world. By contrasting images of emaciated San and obese Westerners and reinforcing popular notions of 'biopiracy' on the part of large pharmaceutical companies, the media captured the public's imagination and embarrassed the CSIR and Phytopharm, and this in turn encouraged the CSIR to enter into high-level negotiations with the San.

For the San, the following three organizations played significant roles throughout the case:

- WIMSA, the San networking and advocacy organization established in 1996 at the request of San groups in the region to lobby for San rights
- The South African San Council, a voluntary association established as part of WIMSA by the three San communities of South Africa (the ≠Khomani, !Xun and Khwe) in November 2001
- The Cape Town-based South African San Institute (SASI), a San service NGO helping San-based organizations access funding and expertise

As a South African state institution, the CSIR was reluctant to negotiate with parties outside the country, so, through WIMSA, the South African San Council was formally mandated to represent the San of Namibia and Botswana as well as those in South Africa in all benefit-sharing negotiations about *Hoodia*. This arrangement recognized the fact that knowledge about the plant crossed national borders, and that the details of sharing benefits among San in different countries needed further consideration. WIMSA and SASI instructed their lawyer to negotiate with the CSIR on behalf of the San, and discussions between the two parties began in earnest.

Early on in the negotiations, the San faced a difficult choice. Should they oppose or even challenge the patent, based on ethical considerations and lack of novelty (the legal argument that the product was not a new invention), or should they adopt a more practical approach and actively negotiate a share of the royalties? This was a critical moral dilemma. As described by Vermeylen in Chapter 10, the sharing of knowledge is a culture-defining attribute of communities such as the San and basic to their way of life. Traditional knowledge of plants is viewed as collective and the idea of 'owning' life is abhorrent. The patenting of active compounds of *Hoodia* by the CSIR ran counter to this belief, yet brought with it lucrative opportunities.

Ultimately, however, the principle of 'no patents on life' was considered 'too expensive' (Chennells 2003) and the poverty-stricken San opted for a share of royalties. Writing to the CSIR president in 2001, the San lawyers stated that a legal challenge of any nature did 'not form part of our clients' plans', but emphasized that the San looked on their traditional knowledge regarding *Hoodia*, as well as other plant uses, as collective San intellectual property that it should not morally be possible for any individual or entity to own (Chennells 2001).[7]

6.5.2 Reaching a Memorandum of Understanding

In February 2002, three months after the formal commencement of negotiations, a memorandum of understanding was reached between the CSIR and the South African San Council including the following key aspects.

---

[7] Of interest is the subsequent appeal against the patent by the European Patent Office, on the basis of it lacking novelty and being based on prior art. The appeal was subsequently overturned.

- The CSIR acknowledged that the San were the 'custodians of an ancient body of traditional knowledge and cultural values, related inter alia to human uses of the *Hoodia* plant', and that such knowledge pre-dated scientific knowledge developed by Western civilization over the past century.
- The CSIR committed itself to recognizing the role of indigenous peoples as custodians of their own knowledge, innovations and practices, and to providing for fair and equitable benefit sharing.
- The San acknowledged and accepted the CSIR's explanation of the 'context' in which it first registered the P57 patent, without having first engaged the San in negotiations with respect to material transfer, information transfer and associated benefit sharing.
- The CSIR recognized the San as originators of the body of traditional knowledge associated with human uses of *Hoodia*.
- Any intellectual property arising from the traditional use of *Hoodia* and related to the CSIR patents for P57 remained vested exclusively with the CSIR. The South African San Council had no right to claim any co-ownership of the patents or products derived from the patents.
- The CSIR and the San committed themselves to negotiating in good faith in order to arrive at a comprehensive benefit-sharing agreement.

The parties agreed to disclose fully to each other any 'matters of significance' relating to the agreement, and that all relevant disclosable information held by the CSIR relating to the P57 patent and subsequent licensing agreements would be made available to the San.

An additional understanding considered the San and the CSIR to be the primary parties with regard to benefit sharing. This point is especially significant because it effectively excluded other groups – genuine or opportunist – from claiming benefits through prior knowledge about *Hoodia*. While this helped to address concerns expressed earlier by the CSIR and Phytopharm regarding the need to identify genuine holders of traditional knowledge about the plant, it also raised new concerns from some commentators about excluding non-San groups, such as the Nama, Damara and Topnaar, who historically occupied, and still occupy, areas where *Hoodia* grows, and undoubtedly used the plant as a medicinal remedy and as a food and water substitute.

6.5.3 *Developing Positions and Identifying Key Issues of Concern*

While the memorandum of understanding represented an important first step, a concrete benefit-sharing agreement was still some way off. At a series of CSIR-funded workshops and meetings, representatives of the San, the CSIR and, in some cases, government departments and NGOs were brought together to further articulate concerns and positions (e.g. Spies 2002). Key issues arising from these discussions focused on three main themes:

1. Building trust between the parties
2. Identifying genuine holders of traditional knowledge about *Hoodia* and potential beneficiaries
3. Ensuring the broader protection and promotion of San cultures and knowledge

6.5.3.1 Building Trust

The development of trust between the CSIR and the San emerged initially as a major concern (e.g. Spies 2002), more especially given the CSIR's history as an institution shaped by the apartheid regime and serving the interests of a repressive government for nearly 40 years. While transformation of this state institution is now well under way, its initial inertia in drawing the San into the project created mistrust and negative impressions amongst the San: how could they be sure that they would receive appropriate royalties and other benefits, and access to all the necessary information? At an early stage in the negotiations the South African San Council referred in writing to the CSIR's alleged collusion with the apartheid regime as a potential problem in building trust. This outraged the CSIR board, but the frank exchanges that ensued cleared the air and enabled the parties to develop a more trusting relationship as they moved towards a final agreement (Chennells 2004).

6.5.3.2 Identifying Holders of Traditional Knowledge and Beneficiaries

The San immediately commenced a process amongst communities represented by WIMSA to establish the extent to which *Hoodia* was known and used. Responses from far-flung communities in South Africa, Namibia and Botswana confirmed published records that *Hoodia*, known as *!Xhoba* to the San, was still well known and used for a number of purposes, chiefly as a sustaining veld[8] food that also reduced hunger and thirst (R. Chennells, private notes). Some informants advised against feeding the plant to small children for sustained periods, but otherwise it was confirmed to have a safe and ancient history. This bolstered the belief of the San, as the first peoples on the subcontinent, that their traditional knowledge of *Hoodia* predated that of pastoralists who had subsequently entered and settled in Southern Africa. The San view was that they had shared knowledge with all subsequent migratory groups and were thus the primary holders of traditional knowledge relating to *Hoodia*.

Despite this opinion, parties were anxious about the conflict that could arise between the San and other groups such as the Nama and Damara. Because both the plant and traditional knowledge about its use extend across Namibia, South Africa and Botswana, this matter was potentially especially complex and fraught. How

---

[8]An Afrikaans word meaning 'uncultivated lands or grassland'.

could a system be created that ensured fairness and equity across three countries and among the relatively new organizational structures set up by different San groups in those countries?

The restricted distribution of *Hoodia* suggested that not all San groups had utilized the plant within living memory (Fig. 6.2). But identifying groups that did have a clear record of historical use was near impossible, given the San's background of resettlement and dislocation over millennia, and also the manner in which the San have moved about the landscape over the centuries, aggregating and dispersing according to season and resource availability (Hitchcock and Biesele 2001). Moreover, thousands of people in southern Africa claim San descent and a recent history of using *Hoodia*. Knowledge about the appetite-suppressant properties of *Hoodia* is shared among a broad spectrum of communities in the region, including the Nama, Damara and other Khoe-speaking peoples, who share their linguistic roots with the San and have suffered a similar history of persecution and marginalization.

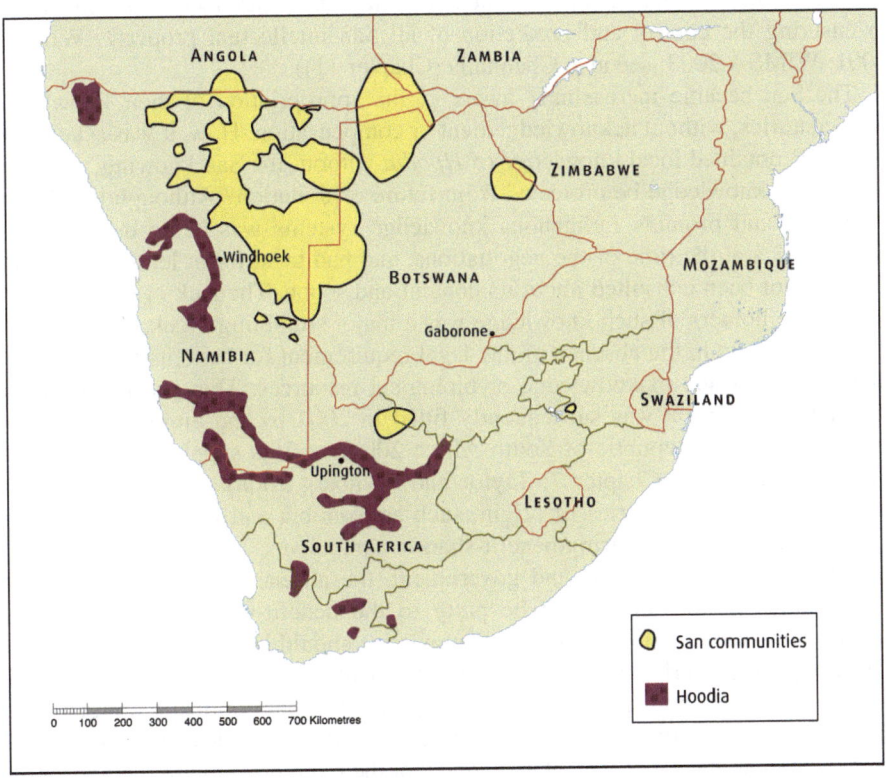

Fig. 6.2 The Distribution of *Hoodia* Species and Occurrence of the San in Southern Africa (Sources: *Hoodia* distribution from data provided by the National Herbarium Pretoria Computerised Information System PRECIS (South African National Biodiversity Institute); San data from Suzman (2001), http://www.san.org.za; after Wynberg (2006))

Resolving these uncertainties was difficult, but there was agreement amongst the San that a nit-picking exercise to link benefit sharing to specific communities using *Hoodia* would be futile and potentially divisive. WIMSA took a binding decision at an annual general meeting in 2001, after years of discussions, to the effect that heritage was indivisible, and that benefits resulting from shared heritage, such as *Hoodia*, should thus be shared equally amongst all San peoples. This decision led to a formula, arrived at collectively by the San during the negotiation process, for the equal division of financial benefits among the countries that WIMSA represented.

6.5.3.3 Protecting San Culture and Knowledge

More generally, the San sought further clarity about how they could more effectively protect their cultural heritage, including their world-renowned rock art, as well as their rich ethnobotanical and environmental knowledge. In the years preceding the benefit-sharing agreement, the San-affiliated NGO SASI had begun to assist WIMSA in establishing a code of conduct for research and researchers, and in ensuring the control and protection of all San intellectual property (WIMSA 2001; WIMSA 2003; see also Chennells (Chapter 11)).

The San became increasingly aware of the appropriation of their knowledge over centuries, without acknowledgement or compensation. How, it was asked, had the CSIR obtained local knowledge of *Hoodia* without the San knowing, and how could such knowledge be protected from future exploitation? Although legislation to protect and promote indigenous knowledge systems was being developed in South Africa at the time of the negotiations, and had been for at least 5 years, the San had not been consulted about its content and scope. The lack of legislation to protect the holders of such knowledge was a major stumbling block, requiring the San to negotiate in the absence of any legal requirement for benefit-sharing agreements with owners of knowledge or biological resources. This gap in the South African statute book was subsequently filled in 2004 by the introduction of the Biodiversity Act (Republic of South Africa 2004), and its supplementary regulations (see Wynberg, Chapter 7; Taylor and Wynberg 2008). A similar situation pertained in other countries of origin, such as Namibia and Botswana, where no law was yet in place requiring benefit-sharing agreements.

On the part of the CSIR and government, the absence of legislation created uncertainties as to who should be party to the benefit-sharing agreement and exactly how traditional or indigenous knowledge should be obtained or used. The CSIR stepped gingerly, unsure (and doubtless unenthusiastic) about 'shedding their white coats' and entering into protracted negotiations, but politically obliged to do so. A primary concern for the CSIR was to ensure that the San leaders they engaged with were genuine and representative, and that their agreement with the San would not lead to a flurry of claims to ownership of the knowledge from third parties.

Represented by Petrus Vaalbooi, chair of the South African San Council, with Roger Chennells, one of the authors of this chapter, acting as legal representative, a series of meetings ensued between the San and the CSIR. In March 2003, less than 2 years after they had commenced, negotiations concluded on the specifics of a mutually acceptable benefit-sharing agreement. Announcing the deal, Ben Ngubane, South African Minister of Arts, Culture, Science and Technology, referred to its historical significance in 'symbolising the restoration of the dignity of indigenous societies' and unleashing benefits by joining together owners of traditional knowledge and local scientists to add value to the biodiversity and indigenous knowledge systems of southern Africa. It was the 'right thing' to do, he said (Ngubane 2003).

6.5.4 The CSIR-San Benefit-Sharing Agreement

The parties negotiated at arm's length for 18 months, the San initially claiming 10% of the royalties in response to the CSIR's early offer of 3%. Both parties argued strongly in favour of their positions, each listening to the other's position, considering and reconsidering implications, moving steadily to ensure progress and finally, reluctantly, settling on the agreed amounts set out below.

In terms of the agreement (CSIR and South African San Council, 2003) the San would receive 6% of all royalties received by the CSIR from Phytopharm as a result of the successful exploitation of products (Fig. 6.3). This would be for the duration of the royalty period or for as long as the CSIR received financial benefits from commercial sales of the products (Provisions 1.5 and 2). The San would also receive 8% of the milestone income received by the CSIR from Phytopharm when certain performance targets were reached during the product development period. In the event of successful commercialization, these monies would be payable into a trust set up jointly by the CSIR and the South African San Council to raise the standard of living and well-being of the San peoples of southern Africa[9] (Fig. 6.3). Both the CSIR and the San Trust were required to put clear and transparent accounting procedures in place with regard to financial benefits paid by the CSIR and used by the San Trust. The trust would include representatives of the CSIR, the ≠Khomani, !Xun and Khwe, other San stakeholders in southern Africa, WIMSA, a South African lawyer nominated by the South African San Council and the Department of Science and Technology, with strict rules determining the distribution of funds to beneficiaries. Payments would not be made to individuals and would need to be used to attain the aims and objectives of the trust. No distribution of funds would be made to a beneficiary community or institution unless a request, approved formally by the trust, set out a detailed budget and coherent plan, identified a bank account opened by elected representatives with a proper constitution, and indicated the capacity to account fully for the proper expenditure of funds (see also Wynberg et al. Chapter 12, for a further account of the trust's operation).

---

[9] Deed of Trust of the San *Hoodia* Benefit-Sharing Trust.

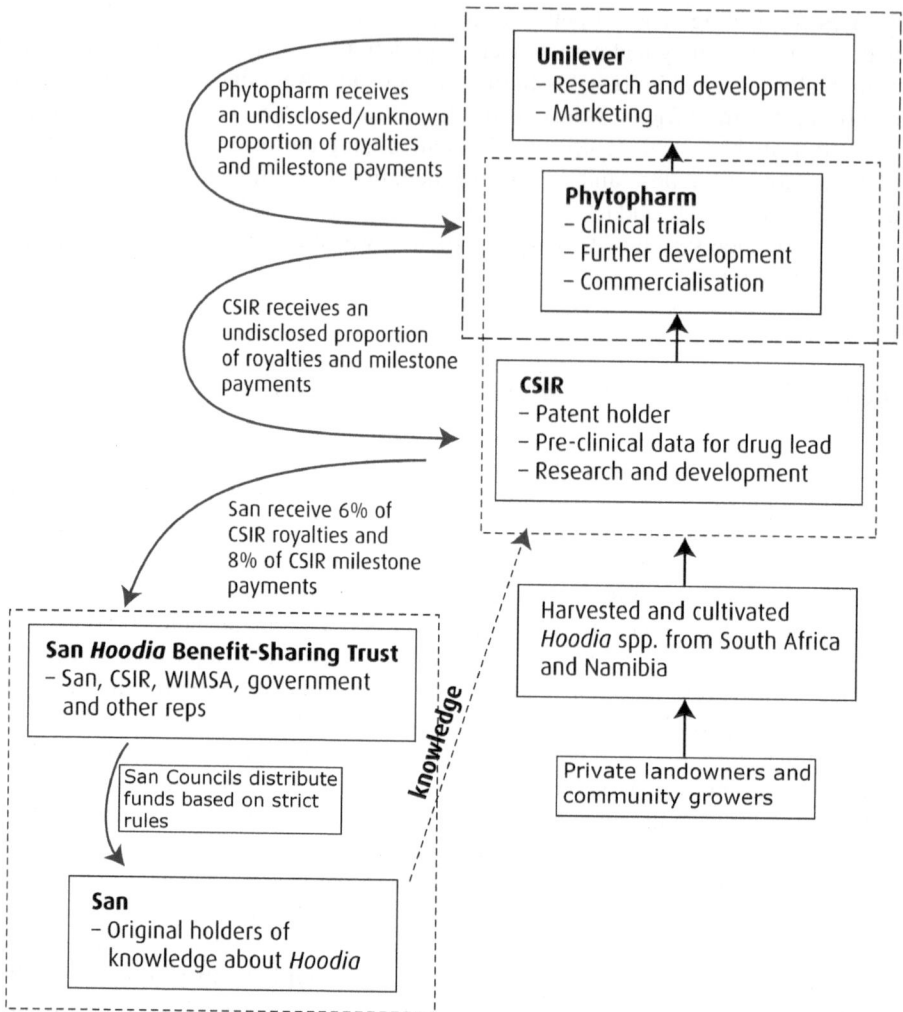

Fig. 6.3 Benefit Sharing and Value-Adding Under the San-CSIR-Phytopharm-Unilever Agreements

The benefit-sharing agreement also committed the parties to conserving biodiversity and undertaking best-practice procedures for plant collection (Provision 3.6), required the CSIR to grant the San access to existing study bursaries (Provision 3.7) and, significantly, laid the groundwork for further collaboration in bioprospecting (Provision 3.8).

In addition to spelling out the details with respect to benefit sharing and administrative aspects such as accounting, the agreement also broadly covered intellectual property issues and, importantly, set out comprehensive measures to protect and indemnify the CSIR. 'Knowledge' was defined as 'the traditional knowledge on the uses of the *Hoodia* plant that occurs in Southern Africa, originally in the hands of the San people'.

Provision 4 of the agreement specified that 'any intellectual property that may be developed or created by the CSIR, including any patent, trade mark or plant breeder's right, as a result of any use of the traditional knowledge, shall be and remain vested in the CSIR'. Moreover, the South African San Council had no right to claim any co-ownership of the patents or products derived from the patents.

Provision 6, 'Warranties and Indemnity', included an undertaking and warranty by the San that, *inter alia*, it was the legal custodian of traditional indigenous knowledge on the use of *Hoodia*; that it would not assist or enter into an agreement with any third party for the development, research and exploitation of any competing products or patents; that it would not approach Phytopharm or Pfizer to obtain additional financial benefits; and that it would not contest the enforceability or validity of the CSIR's right, title and interest in the P57 patent and related products.

A further provision on third-party claims (Provision 9) set out various measures to protect the CSIR against claims by any third party for intellectual property infringement and stipulated that a successful third-party claim against the CSIR could lead to a review of the agreement to accommodate claimants in the sharing of financial benefits. It also required the South African San Council to share financial benefits with a third party if the latter were successful in proving a claim.

In February 2005, the San Trust, formally named the San *Hoodia* Benefit-Sharing Trust, was registered. The content of the trust document was discussed over several meetings, including a consultative conference at Upington, South Africa, in October 2003, during which San delegates from South Africa, Namibia and Botswana debated issues and agreed upon guiding principles relating to benefit sharing. There was unanimous agreement that 75% of all trust income would be equally distributed to the then constituted San councils of Namibia, Botswana and South Africa; and that 25% would be retained by the trust for internal and administration purposes and for allocation to WIMSA. Priorities within the region, such as education, leadership empowerment and land security, were agreed upon as non-binding recommendations to the councils. Principles for benefit sharing that would bind the trust were unanimously endorsed by the WIMSA annual general meeting in December 2003 (WIMSA 2004). The trust began its work in earnest, electing a chair, secretary and treasurer, and started engaging with the practical challenges of distributing milestone income received from the CSIR, at that time a total of some R569,000 (see Wynberg et al. Chapter 12). The derivation of this amount was from two milestone payments to the CSIR, from Pfizer and Unilever respectively, from which 8% was allocated to the San *Hoodia* Trust (Table 6.2).

6.6 *Hoodia* Booms and Busts: 2001–2006

At the same time as institutional arrangements were being established to share benefits arising from *Hoodia* commercialization, a swathe of opportunistic *Hoodia* growers and traders were emerging outside the context of the CSIR-Phytopharm-Unilever agreements. The publicity generated by the agreements, the marketing

Table 6.2 Benefit-Sharing Payments to the San-*Hoodia* Trust from the CSIR, Paid into the Trust Bank Account on 11 May, 2005

Date	Payments received by CSIR	Foreign currency	ZAR amount	San portion
02/03/2000	First milestone Pfizer licence	US$500,000	3 245 750.00	259 660.00
14/03/2005	Unilever licence with Phytopharm milestone payment	350 020	3 867 791.00	309 423.28
	Total		7 113 541.00	569 083.28

opportunities presented by traditional San use of the plant and the patent awarded to the CSIR had led to frenzied interest in *Hoodia* amongst plant traders. By 2004 concerns about the threats posed to natural populations through unregulated collection led to the inclusion of *Hoodia* species in Appendix II of CITES.

By 2006 trade had escalated exponentially—and, in many cases, illegally—from just 25 tons in 2004 to more than 60 tons of wet, harvested material per year, sold as ground powder for incorporation into non-patented dietary supplements (see Fig. 6.4). In North America in particular, dozens of *Hoodia* products were being advertised on the Internet and sold in drugstores and pharmacies as diet bars, pills, drinks and juice, all traded by a myriad of companies 'free-riding' on the publicity and clinical trials of Phytopharm and Unilever. The CSIR patent was focused on the *Hoodia* extract, and nothing prevented other companies from simply selling the raw material for incorporation into herbal and dietary supplements. Many products were of dubious authenticity, contained unsubstantiated quantities of *Hoodia*, made unfounded claims and implied association with the San, who received no benefits (e.g. FDA 2004).

For example, an advertisement by the US-based BioMed Pharmaceuticals promoted Trimphetamine as the 'first commercially available product containing the revolutionary *Hoodia gordonii* cactus plant', based on a standardized natural extract of the plant, and another US-based company, Hi-Tech Pharmaceuticals, marketed a similar *Hoodia*-based product, Lipodrene, citing use of *Hoodia* as an appetite suppressant by the San. A rather barefaced advertisement for the Hoodoba '*Hoodia* gordonii diet pill' described the 'push by western drug companies' to 'sideline the indigenous people and turn this remarkable plant into a synthetic prescription drug', and then went on to do the same, by using the image and knowledge of the San to market the product as a natural extract (see www.*hoodia*-dietpills.com). An Internet advertisement (since removed) for Aloe *Hoodia* described how Pfizer had decided to invest 'millions' to research the benefits of the plant as a new anti-obesity drug and an advertisement for Pure *Hoodia* referred to the success of clinical trials for *Hoodia* (see www.purehoodia.com). These and related products raised important ethical and legal issues, more significantly in their neglect of the San and countries of origin as beneficiaries of commercialization, but also in the extent to which they free-rode on the research done by the CSIR and Phytopharm to demonstrate safety and efficacy.

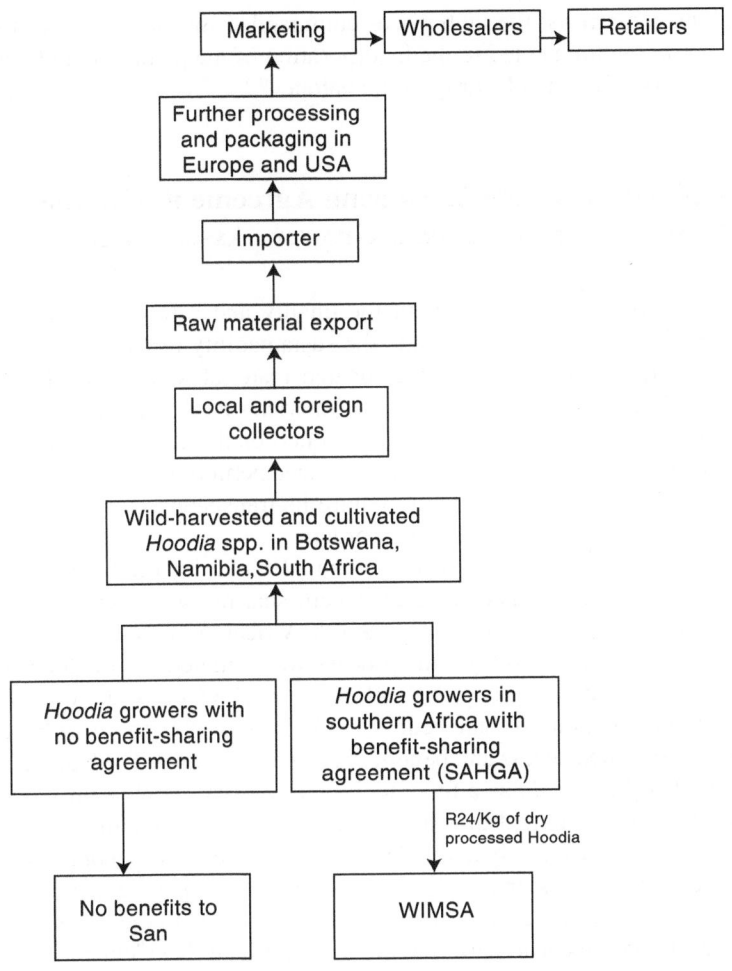

Fig. 6.4 Benefit Sharing through SAHGA and the Hoodia Value Chain Based on Trade of Raw Material

Concerns led to the closer analysis of products by the US Food and Drug Administration (FDA), which revealed that many had little or no *Hoodia* content and lacked adequate evidence of safety (e.g. FDA 2004). The US Federal Trade Commission (FTC) also brought action against spammers sending e-mail messages about *Hoodia* weight-loss products, alleging that the claims made for the products were false and unsubstantiated (FTC 2007). Along with this boom, poaching and illegal harvesting of wild *Hoodia* was widespread and unregulated, and farmers planted hundreds of hectares in the expectation of the boom to come. In South Africa and Namibia, illegal trade and harvesting of *Hoodia* resulted in a number of prosecutions and arrests; the high prices commanded for the dry product of up to US$200 per kilogram had led to the incorporation of the plant into a global underground network of diamonds, drugs and abalone.[10]

6.7 Negotiating a Benefit-Sharing Agreement with the Southern African *Hoodia* Growers Association

From 2006, increasing concern about the quality and safety of material sold as *Hoodia*, and about over-harvesting and the sustainability of *Hoodia* supply, led to a more regulated industry based on cultivated material. Greater vigilance on the part of the FDA and FTC as well as the American Herbal Products Association rapidly reduced the number of illegitimate products on the US market, and regulators in South Africa, Namibia and Botswana introduced permitting procedures to prohibit the harvesting of *Hoodia* in the wild, require its transparent cultivation and set in place mechanisms to track trade across borders.

In South Africa, those involved in growing *Hoodia* for the herbal and dietary supplement market negotiated another benefit-sharing agreement with the San, based on a levy on processed *Hoodia* (South African San Council and Southern African *Hoodia* Growers 2006). This process was initiated in late 2005 when the San were approached by a group of South African *Hoodia* growers who were cognizant of their obligations to share benefits with the San under the 2004 Biodiversity Act and its anticipated access and benefit-sharing regulations. The San realized that the new market for *Hoodia* as a food additive or dietary supplement was likely to grow over the years, and that they had a right to share the benefits. Because these products did not relate directly to the P57 patent and the use of *Hoodia* extracts, the San were legally able to sign an additional benefit-sharing agreement with *Hoodia* growers that was not in breach of their prior agreement with the CSIR.

Negotiations commenced between the South African San Council (again acting on behalf of WIMSA) and the SAHGA, which represented the interests of some commercial growers of *Hoodia* in South Africa who had agreed to comply with certain standards of best practice, safety, fair trade and benefit sharing. In March 2006 a preliminary benefit-sharing agreement was concluded with the SAHGA. In terms of the agreement 6% of the gross value of *Hoodia* sold would be allocated to WIMSA – 4% into a trust for the San and 2% to WIMSA or the South African San Council. No member was permitted to sell to vendors engaged with the production or marketing of illegal *Hoodia* products.

Royalties of R176,000 (US$22,000) trickled in from this agreement, but it was soon replaced with another more comprehensive initiative that included the majority of South African *Hoodia* growers as well as South African provincial environmental

---

[10] An endangered marine mollusc, highly sought after as a cultural delicacy in the East and subject to high levels of illegal trade.

government agencies responsible for ensuring sustainable use of *Hoodia* and administering permits (see Wynberg, Chapter 7). After a year of negotiations, during which the different realities and negotiating positions of the respective parties emerged in an increasingly mature climate of transparency, a benefit-sharing agreement was concluded in March 2007 between the San and the newly formed SAHGA. This had been preceded by the signing of a memorandum of understanding in January 2007 between the San (represented by WIMSA), *Hoodia* growers and the Western Cape and Northern Cape environmental departments[11] which captured the intention of the parties as they entered negotiations.

The benefit-sharing agreement (WIMSA and the Southern African *Hoodia* Grower's Association 2007), drafted to be compliant with the provisions of the Biodiversity Act, acknowledged the San to be the primary holders of traditional knowledge about *Hoodia*, having a legal right to share benefits arising from its harvesting, growing and marketing. It also recognized the urgent need for regulation to minimize impacts on wild populations and to ensure the attainment of standards of legality, safety and fair trade. The stated objectives of the non-profit SAHGA included:

- To regulate the legal production and harvesting of *Hoodia* by its members, in compliance with the CBD
- To promote a sustainable *Hoodia* industry in southern Africa
- To liaise with all role players
- To gather and exchange relevant information relating to permits, quality control, sales and compliance
- To promote research

Two San representatives were elected to be members of the board of directors and another two were designated as observers. WIMSA in turn was to ensure the proper administration of financial benefits, and to further the objectives of SAHGA and help with effective marketing of *Hoodia*. Although the stated intention of the parties was to create an exclusive joint venture and benefit-sharing agreement, WIMSA was entitled, on good cause, to motivate to SAHGA for the signing of another, separate agreement. Parties additionally agreed to promote SAHGA as the only legitimate source of *Hoodia* for the food, food additive and dietary supplement market, outside of the CSIR-Unilever agreement and to 'inform the world' that *Hoodia* products outside of the two benefit-sharing agreements were illegal under the CBD. The agreement also, significantly, acknowledged other groups holding traditional knowledge of *Hoodia*, such as the Nama and Damara, and provided an opening for further discussions and possible agreements with such groups.

Financial benefits for the San were formulated based on a ZAR 24 levy charged on each kilogram of dry, processed *Hoodia*, paid prior to the issuing of CITES export permits and to be revisited on an annual basis. Calculation of the levy was

---

[11] Unpublished signed legal agreement.

based on a number of factors including the previous SAHG levy of 6% of the sale from the farm, as well as conditions in the world *Hoodia* market – recognizing its high levels of fluctuation, the need for the levy to be affordable for growers and other equity considerations. The agreement also provided for re-evaluation after 1 year, taking into account the need for the eventual amount to be fair to both sides. Parties were fully aware that the original figure of 6% had been agreed upon with SAHG without the benefit of adequate knowledge about trade volumes, without extensive calculation of the likely implications of percentages for all parties, and without sufficient reliable information to fix an appropriate percentage with certainty. Conflict resolution was proposed through mediation or, failing this, through arbitration. The agreement, whilst negotiated in South Africa, was drafted in such a way as to welcome and enable the participation of *Hoodia* growers from neighbouring Namibia and Botswana in due course.

At the time of going to press, the SAHGA benefit-sharing agreement had failed to deliver any of the promised payments to the San, largely because the Minister had not endorsed the agreement, thereby rendering it unenforceable by government in terms of the 2008 Biodiversity, Access and Benefit-Sharing Regulations (see Wynberg, Chapter 7). The agreement is currently being renegotiated and redrafted in such a way that compliance at all levels of government will give effect to the primary intention of the parties, namely that benefits from growing *Hoodia* be shared with the San.

6.8 Implementation Challenges

The conclusion of two benefit-sharing agreements is a major achievement. Indeed, these agreements are very rare examples indeed of the much-touted benefits from bioprospecting having practical realization. Nonetheless, implementation poses a number of challenges to the San, to those involved in the *Hoodia* industry and to regulators and policymakers.

6.8.1 Decision-Making and the Distribution of Benefits

One of the key challenges concerns the way in which decisions will be made about the sharing of existing and, hopefully, future benefits. The CSIR-San agreement will pay 6% of royalties into the San Trust, which, as described above, has begun preparing the policies and structures necessary to distribute anticipated flows of money. The fair and equitable distribution of large sums of money to beneficiaries in three different countries would be an enormous challenge for any organization. The fact that these beneficiaries are impoverished indigenous peoples, wrestling with problems of organizational cohesion and underdevelopment as described in Chennells et al. (Chapter 9) and Wynberg et al. (Chapter 12), makes this challenge even more complex.

The SAHGA benefit-sharing agreement also promises to deliver millions of rands within the next few years directly to the San regional organization WIMSA. This money has no prior allocations earmarked, so distributing it wisely will present the relatively inexperienced board with major challenges.

The responsibility on San individuals on the San Trust, as well as on the WIMSA board, to meet heightened expectations and to act wisely and transparently in the eyes of the watching world will be onerous indeed. NGOs entrusted with providing support will be expected to shoulder part of this burden. The objective will be to minimize the negative social and economic impacts, and the intracommunity conflicts that may arise following the introduction of large sums of money into San communities.

There is limited international and local experience in the administration and implementation of such agreements, and few, if any, cases address the sharing of benefits within communities. As Barrett and Lybbert (2000) point out, benefit-sharing questions have thus far remained issues of distribution between the community in aggregate and outsiders, with little practical experience at a local and intracommunity level. There have been some early indications, however, of the divisive impact that natural product trade can have in indigenous communities. In India, for example, the commercialization of Jeevani *(Trichopus zeylanicus)*, a wild plant with anti-fatigue properties, has led to divisions amongst the tribal community, the Kanis, as to how their knowledge should be used (Tobin 2002; Gupta 2004; Chaturvedi, Chapter 13). In Peru, a 1996 agreement of the International Cooperative Biodiversity Group also led to conflict between organizations representing local Aguarana communities, as well as at a national level (Tobin 2002. Greene 2004).

In the case of the San, intracommunity issues are especially complex. The organizations set up to represent the San politically are relatively new, and the introduction of Western values and economies into supposedly traditional communities, already fractured and 'hybridized', presents a set of diverse social and economic problems. Robins (2002) describes the social complexities of contemporary San identity, knowledge and practice, and charts the intracommunity divisions and conflict that emerged between self-designated 'traditionalists' and 'Western bushmen' when San land claims were lodged in the Northern Cape province of South Africa. While these claims resulted in significant benefits for the San, they also had unintended consequences in the form of conflict. Robins (2002) points out the contradictions between San 'cultural survival' and the promotion of the values of 'civil society' and 'liberal individualism', a conclusion that holds particular resonance for the *Hoodia* case, contextualized as it is within the international discourse of indigenous peoples, a vigilant NGO community alert to biopiracy cases, and a new policy framework that requires fair and equitable benefit sharing for the use of traditional knowledge.

The possible compensation of other groups that use *Hoodia* and have traditional knowledge of the plant, such as the Nama, Damara and Topnaar, also represents a major challenge that will have to be resolved, especially once *Hoodia* markets mature and significant profits begin to flow. Already, Namibia has articulated a position that supports the inclusion of the Nama and other groups in benefit-sharing arrangements, particularly relating to participation in *Hoodia* growing projects (Ministry of Environment and Tourism, 2007). This position is bolstered by the fact that *Hoodia* wild and cultivated populations occur in areas occupied by Nama communities. A 2008 meeting between *Hoodia* growers from South Africa and Namibia recognised the need for an alignment of approaches on both benefit sharing as well as marketing (University of Cape Town and University of Central Lancashire, 2008), and led to the San agreeing to commence negotiations with Nama traditional leaders in Namibia. However, Nama communities, even more than the San, lack organizational structures and cohesion and have required substantial support to get to the point at which they can negotiate their rights, as well as manage and disburse incoming funds. In the interim, structures have emerged through the *Hoodia* Growers Association of Namibia to raise and manage funds for the inclusion of the Nama and other indigenous groups in the *Hoodia* industry, with the intention of building their organizational and technical capacity in the medium to long term. The objective is that these two important indigenous groupings, both holders of traditional knowledge relating to *Hoodia*, will formalise a practical agreement about how benefits from the growing of *Hoodia* are to be shared between their respective communities.

6.8.2 *Regional Differences in Benefit-Sharing Policies*

One of the more interesting aspects of the case lies in its regional implications. *Hoodia* is a biological resource that is shared across national political boundaries, and knowledge of the plant is similarly shared by communities straddling these boundaries. Thus far, however, South Africa has played a leading role: in lodging the patent, developing commercial partnerships with multinational companies, negotiating benefit-sharing arrangements with the San and facilitating legal trade in the plant. Botswana and Namibia, by comparison, although involved in harvesting and cultivating *Hoodia*, have not yet legalized trade in the plant nor developed commercial partnerships.

Moreover, as described in Wynberg (Chapter 7), South Africa has adopted access and benefit-sharing (ABS) legislation and supports recognizing the San as a community with clear rights to benefit from *Hoodia*, but Namibian and Botswanan policies have been more ambivalent. Neither Namibia nor Botswana has ABS legislation and in both countries benefits from *Hoodia* are considered to belong to the state,[12] rather than the San or other traditional knowledge holders. Unsurprisingly, these divergent policy approaches have led to concerns.

---

[12]The CBD regulates relationships between states and affirms that countries have national sovereignty over their genetic resources. The distribution of such benefits is left to national discretion, within the requirements of article 15 and article 8j, which declare that holders of traditional knowledge have rights over their knowledge (see also Wynberg and Laird, Chapter 5).

A central concern relates to the difficulties of controlling trade. There have been many reports of illegal material entering South Africa from Namibia and being exported from South Africa under permit. The areas in which the plant occurs are typically very remote and illegal harvesting is difficult to monitor and combat. Steps could be taken to address these concerns, but their efficacy would be questionable without a regionally coherent position on *Hoodia* use. Strategic approaches to value-adding and the use of marketing tools such as geographical indications would also be undermined in the absence of strong regional collaboration – needed at government, industry, farmer and community level.

Although the San Trust, which was set up to disburse benefits, already implements benefit sharing across regional boundaries, based on an acknowledgment of the shared nature of *Hoodia* knowledge, there is clearly a need for benefit-sharing strategies to be developed at regional and national levels in cases where genetic resources are shared across boundaries.

6.8.3 Hoodia Trade and Markets

Without the development of a sustainable and viable industry, no benefits will emerge, and a set of complex challenges also confronts those involved in trading and growing *Hoodia*. As with other agricultural commodities, *Hoodia* markets follow the law of supply and demand, which determines the prices, quantities and allocation of resources (Wall 2001). In line with the classical model described by Homma (1992), *Hoodia* has moved through a rapid expansion phase, followed by a stabilization phase, where an equilibrium has been reached between supply and demand, supposedly close to the maximum capacity of extraction of the product. Prices have consequently risen because of the inability to meet a growth in demand, which, as Wynberg (Chapter 7) describes, has led to the adoption of policies to protect the sector or stimulate sustainable production. The shrinking of the resource, restrictive policies on wild harvesting and incentives to cultivate have stimulated a substantial increase in *Hoodia* cultivation, with the challenge now to secure markets for this material. Similarly, the recent withdrawal of Unilever from *Hoodia* development has led to an unstable market and questions as to whether a product can be developed that is safe, efficacious and desirable to consumers.

Further challenges lie in the monitoring of compliance with the benefit-sharing agreements. While this is relatively straightforward and effective for the CSIR-San benefit-sharing agreement, which has clear milestones and reporting mechanisms, it is less so for the SAHGA benefit-sharing agreement. Many *Hoodia* traders wish their trade volumes to remain confidential, yet the agreed levy to the San cannot be calculated without this information. The SAHGA agreement depends largely on good faith and the proactive declaration by growers of volumes traded and monies owed. As already noted, however, there is no government endorsed benefit-sharing agreement to date and many growers have proved reluctant to provide the necessary information and levies. It is anticipated that the redrafted agreement will assist with enforcing compliance by *Hoodia* growers and traders.

Hoodia sales are also currently severely depressed as a result of an increased crackdown by compliance institutions on new and unregulated products. The environmental government agencies responsible for issuing permits are not legally required to provide SAHGA with this vital information, but with the promulgation of the regulations and an intended amendment of the SAHGA constitution, it is anticipated that the intended benefit-sharing payments will flow to the San within the next year.

Some of the greatest threats to benefit sharing lie outside the region. Although no conclusive figures exist, it is well known that extensive *Hoodia* populations have been established elsewhere in the world. Some of this genetic material may have been acquired before the entry into force of the CBD, and some could just as easily have been smuggled out of the region without the required permission. It is therefore possible that a *Hoodia* industry could thrive outside of southern Africa, without channelling benefits to the original knowledge holders.

6.9 Conclusion

The *Hoodia* case study tells a complex story with many strands, and from it a number of important lessons and conclusions can be drawn that ought to be integrated into ongoing debates about ways in which benefit sharing for communities can be made more equitable. One of the most crucial lessons is the need to get it right from the start. Obtaining the prior informed consent of communities holding knowledge about biodiversity from the very outset of a project – and engaging them as active partners – is an absolutely fundamental principle of benefit sharing. The *Hoodia* case study illustrates what can go wrong when this principle is ignored.

The negotiating process between the CSIR and the San has demonstrated the importance of relationship building between role players and of having in place a political climate conducive to fair deliberations. It has also affirmed the importance of community-based institutions through which holders of traditional knowledge can be represented in negotiations and benefits can be channelled. The process has highlighted the prominent role played by NGOs, legal representatives and intermediaries in benefit sharing – in this case not only in helping the San attain their rights, but also in shaping San politics and economic development.

One of the major impacts of the commercialization of *Hoodia* has been the wide-ranging interest it has aroused about the importance of protecting traditional knowledge and ensuring that holders of such knowledge receive fair compensation. Amongst the San, the *Hoodia* case is considered an important empowering tool to enable more informed decisions to be made about their intellectual property and ways to protect it. At government level, the case has led directly to an increased focus and emphasis on biodiversity and its potential value, and, in South Africa, on the inclusion of prior informed consent and benefit sharing in biodiversity legislation and the requirement of disclosure of origin prior to the granting of patents. At the international level, the case is widely considered to have set precedents about the ways in which holders of traditional knowledge should be compensated for their knowledge.

There is clearly an urgent need to introduce new forms of protection for traditional knowledge that not only give communities rights over their knowledge, but also enable the wider preservation and promotion of such knowledge systems. The *Hoodia* case demonstrates the value of an integrated system to protect and promote traditional knowledge and, in addition, the importance of so-called 'defensive protection' to prevent the misappropriation of traditional knowledge.

Some of the lessons are still to be learnt and some are only unfolding. If the San receive significant sums of money, it will be extremely difficult to determine who benefits and how benefits are spread across geographical boundaries and within communities, and to minimize the negative social and economic impacts and conflicts that could follow the introduction of large sums of money into impoverished communities. The due compensation of other communities such as the Nama, Damara and Topnaar will also require careful consideration, including the fact that participation in government-assisted growing schemes is a significant benefit. Above all, beneficiaries will need continued legal, administrative and technical support to claim

---

Box 6.1 What is *Hoodia*?

Species of the genera *Hoodia* and related *Trichocaulon* have long been used as thirst quenchers and appetite suppressants (White and Sloane 1937) (Fig. 6.5). Both genera are members of the Apocynaceae family, succulent perennials adept at storing moisture during the long dry spells of their native habitats (CITES 2004). The unusual flowers are flat and saucer-like in shape and brownish in colour, and form prolifically near the stem tips in summer, when they are often characterized by a distinct carrion smell to attract pollinating flies. The stems are cylindrical, leafless and typically multi-angled, ribbed and spiny. More than 20 species have been recorded from southern Africa, although the species of most interest for their appetite-suppressing properties are *Hoodia gordonii, H. currorii, H. flava, H. lugardii* (now *H. currorii* subsp. *lugardii*), *H. piliferum* (previously *Trichocaulon piliferum*), *H. officinale* (previously *Trichocaulon officinale*) (Van Wyk and Gericke 2000; White and Sloane 1937; patent WO 9846243A2). Vernacular names for the plants include *ghaap* (sometimes spelt *ngaap, ghap, gap* or *gnaap*) and !khobab, |goa.-|, |khowa.b, |goai-|, |khoba, |khoba.b|s, |khowab, |goab, otjinove, !nawa#kharab, sekopane or *seboka* (White and Sloane 1937; Smith 1966; Malan and Owen-Smith 1974; Van Wyk and Gericke 2000; Hargreaves and Turner 2002; CITES 2004).

The genus *Hoodia* was named in 1830 after Van Hood, a keen grower of succulent plants (Barkhuizen 1978). Two types of *ghaap* were previously

(continued)

Box 6.1 (continued)

recognized by colonists and indigenous communities alike: true ghaap (*Trichocaulon* species) and the allied genus *Hoodia,* which was known as *bitterghaap, bobbejaanghaap* (translated from Afrikaans as 'baboon soap', referring to the slimy inner texture of the skins of *Hoodia* and to the fact that it is not suitable for human use), *jakkalsghaap, slangghaap, wildeghaap* or *wolweghaap*, the prefix used to denote worthlessness or inferiority (Smith 1966). *Trichocaulon* species have smaller, more rounded and almost thornless stems with small flowers, whilst *Hoodia* have long, narrow and thorny stems with large showy flowers. However, Bruyns (1993) showed there to be considerable overlap between the two groups and united all *ghaap* species under *Hoodia*.

Fig. 6.5 Flowering *Hoodia gordonii*, Ceres (Karoo), Western Cape, South Africa (Photo: Rachel Wynberg)

what is rightfully theirs, and to do so in a manner that deliberately – though cautiously – brings tangible and effective benefits to the original holders of *Hoodia* knowledge.

References

Barkhuizen, B. P. (1978). *Succulents of Southern Africa*. Cape Town: Purnell.
Barnard, A. (1996). Lourens van der Post and the Kalahari debate. In P. Skotnes (Ed.), *Miscast: negotiating the presence of the Bushmen*. Cape Town: UCT Press.
Barnett, A. (2001). In Africa the *Hoodia* cactus keeps men alive. Now its secret is "stolen" to make us thin. *The Observer*, 17 June. http://www.guardian.co.uk/world/2001/jun/17/international-educationnews.businessofresearch. Accessed 5 April 2008.
Barrett, C. B., & Lybbert, T. J. (2000). Is bioprospecting a viable strategy for conserving tropical ecosystems? *Ecological Economics, 34*, 293–300.
Bloch, A., & Thomson, C. (1995). Position of the American Dietetic Association: phytochemicals and functional foods. *Journal of the American Dietetic Association, 95*(4), 329–334.
Boonzaier, E., Malherbe, C., & Smith, A. (1996). *The Cape Herders, a history of the KhoiKhoi of South Africa*. Cape Town: David Philip.
Bruyns, P. (1993). A revision of *Hoodia* and *Lavranea* (*Asclepiadaceae–Stapelieae*). *Botanische Jahrbücher für Systematik, Pflanzengeschichte und Pflanzengeographie*, 115.
Cape Times (1997). Plant helps in fighting obesity, 26 June.
CBD (2002). Bonn guidelines on access to genetic resources and fair and equitable sharing of the benefits arising out of their utilization. Decision VI/24, 2002, Secretariat of the Convention on Biological Diversity, Quebec. www.cbd.int/doc/publications/cbd-bonn-gdls-en.pdf. Accessed 22 March 2008.
Chennells, R. (2001). Letter to the executive president of the CSIR, 5 July.
Chennells, R. (2003). Ethics and practice in ethnobiology, and prior informed consent with indigenous peoples regarding genetic resources. Paper presented at a conference on biodiversity, biotechnology and the protection of traditional knowledge, St Louis, MS, 4–6 April.
Chennells, R. (2004). Attorney's notes relating to San *Hoodia* case.
CITES (2004). Amendments to Appendices I and II of CITES. Proposal to the thirteenth meeting of the Conference of the Parties, Bangkok, Thailand, 2–14 October. www.cites.org/common/cop/13/raw_props/BW-NA-ZA-*Hoodia*.pdf. Accessed 7 April 200.
Colchester, M. (2003). Salvaging nature: indigenous peoples, protected areas and biodiversity conservation. World Rainforest Movement, Montevideo. www.wrm.org.uy/subjects/PA/texten.pdf. Accessed 5 April 2008.
CSIR (1999). Annual Report 1998, Council for Scientific and Industrial Research. Pretoria, South Africa.
CSIR (2001). Adding value to South Africa's biodiversity and indigenous knowledge through scientific innovation. www.csir.co.za. Accessed 4 February 2002.
CSIR and the South African San Council (2003). Benefit-Sharing Agreement, 24 March 2003.
Deacon, H. J., & Deacon, J. (1999). *Human beginnings in South Africa: uncovering the secrets of the Stone Age*. Cape Town: David Phillip Publishers.
Department of Trade and Industry (2008). Address by the Minister of trade and industry. Mandisi Mpahlwa to the National Council of Provinces, 29 May 2008. http://www.thedti.gov.za/article/articleview.asp?current=1&arttypeid=2&artid=1575.
Diamond, J. (1998). *Guns, germs and steel: a short history of everybody for the last 13, 000 years*. London: Vintage.
Dixey, R. (2004). Press conference to discuss the announcement of a licence and joint development agreement between Phytopharm and Unilever for *Hoodia gordonii* extract.

Douglas, J. (2008). Phytopharm, unilever scrap *Hoodia* development, 14 November. http://www.easybourse.com/bourse-actualite/marches/update-phytopharm-unilever-scrap-*hoodia*-development-562184. Accessed 30 March 2009.

Dutfield, G. (2002). Indigenous peoples' declarations and statements and equitable research relationships. In S. A. Laird (Ed.), *Biodiversity and traditional knowledge: equitable partnerships in practice*. London: Earthscan.

Ekpere, J. A. (2001). *The OAU's model law: the protection of the rights of local communities, farmers and breeders, and for the regulation of access to biological resources. An explanatory booklet.* Organisation of African Unity, Scientific, Technical and Research Commission, Lagos, Nigeria.

FDA (2004). 75-day premarket notification of new dietary ingredients. U.S. Food and Drug Administration, Washington, DC. http://www.fda.gov/ohrms/dockets/dockets/95s0316/95s-0316-rpt0218-vol156.pdf.

FTC (2007). FTC stops international spamming enterprise that sold bogus *Hoodia* and human growth hormone pills', 10 October, Federal Trade Commission, Washington, DC. www.ftc.gov/opa/2007/10/*hoodia*.shtm. Accessed 5 April 2008.

Geingos, V., & Ngakaeaja, M. (2002). Traditional knowledge of the San of Southern Africa: *Hoodia gordonii*. Presentation to the Second South-South Biopiracy Summit: 'Biopiracy – Ten Years Post Rio', 22–23 August 2002. Johannesburg, South Africa.

Greene, S. (2004). Indigenous people incorporated? Culture as politics, culture as property in pharmaceutical bioprospecting. *Current Anthropology, 45*(2), 211–238.

Gupta, A. K. (2004). WIPO-UNEP study on the role of intellectual property rights in the sharing of benefits arising from the use of biological resources and associated traditional knowledge. WIPO publication number 769, World Intellectual Property Organisation and United Nations Environment Programme.

Hargreaves, B. J., & Turner, Q. (2002). Uses and misuses of *Hoodia*. *Asklepios, 86*, 11–16.

History of the Bosjesmans, or Bush People (1847). The Aborigines of Southern Africa. London: Chapman, Elcoate & Co.

Hitchcock, R. K., & Biesele, M. (2001). San, Khwe, Basarwa, or Bushmen? Terminology, identity, and empowerment in southern Africa. http://www.khoisanpeoples.org/indepth/ind-identity.htm. Accessed 30 March 2008.

Hitchcock, R., Ikeya, K., Biesele, M., & Lee, R. (2006). *'Introduction', updating the San: image and reality of an African People in the 21st century*, Senri Ethnological Studies, vol 70. Osaka: National Museum of Ethnology.

Homma, A. K. O. (1992). *The dynamics of extraction in Amazonia: a historical perspective. Advances in Economic Botany, vol 9*. New York: New York Botanical Garden.

KFO (2006). Annual Report. Kuru Family of Organisations, Ghanzi, Botswana. www.kuru.co.bw.

Khoisis (1983). Occasional papers of the Departments of Anthropology and Archaeology, vol 4. University of Stellenbosch, South Africa.

Lee, R. B. (1976). *Introduction–Kalahari hunter gatherers: studies of the 'Kung San and their neighbours'*. Cambridge, MA: Harvard University Press.

Lee, R. B., Hitchcock, R., & Biesele, M. (2002). Foragers to first peoples. *Cultural Survival Quarterly, 26*(1), 18–20.

Malan, J. S., & Owen-Smith, G. L. (1974). The ethnobotany of Kaokoland. *Cimbebasia Series B, 2*(5), 131–178.

Marloth, R. (1932). *The Flora of South Africa, vol 3*. Cape Town/London: Darter Bros/Wheldon & Wesley.

Masson, F. (1796). *Stapelia Nova: or, a collection of several new Species of that genus discovered in the interior of Africa*. London: W. Bulmer & Co.

Ministry of Environment and Tourism (MET) (2007). Conference on *Hoodia*, 22–23 February 2007. Schützenhaus, Keetmanshoop, Namibia.

Ngubane, B. (2003) Address by the Minister of Arts, Culture, Science and Technology at the signing of a benefit-sharing agreement between the CSIR and the San, 24 March, Molopo Lodge, South Africa, www.info.gov.za/Speeches/2003/03032410461009.htm. Accessed 5 April 2008.

Penn, N. (1996). Fated to perish: the destruction of the Cape San. In P. Skotnes (Ed.), *Miscast: negotiating the presence of the Bushmen*. Cape Town: UCT Press.

Phytopharm (1997). Phytopharm to develop natural anti-obesity treatment, Press release, 23 June. www.phytopharm.com/news/newsreleases/?page=10&id=1814. Accessed 5 April 2008.

Phytopharm (2001). Successful completion of proof of principle clinical study of P57 for obesity, Press release, 5 December. www.phytopharm.com/news/newsreleases/?page=7&id=1749. Accessed 5 April 2008.

Phytopharm (2002). Future development of P57, Press release, 30 July. www.phytopharm.co.uk/news/newsreleases/?page=6&id=1726. Accessed 5 April 2008.

Phytopharm (2003). Annual report and accounts for the year ended 31 August 2003. Cambridge, UK: Phytopharm plc.

Phytopharm (2004). Interim results for the period to 29 February 2004, Press release, 5 May. www.phytopharm.com/files/news/1588/Interim2004.pdf. Accessed 5 April 2008.

Phytopharm (2007). Phytopharm initiates stage 3 activities of joint development agreement for *Hoodia* extract with Unilever, 24 September. www.phytopharm.com/news/newsreleases/?id=7324. Accessed 5 April 2008.

Phytopharm (2008). Preliminary results for the period ended 30 September 2008, 27 November 2008. http://www.phytopharm.co.uk/news/newsreleases/?id=16517.

Phytopharm (2008). Unilever returns rights to *Hoodia* extract, 12 December. http://www.phytopharm.com/news/newsreleases/?id=16553. Accessed 30 March 2009.

Phytopharm (2009). Progression of the *Hoodia* project under Phytopharm plc, 4 March. Letter to South African government departments.

Povey, K. (2007). Unilever, personal communication, October.

Republic of South Africa (2004). National Environmental Management: Biodiversity Act (Act 10 of 2004). *Government Gazette*, Pretoria. www.info.gov.za/gazette/acts/2004/a10-04.pdf. Accessed 5 April 2008.

Robins, S. (2002). NGOs, "bushmen", and double vision: the ≠Khomani San land claim and the cultural politics of "community" and "development" in the Kalahari. In T. A. Benjaminsen, B. Cousins, & L. Thompson (Eds.), *Contested resources: challenges to the governance of natural resources in South Africa*. Cape Town: programme for Land and Agrarian Studies, School of Government, University of the Western Cape.

SASI (2007). Personal communication with A. Thoma and other members of the South African San Institute, November.

Silvain, R. (2006). Drinking, fighting and healing: San struggles for survival and solidarity in the Omaheke region. In R. Hitchcock, K. Ikeya, M. Biesele, & R. Lee (Eds.), *Updating the San: image and reality of an African people in the 21st century. Senri Ethnological Studies, vol 70*. Osaka: National Museum of Ethnology.

Skotnes, P. (1996). *Miscast: negotiating the presence of the Bushmen*. Cape Town: UCT press.

Smith, C. A. (1966). Common names of South African Plants. Botanical Survey Memoir No. 35, Department of Agricultural Technical Services, Pretoria.

Soodyall, H. (2006). *The prehistory of Africa. Tracing the lineage of modern man*. Cape Town: Jonathan Ball Publishers.

South African San Council and Southern African *Hoodia* Growers (2006). Benefit-sharing agreement and joint venture, March 2006. Unpublished signed legal agreement.

Spies, C. (2002) Report on workshop on benefit sharing between South African San Council and the CSIR on the *Hoodia* P57 project, 13–14 June, Molopo Lodge, Northern Cape province, South Africa.

Stafford, L. (2009). After another cancelled partnership, the future of *Hoodia* remains unclear, *HerbalEGram*, 6(2). http://cms.herbalgram.org/heg/volume6/02%20February%20/Hoodia_Nixed.html. Accessed 30 March 2009.

Stephenson, D.J. (Ed.) (2003). Case Study of the Patenting of P57 and Related Benefit-Sharing and Licensing Agreements with Regard to the Intellectual Property Rights of the San Peoples of Southern Africa. First Peoples Worldwide, Virginia, USA.

Steyn, H.P., & du Pisani, E. (1985). Grass-seeds, game and goats: an overview of Dama subsistence. *SWA Wissenschaftliche Gesellschaft*, Journal XXXIX.

Suzman, J. (2001). *Regional assessment of the San in Southern Africa*. Windhoek, Namibia: Legal Assistance Centre.

Taylor, M., & Wynberg, R. (2008). Regulating access to South Africa's biodiversity and ensuring the fair sharing of benefits from its use. *South African Journal of Environmental Science and Policy, 15*(2), 217–243.

Theal, G. M. (1892–1919). *The history of South Africa (Vol. 4)*. London: George Allen & Unwin.

Tobin, B. (2002). Biodiversity prospecting contracts: the search for equitable agreements. In S. A. Laird (Ed.), *Biodiversity and traditional knowledge: equitable partnerships in practice*. London: Earthscan.

Unilever (2009). Termination of the Unilever *Hoodia* Project, 3 March, Letter written to South African government departments.

University of Cape Town and University of Central Lancashire (2008). San-!Khoba: indigenous peoples, consent and benefit sharing. *Hoodia* Information Workshop, Protea Hotel, Upington, 8 May.

Van den Eynden, V., Vernemmen, P., & Van Damme, P. (1992). *The ethnobotany of the Topnaar*. Belgium: Universiteit Ghent.

Van Wyk, B. E., & Gericke, N. (2000). *People's plants: a guide to useful plants of Southern Africa*. Pretoria: Briza Publications.

Vermeylen, S. (2007). Contextualising 'fair' and 'equitable': the San's reflections on the *Hoodia* benefit-sharing agreement. *Local Environment, 12*(4), 1–14.

Von Koenen, E. (2001). *Medicinal, poisonous, and edible plants in Namibia (4th ed)*. Windhoek, Namibia: Klaus Hess Publishers.

Wall, N. (2001). *The complete A–Z economics handbook*. London: Hodder & Stoughton.

Watt, J. M., & Breyer-Brandwijk, M. G. (1962). *The medicinal and poisonous plants of Southern and Eastern Africa (2nd ed)*. London: Livingstone.

White, A., & Sloane, B. L. (1937). *The Stapelieae III (2nd ed)*. CA: Pasadena.

Wilmsen, E. N. (1989). *Land filled with flies: a political economy of the Kalahari*. IL: University of Chicago Press.

WIMSA. (2001). *Media and research contract of the San of southern Africa, approved by the WIMSA Annual General Assembly on 28 November*. Windhoek, Namibia: Working Group for Indigenous Minorities in Southern Africa.

WIMSA. (2003). *The San of Southern Africa: heritage and intellectual property*. Windhoek, Namibia: Working Group for Indigenous Minorities in Southern Africa.

WIMSA. (2004). *WIMSA Annual Report, April 2003 to March 2004*. Windhoek, Namibia: Working Group for Indigenous Minorities in Southern Africa.

WIMSA and the Southern African *Hoodia* Grower's Association (2007). Benefit-sharing agreement and joint venture, as contemplated by the South African Biodiversity Act 10 of 2004.

Wynberg, R. (2004). Rhetoric, realism and benefit sharing – use of traditional knowledge of *Hoodia* species in the development of an appetite suppressant. *World Journal of Intellectual Property, 6*(7), 851–876.

Wynberg, R. (2006). Identifying pro-poor, best practice models of commercialisation of southern African non-timber forest products, PhD thesis, University of Strathclyde, Glasgow.

Now today, I hear that some people are saying that there are laws that have stopped the hunting of animals and I don't know who told that person to create this law that has stopped us from killing animals. We really don't know how someone can come home and start to pass laws on what we used to know as our land, and these things are being decided in our absence (anonymous San community member, South Africa)

Chapter 7
Policies for Sharing Benefits from *Hoodia*

Rachel Wynberg

Abstract This chapter provides an analysis of the policies and laws that have emerged in southern Africa to regulate the harvesting, trade, and commercial development of *Hoodia*. Many of these policies have evolved rapidly, alongside the commercialisation of *Hoodia* and the increasing prominence of access and benefit sharing as a policy issue. However, policy implementation has been challenging, complicated by the fact that both the traditional knowledge that was used to develop *Hoodia* and the species involved cross national borders and involve a number of distinct indigenous communities. Each of the three countries with which *Hoodia* and its knowledge are associated has evolved a distinct regulatory approach towards the plant's conservation and use, and to the way in which ABS issues are framed. Moreover, southern African countries are at very different points of legislating for ABS, hold inconsistent understandings of the role of traditional knowledge holders, and also have varied approaches and capacities for bioprospecting and natural product development. While these more slippery political issues of benefit sharing and indigenous peoples remain disconnected and incoherent between southern African countries, those countries have increasingly collaborated to design joint policies for *Hoodia* management, with steps put in place to collaborate more strongly on poaching, trade and the transport of illegally harvested material. This bodes well for future cooperation and suggests a positive environment within which policy resolutions can be found.

Keywords policy • law • access and benefit sharing • southern Africa • indigenous peoples

---

R. Wynberg
Environmental Evaluation Unit, University of Cape Town, Private Bag X3, Rondebosch, 7701, Cape Town, South Africa
e-mail: rachel@iafrica.com

7.1 Introduction

A bewildering complexity of policies and laws has emerged in southern African countries to regulate the harvesting, trade, and commercial development of *Hoodia*. These exist at a convoluted interface between biodiversity conservation, access and benefit sharing (ABS), intellectual property rights, science and technology, and traditional knowledge. As this chapter illustrates, the manifold laws that regulate each of these components typically have little coherence, at best, or are contradictory, at worst. Additionally, they are administered in substantially different ways by a range of government institutions, with overlapping mandates and unclear roles and responsibilities.

These complexities are exacerbated for *Hoodia* regulation because both the traditional knowledge that was used in the commercial development of *Hoodia* and the species involved cross national borders, involving the governments of South Africa, Namibia and Botswana, as well as indigenous communities of the San, Nama, Damara and other groups. However, each of the three countries with which *Hoodia* and its knowledge are associated has evolved a distinct regulatory approach towards the plant's conservation and use, and to the way in which ABS issues are framed. The case thus raises the important questions of how benefits can be equitably shared across communities and regions in such situations, what policies best serve the interests of indigenous communities and national governments, and how such policies can be coherently implemented at a regional level.

Through the lens of the *Hoodia* case study, this chapter explores the variety of policy tools that governments in southern Africa have used to manage and implement requirements of the Convention on Biological Diversity (CBD) for conservation, sustainable use and equitable benefit sharing, the pitfalls encountered along the way, and likely challenges arising in the future.

7.2 ABS Regulation in Southern Africa

Biodiversity conservation, traditional knowledge and intellectual property protection are increasingly under the legal spotlight, and many countries today, including South Africa, Namibia and Botswana, have adopted national and regional laws to comply with international agreements and policies governing these issues. Some of the key international treaties are the CBD, the Convention on Trade in Endangered Species (CITES), the International Treaty on Plant Genetic Resources for Food and Agriculture, and the Trade-Related Intellectual Property Rights (TRIPS) Agreement of the World Trade Organization. These treaties have been complemented by policy statements such as the United Nations Declaration on the Rights of Indigenous Peoples, which represents a crucial advance in furthering the rights of indigenous peoples (UN 2007). National and regional experiences of implementing these various agreements are extremely varied but have in common a certain degree of legal novelty and fluidity because of the untested nature and newness of the issues.

As in many other countries, ABS is a relatively new legal domain for South Africa, Namibia and Botswana. Each of these countries, however, has a diverse suite of policies and laws of relevance to ABS and the use and trade of biological resources, including *Hoodia*. Table 7.1 summarizes key laws and policies in each country, illustrating that two broad regulatory approaches to *Hoodia* have emerged: the first based on established protocols for species management, conservation, sustainable use and trade, and the second – the focus of this chapter – concentrating on the more recent issues of benefit sharing, prior informed consent and traditional knowledge protection.

What is clear is that although ABS policy frameworks have been under development in southern Africa since the mid 1990s, their adoption has been erratic and their implementation weak. This embryonic state of ABS policy and law in the region, the general confusion that has resulted from the overlapping mandates of different government bodies and research institutions, and the multiplicity of only partially relevant laws (see Table 7.1) have led to an extremely incoherent policy climate for *Hoodia* regulation. In fact, most policy interventions in southern African countries to regulate access to *Hoodia* genetic resources, protect traditional knowledge associated with the plant and ensure the fair sharing of benefits from its use have emerged 'after the fact' or, in some cases, not at all. The initial acquisition of traditional knowledge about the appetite-suppressing properties of *Hoodia*, obtained by the Council for Scientific and Industrial Research (CSIR) from botanical accounts (Marloth 1932) in the 1960s, the CSIR's patenting of these properties in 1997 without the consent of the San, the traditional knowledge holders, and the CSIR's subsequent licensing agreement with Phytopharm to commercially develop a product elicited little, if any, policy response from any southern African government at the time. Only after considerable media attention in 2001 did the CSIR consent to negotiations with the San to develop a benefit-sharing agreement, but this was largely done in a legal vacuum. It was partly the unfolding of these experiences and the high-profile nature of the case that gave impetus to the development of binding laws in South Africa and elsewhere.

In South Africa, this has been encapsulated by the National Environmental Management: Biodiversity Act (10 of 2004) ('the Biodiversity Act') and the 2008 promulgation of ABS regulations to give effect to the Act. As described in Box 7.1, this regulatory framework for the first time addresses the need for bioprospectors to obtain prior informed consent from custodians of biodiversity and holders of traditional knowledge before initiating any project. It also requires a benefit-sharing agreement to be developed between different stakeholders to ensure that holders of traditional knowledge or custodians of biodiversity are fairly compensated.

The inclusion of prior informed consent and benefit sharing in South African legislation represents a major step forward in redressing past imbalances in the way in which biodiversity and traditional knowledge have been exploited. Yet the implementation of these laws presents major challenges (Crouch et al. 2008; Taylor and Wynberg 2008). Aside from the fact that the Act fails to vest ownership of genetic resources in the state, (due to a concern that to do so may infringe constitutionally protected property rights), and thus limits the extent to which wider community

Table 7.1 Key Laws and Policies Pertaining to ABS and the Use, Trade and Conservation of *Hoodia* in South Africa, Namibia and Botswana

South Africa

Policy/law	Relevant provisions/content
South Africa	
Constitution of the Republic of South Africa (Act 108 of 1996)	Conservation and ecological sustainability are given prominence in the Bill of Rights. Does not vest ownership of genetic resources in the state
White Paper on the Conservation and sustainable Use of South Africa's Biological Diversity (1997)	Goal 3 aims to 'ensure that benefits derived from the use and development of South Africa's genetic resources serve national interests'
Indigenous Knowledge Systems Policy (2004)	Is an enabling framework to stimulate and strengthen the contribution of indigenous knowledge to social and economic development in South Africa
Policy Framework for the Protection of Indigenous Traditional Knowledge through the Intellectual Property System and the Intellectual Property Laws Amendment Bill (2008)	Provides that the law of trademarks/geographical indications may be able to provide protection of certain names/features associated with traditional knowledge, that a national council consisting of experts on traditional knowledge can advise the Minister of Trade and Industry and the registrar of intellectual property, and that communities may form business enterprises to administer and commercialize their traditional intellectual property
National Environmental Management Act (107 of 1998)	Gives legal effect to the Constitution and to the White Paper on Environmental Management Policy. Sets in place procedures and mechanisms for cooperative governance and regulates environmental impact assessments
National Forests Act (84 of 1998)	Overall purposes include the sustainable use, management and development of forests, the restructuring of state forestry, the protection of certain forests and trees, and the promotion of community forestry. Certain activities may be licensed in state forests, including the collection of biological resources
National Environmental Management: Biodiversity Act (10 of 2004)	Provides for the management and conservation of biological diversity, the use of indigenous biological resources in a sustainable manner, and the fair and equitable sharing among stakeholders of benefits arising from bioprospecting involving indigenous biological resources
National Environmental Management: Biodiversity Act (10 of 2004): Threatened or Protected Species Regulations	List *Hoodia gordonii* and *Hoodia currorii* as protected species
National Environmental Management: Biodiversity Act (10 of 2004): Bioprospecting, Access and Benefit Sharing Regulations	Make a distinction between the 'discovery' and 'commercialization' phase of a bioprospecting project, but regulate the two phases identically. National minister responsible for environment issues permits for bioprospecting and export for bioprospecting purposes. Foreigners may only apply for permits jointly with a South African collaborator. Export must be in the public interest. Benefit-sharing agreement may be refused if there is no provision for enhancing scientific and technical capacity to conserve, use and develop biodiversity or to promote conservation

7 Policies for Sharing Benefits from *Hoodia*

Agricultural Pests Act (36 of 1983)	Provides for the prevention and combating of agricultural pests, and regulates the importation of controlled goods, including plants, pathogens and insects. Prohibits any person from importing into South Africa any plant without a permit. The Minister of Agriculture has imposed a number of controls concerning the import of seeds, for example by requiring phytosanitary certificates
Patents Act (57 of 1978)	Governs the registration and granting of patents for inventions. Provides for the patenting of microorganisms and microbiological processes but prohibits the patenting of plants and animals. The 2005 amendment requires an applicant for a patent to furnish information on the use of indigenous biological resources or traditional knowledge in an invention
Traditional Health Practitioners Act (35 of 2004)	Provides a regulatory framework to ensure the efficacy, safety and quality of traditional health care services
Various Provincial Ordinances and Acts	The provinces have, in all, 28 legal instruments for nature conservation. In general they allow for the establishment and protection of nature reserves, for the conservation of threatened species, and for fishing and hunting. Many of these laws are outdated and the nine provinces are at different stages of phasing out old laws and developing and implementing new ones *Hoodia* is listed as a protected genus in the Northern Cape, through the Environmental Conservation Ordinance (19 of 1974), and in the Western Cape and Free State provinces through similar legislation, and a permit is required from provincial authorities to collect, cultivate, transport or export *Hoodia* spp.
Namibia	
Namibian Constitution (1990)	Obliges the government to adopt policies aimed at the 'maintenance of ecosystems, essential ecological processes and biological diversity of Namibia and utilization of living natural resources on a sustainable basis for the benefit of all Namibians, both present and future'. Recognizes the existence and importance of customary law, declaring it to be of the same value as common law. Vests ownership of all non-privately owned land and natural resources in the state
Environmental Management Act (7 of 2007)	Requires benefit sharing. Sets up an advisory council which includes access to genetic resources
Environmental Investment Fund of Namibia Act (13 of 2001)	Establishes a fund to support environmental and natural resource management in Namibia
Nature Conservation Ordinance (4 of 1975)	Primary legislation governing nature conservation in Namibia. Sets in place a permitting system for protected species, including *Hoodia* spp., requiring prior authorization for harvesting and trade. Requires a permit for the picking and transport, sale, donation, export and removal of protected plants. Requires the written permission of landowners before any indigenous plant is picked. The 1996 amendment gives rights over wildlife and tourism to communal farmers

(continued)

Table 7.1 (continued)

Policy/law	Relevant provisions/content
National Agriculture Policy (1995)	Aims to achieve growth in agricultural production and profitability, ensure food security, improve living standards for farmers and farm workers, and promote sustainable use of land and natural resources. Aims to promote diversification of rural livelihoods
Traditional Authorities Act (17 of 1995)	Requires traditional authorities to ensure that members of their communities use natural resources sustainably and in a manner that conserves the environment
Research, Science and Technology Act (23 of 2004)	Establishes the National Commission on Research, Science and Technology
Plant Quarantine Act (7 of 2008)	Requires phytosanitary certificates to accompany exports of raw material
Botswana	
Wildlife Conservation and National Parks Act (28 of 1992)	Governs the use of resources in national parks, protected areas and game reserves, as well as procedures to access biological resources. Implements CITES
Agricultural Resources (Conservation) Act (2006)	Provides for regulations to control access to biological resources and sets in place permitting requirements for the harvesting, export and trade of *Hoodia*, except when for domestic use. Includes no specifications for benefit sharing and prior informed consent
Forest Act (38 of 2004)	Protects forests and regulates the use of forest resources
Tribal Land Act (32 of 2002)	Recognizes that a community collectively owns the land as well as the resources on it, but gives decision-making power over those resources to tribal land boards. Has relevance for the provision of prior informed consent and the negotiation of benefits
National Biodiversity Strategy and Action Plan (2004)	Recommends the development of a strategy on ABS, traditional knowledge and property rights
Community-Based Natural Resource Management Policy (2006)	Includes traditional knowledge protection and benefit sharing and aims to protect the intellectual property rights of communities

benefits can be secured (see Taylor and Wynberg 2008), its permitting requirements are unduly onerous and complex, especially for applicants engaged in the exploratory or research phases of a project (Crouch et al. 2008). As described below, even companies simply wanting to trade biological material such as sliced and dried *Hoodia* now face a labyrinth of permitting procedures that are poorly aligned between multiple layers of government bureaucracy. The confusion that results has direct impacts on the ability of communities to obtain concrete benefits from biodiversity.

Box 7.1 The Legal and Institutional Framework for Bioprospecting and ABS in Southern Africa

South Africa

A diverse suite of policies and laws of relevance to ABS is in place in southern African countries (Table 7.1), but these are most developed in South Africa, which has actively engaged in bioprospecting for decades. Until recently the commercial development of South Africa's biological resources took place in a legislative vacuum, but now this has been filled by a specific regulatory ABS framework, articulated through the National Environmental Management: Biodiversity Act (10 of 2004) and regulations passed under that Act in 2008. A key objective of the Act is to provide for 'the fair and equitable sharing among stakeholders of benefits arising from bioprospecting involving indigenous biological resources'.

The Act requires bioprospectors to obtain a permit for bioprospecting involving indigenous biological resources, and for the export of these resources. Prior informed consent is required from all stakeholders before a permit is issued, and this must be reflected in a benefit-sharing agreement. Stakeholders include those who give access to indigenous biological resources (e.g. private landowners, the state or a community that communally owns land) and indigenous communities who use the resource traditionally or have knowledge of its properties. Benefit-sharing agreements must be entered into with both categories of stakeholders and, in addition, a material transfer agreement must be entered into with those who give access to resources (see Fig. 7.1). These must be approved by the minister responsible for the environment who may take steps to ensure that the negotiations around the agreement take place on an equal footing and that the resultant agreement is fair and equitable. The minister may also refuse to approve a benefit-sharing agreement if the agreement does not allow for enhanced scientific knowledge and technical capacity to conserve, use and develop indigenous biological resources. The Act is prescriptive about what must be included in benefit-sharing and material transfer agreements.

(continued)

Box 7.1 (continued)

The Act establishes a Bioprospecting Trust Fund, managed by the director-general of environmental affairs, into which all money arising from benefit-sharing agreements must be paid, and from which all payments to stakeholders will be made. This applies equally in the case of *Hoodia*, despite that agreement having been negotiated prior to the promulgation of the Biodiversity Act. The Fund, however, involves no discretionary powers and will simply act as a bank account through which any incoming funds will be channelled to the existing San-*Hoodia* Trust or other identified beneficiaries.

Importantly, the regulations prescribe that existing bioprospecting projects, which include those relating to *Hoodia*, must, within six months of the regulations coming into effect (i.e. by 1 October 2008), submit an application to the minister for a bioprospecting permit, together with a benefit-sharing agreement or a written request for assistance to negotiate such an agreement. At the time of writing, these had been submitted to the department responsible for environmental affairs but not yet formally approved.

Namibia

Although Namibia's legal and institutional framework for ABS is not as advanced as South Africa's, it has adopted a progressive and proactive policy approach to ensuring access to genetic and biological resources and the fair sharing of benefits derived from these resources, commencing as early as 1997. Although national legislation is not yet in place, ABS has been effected in practice through bilateral agreements, existing laws that facilitate ABS more broadly (Table 7.1) and the active engagement of key government departments, research institutions and non-governmental organizations through the high-level Bioprospecting Committee. In contrast to South Africa, Namibia has decided to delay promulgation of its draft legislation entitled *Access to Genetic Resources and Associated Traditional Knowledge* until the international ABS regime is adopted and finalized (Nott and Wynberg 2008).

Botswana

Botswana, like Namibia, does not have any laws that explicitly regulate bioprospecting activities, ABS and traditional knowledge protection, although the proposed National Environmental Management Act will likely include provisions to regulate these activities. Some sectoral laws include provisions of relevance to ABS, but most natural resource laws have been designed to meet the objectives of conservation and sustainable use and do not explicitly address benefit sharing (UNU–IAS 2008). Despite broad recognition of the importance of traditional knowledge, no laws or policies set in place requirements for informed consent prior to the use of this knowledge. Similarly, requirements for prior informed consent are not articulated in any law. In practice, however, a comprehensive permitting procedure is in place for foreign researchers wishing to do any form of research in Botswana.

7 Policies for Sharing Benefits from *Hoodia*

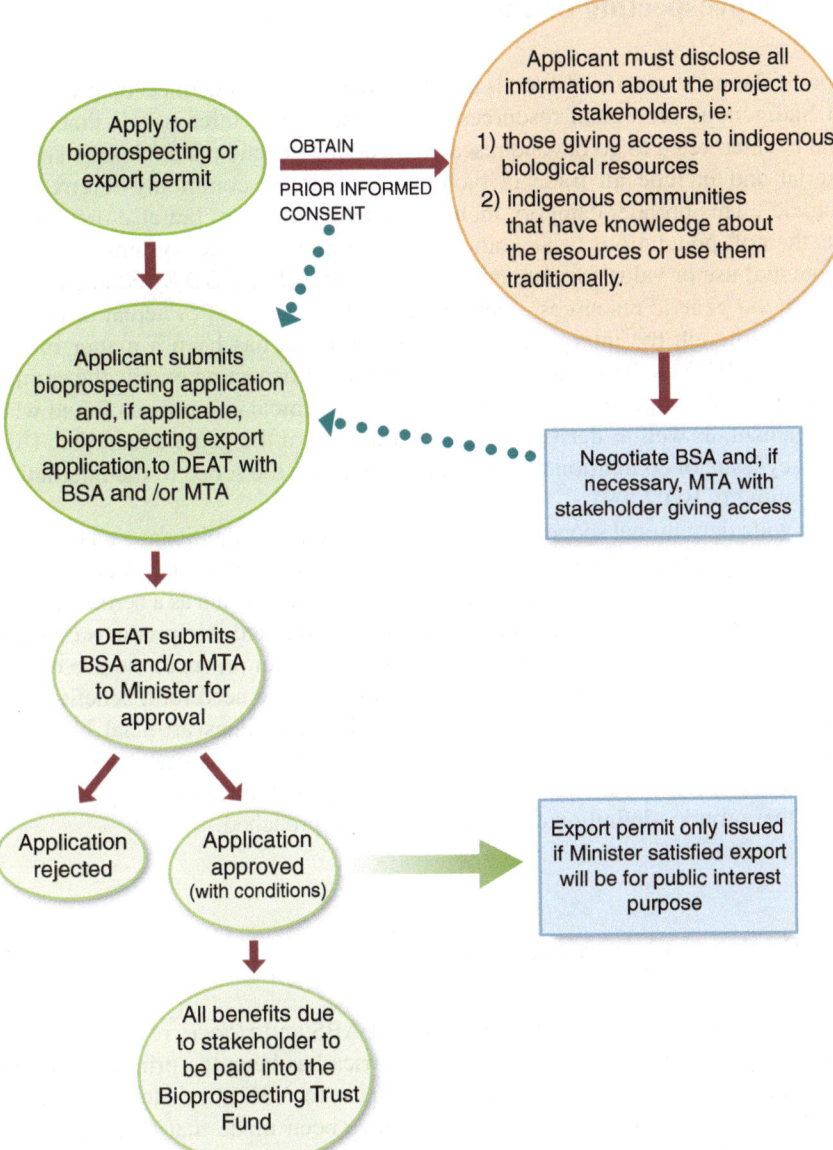

Fig. 7.1 Process Prescribed by the ABS Regulations to Obtain a Bioprospecting Permit or Bioprospecting Export Permit

7.3 Bioprospecting or Biotrade?

Of particular relevance for *Hoodia* is that the Biodiversity Act defines 'bioprospecting' and 'indigenous biological resources' very widely. The inference is that the Act could be interpreted to go beyond research involving genetic material or biochemical material and include all trade in biological resources, commonly referred to as 'biotrade'. This broader category includes genetic resources, but also organisms or parts thereof, populations or any other biotic component of ecosystems with actual or potential use or value for humanity (UEBT 2007). The CBD by contrast focuses narrowly on genetic resources – defined as 'genetic material of actual or potential value' – although the interpretation of this definition has been a matter of some dispute (see IFPMA 2006; Rosenberg 2006) as bioprospecting entails the commercial use not only of genetic material, but also of chemical compounds found within the organism, as well as derivatives and products from the genetic material. This is also a key issue of contention in negotiations to define the scope of the international ABS regime (CBD 2009).

The wide definitional scope of the Biodiversity Act has significant – albeit unclear and complex – implications for *Hoodia*. For example, *Hoodia* has been developed both as a genetic resource, to be included in patented extracts, and as a herbal medicine, where the raw material is simply dried, cut and incorporated into products (Wynberg and Chennells, Chapter 6; Wynberg and Taylor 2009). In practice, the use of traditional knowledge for both types of products prescribes the need for a benefit-sharing agreement, but there are clearly overlapping and sometimes artificial boundaries between trade in genetic resources and that in biological organisms. Experiences of regulating ABS for *Hoodia* are thus likely to set important international precedents as to how these murky definitional questions can be approached.

7.4 Regional Coherence in ABS

Other debates have also arisen in the context of ABS and *Hoodia* regulation. As the economic powerhouse of the region, South Africa has been the primary developer of *Hoodia*, and the furthest advanced with respect to ABS laws. However, this has not been without regional conflict. Claims have been made that both the genetic resources and knowledge associated with *Hoodia* were originally collected in Namibia, and, as has been described in other chapters (Wynberg and Chennells, Chapter 6; Vermeylen, Chapter 10), the recognition and compensation of indigenous groups other than the San that have traditional knowledge of *Hoodia* have surfaced as important issues requiring resolution. Steps have now been taken to include other indigenous groups in initiatives to share benefits from *Hoodia*, but the initial conflicts point to a wider problem of traditional knowledge use and protection not being adequately incorporated into ABS frameworks. In South Africa, for example, the Department of Arts and Culture has been a strong leader in developing policy and, recently, laws to promote indigenous knowledge systems (Table 7.1). However, the Department of Water and Environmental Affairs – which has little first-hand knowledge of working with communities – remains responsible for ensuring that holders of traditional knowledge of biodiversity are fairly compensated for the use of this knowledge. Although government departments recognize these overlaps, in practice issues associated with traditional knowledge and biodiversity often fall between two stools, aggravated by the seemingly insurmountable challenges of identifying knowledge holders and beneficiary communities.

At the regional level this is even more complicated. Although all three countries are signatories to the United Nations Declaration on the Rights of Indigenous Peoples (UN 2007), there have been differences of opinion among them about the right to self-determination of indigenous peoples and the control that indigenous people should have over natural resources on traditional lands. While South African policy has placed particular emphasis on affirming the rights of indigenous peoples, both Namibia and Botswana have adopted policy approaches that do not necessarily support specific indigenous groups, but rather see indigeneity as a characteristic of all citizens. This reflects the current unease in many African States with the United Nations definition of 'indigenous peoples', regarded as potentially confusing in a continent where the majority of citizens consider themselves indigenous. In neither Botswana nor Namibia are the San referred to as a separate linguistic or ethnic group, and in Botswana they are euphemistically referred to in development policies as 'remote area dwellers'. This has played out directly in the *Hoodia* case, evidenced by disagreements between countries about the extent to which the San should be recognized as the original holders of traditional knowledge about *Hoodia*, and whether or not they should be primary beneficiaries of benefit-sharing agreements.

Ideally, common regional policies should govern the sharing of benefits arising from the use of strategic resources such as *Hoodia* and associated traditional knowledge, but in practice the complexity and diversity of legal and institutional mechanisms across countries, and the multiple jurisdictions and cross-cutting nature of conservation, trade, traditional knowledge, intellectual property and benefit sharing, mean that governments have found it difficult to fully streamline policies.

7.5 Linkages Between *Hoodia* Conservation, Trade and ABS

Some progress has been made in bringing regional policy coherence to *Hoodia* conservation, but the interface between ABS requirements and regulations for *Hoodia* trade and use has emerged as one of the most neglected issues in policy implementation. As Table 7.1 illustrates, a fairly comprehensive body of legislation has developed to regulate the use and conservation of *Hoodia* species, mostly embedded within nature conservation laws in South Africa, Namibia and Botswana. The genus *Hoodia* has also been included on Appendix II of CITES, due to increasing interest in the commercial application of *Hoodia* spp. and concomitant concerns about the threats posed to natural populations through unregulated collection (CITES 2004).

Included in this CITES listing is an annotation requiring CITES permits for all parts and derivatives of *Hoodia* species except those bearing a label 'Produced from *Hoodia* spp. material obtained through controlled harvesting and production in collaboration with the CITES Management Authorities of Botswana/Namibia/South Africa under agreement no. BW/NA/ZA xxxx'. The inclusion of this annotation is significant as its intent is to ensure that countries in which *Hoodia* naturally occurs (the so-called 'range states') capture the economic benefits that accrue from commercialization. Unusually for CITES, this signifies an attempt to link trade and benefit sharing, although in practice no exemptions have been granted and CITES permits are required for trade in all parts and derivatives, including seeds.

CITES is implemented in all three countries through environmental departments, based on existing conservation laws. To varying extents a comprehensive and relatively coherent permitting system thus exists to protect wild *Hoodia* populations, comply with CITES 'non-detriment requirements' (essentially showing that harvesting has been conducted in accordance with sustainability guidelines) and provide information about harvesting pressures, illegal activities and resource status (Mupetami 2007). Permits are also required in each country for the collection of seeds, the establishment of cultivated areas, the transport and export of material, and in some cases manufacturing activities associated with *Hoodia*. Wild harvesting is generally discouraged, or permitted only under stringent conditions, and, in response to the initial exponential growth in *Hoodia*, trade authorities are increasingly collaborating to set in place mechanisms to track trade across borders.

ABS, CITES and the wider trade in species are integrally linked in many ways (INA 2004; Ruiz Miller and Lapeña 2007). However, the dovetailing of permitting requirements for each of these activities has proven extremely difficult, requiring permit applicants to comply with an incessant stream of bureaucratic procedures, administered by different authorities. Within government it is also extremely difficult to keep track of such diverse applications. To overcome some of these problems in South Africa, it has been suggested that the ABS permit system and the provincial research permit system be synchronized with current efforts to develop a uniform and coordinated permitting system for CITES, possibly through a single electronic database, which would include information about the application, its status, and existing permits granted. Early experiences of implementing benefit-sharing agreements for *Hoodia* suggest that such information could help considerably in determining, for example, the volumes of material traded, and thus the benefits due to traditional knowledge holders. As described in Wynberg and Chennells (Chapter 6), the lack of such information has been a major stumbling block preventing compliance with and implementation of existing benefit-sharing agreements.

Across all countries, however, monitoring, enforcement and compliance are key constraints that prevent the effective implementation of the *Hoodia* permitting system. Law enforcement capacity is low, the legal processes are cumbersome and seemingly full of loopholes, and the low penalties do not constitute a sufficient deterrent to transgressors, given the high value of the resource. This is exacerbated by the fact that illegal harvesting typically occurs in remote rural areas, with material quickly transported over borders, especially from Namibia to South Africa. Illegal activities peaked from 2003 to 2006, but have now declined with increasing border control, less market demand and increasing vigilance on the part of Namibian, South African and, increasingly United States regulators concerned about deceptive *Hoodia* advertising and the quality of imported material (Federal Trade Commission 2007; Wynberg and Newton 2008).

7.6 Ownership of *Hoodia* Genetic Resources

Finally, maintaining ownership of *Hoodia* genetic resources is a key area that has received attention from regulators, given the ease with which *Hoodia* can be cultivated. Existing *Hoodia* plantations in the United States, Israel and elsewhere are cause for concern and could undermine regional efforts to develop the *Hoodia* industry and bring benefits to indigenous communities in southern Africa. With this in mind, agreement has recently been reached between Namibia, South Africa and Botswana on developing a common permitting protocol to allow seed exports within these range states, but not outside. The difficulties of enforcing such requirements are enormous, however, given the ease with which seed can be transported out of the country. Increasingly, therefore, traders are looking to innovative mechanisms such as geographical indications or localized branding to secure markets. The intention is that this would be done through labels on products indicating not only that they originate from southern Africa, but also that they play a role in delivering benefits to holders of traditional knowledge. Policy tools such as these could well play a more effective enabling role than the prescriptive tools being set in place through ABS legislation.

7.7 Conclusion

Huge hurdles face regulators and administrators attempting to implement ABS in a coherent and meaningful way in southern African countries. Not only are countries at very different points of legislating for ABS, but they also hold inconsistent understandings of the role of traditional knowledge holders, and have varied approaches and capacities for bioprospecting and natural product development. As this chapter has described, many of these incongruities have played themselves out in the arena of *Hoodia* trade, use and benefit sharing. However, while the more slippery political issues of benefit sharing and indigenous peoples remain disconnected and incoherent between southern African countries, those countries have increasingly collaborated to design joint policies for *Hoodia* management, with steps put in place to collaborate more strongly on poaching, trade and the transport of illegally harvested material. This bodes well for future cooperation and suggests a positive environment within which policy resolutions can be found. Moreover, as ABS processes mature and international mechanisms are set in place, ABS policies and procedures are likely to become more streamlined and workable. But even with good policies and laws in place, it is likely that due to its cross-cutting nature, regulating ABS will always be a challenging process.

References

CBD (2009). Report of the Seventh Meeting of the Ad-Hoc Open-Ended Working Group on Access and Benefit-Sharing of the Convention on Biological Diversity, Paris, 2–8 April. www.cbd.int/doc/?meeting=abswg-07. Accessed 26 May 2009.

CITES (2004). Amendments to Appendices I and II of CITES. Proposal to the thirteenth meeting of the Conference of the Parties, Bangkok, Thailand, 2–14 October. www.cites.org/common/cop/13/raw_props/BW-NA-ZA-*Hoodia*.pdf. Accessed 7 April 2006.

Crouch, N. R., Douwes, E., Wolfson, M. M., Smith, G. F., & Edwards, T. J. (2008). South Africa's bioprospecting, access and benefit-sharing legislation: current realities, future complications, and a proposed alternative. *South African Journal of Science, 104*, 355–366.

Federal Trade Commission (2007). FTC Charges Marketers of '*Hoodia*' Weight Loss Supplements with Deceptive Advertising, 27 April 2009. http://www.ftc.gov/opa/2009/04/nutraceuticals.shtm. Accessed 31 May 2009.

IFPMA. (2006). *Guidelines for IFPMA members on access to genetic resources and equitable sharing of benefits arising out of their utilization. International Federation of Pharmaceutical Manufacturers and Associations, 7 April*. Geneva: Switzerland.

INA (2004). Expert Workshop Promoting CITES-CBD Cooperation and Synergy. International Academy for Nature Conservation, Isle of Vilm, Germany, 20–24 April. www.cbd.int/cooperation/final-report-CITES CBD_Vilm_Workshop_Report.doc.

Marloth, R. (1932). *The flora of South Africa, vol 3*. Cape Town/London: Darter Bros/Wheldon & Wesley.

Mupetami, L. (2007). Resource management and protection, Presentation to a conference on *Hoodia*, 22–23 February. Keetmanshoop, Namibia.

Nott, K., & Wynberg, R. (2008). Millenium challenge account Namibia compact, Volume 4: Thematic Analysis Report – Indigenous Natural Products, Namibia Strategic Environmental Assessment. Task order under the Project Development, Project Management, Environmental and General Engineering ID/IQ Contract no. MCC-06-0087-CON-90, Task Order No. 02.

Rosenberg, D. (7 November 2006). Some business perspectives on the international regime. UK: GlaxoSmithKline.

Ruiz Miller, M., & Lapeña, I. (Eds.) (2007). A moving target: genetic resources and options for tracking and monitoring their international flows. IUCN Environmental Policy and Law Paper No. 67/3, IUCN, Gland, Switzerland.

Taylor, M., & Wynberg, R. (2008). Regulating access to South Africa's biodiversity and ensuring the fair sharing of benefits from its use. *South African Journal of Environmental Science and Policy, 15*(2), 217–243.

UEBT (2007). BioTrade Verification Framework for Native Natural Ingredients – 2007-09-20. Union for Ethical BioTrade, Geneva. www.uebt.ch/dl/Engl-UEBT-Nat-Ingredients-Ver-framework-2007-09-20(rev1)b.pdf. Accessed 25 May 2009.

UN (2007). United Nations Declaration on the Rights of Indigenous Peoples. Adopted by United Nations General Assembly Resolution 61/295 on 13 September. www.un.org/esa/socdev/unpfii/documents/DRIPS_en.pdf. Accessed 29 April 2008.

UNI–IAS (2008). *Access to genetic resources in Africa: analysing ABS policy development in four African countries*. Development Cooperation Ireland, United Nations Environment Programme, United Nations University Institute of Advanced Studies.

Wynberg, R., & Newton, D. (2008). A Policy and Trade Assessment for *Hoodia* spp. in Namibia, Botswana and South Africa. Unpublished report, Environmental Evaluation Unit, University of Cape Town and TRAFFIC.

Wynberg, R., & Taylor, M. (2009). Finding a path through the ABS maze: challenges of regulating access and ensuring fair benefit-sharing in South Africa. In E. C. Kamau & G. Winter (Eds.), *Genetic resources, traditional knowledge and the law: solutions for access and benefit sharing*. London: Earthscan.

We lost our culture, because the colonial governments came in and took over everything, now the younger generation don't know our forefathers' culture and traditions. It's only a few of us elders who know about it. I am telling you I feel really sad, I just want to cry because we have lost our whole life.
(Rosa #Gaeses, Etosha Poort, Outjo, Namibia)

Chapter 8
The Struggle for Indigenous Peoples' Land Rights: The Case of Namibia

Saskia Vermeylen

Abstract The enclosure of commons is a historical event not limited to homelands of developed nations. Instead it also characterized their colonialization of other nations. Obtaining additional land was one of the motives of colonialism, but – for indigenous peoples – it meant more than the loss of tangible resources. This chapter, based on fieldwork with the Namibian San, indicates that the enclosure of land led to a loss of social relations that had sustained their culture and identity. Despite the fact that most San live in circumstances far different from their hunter-gatherer days, they are compelled to choose between identities defined by others, in which they are seen as either 'backward' or living 'in harmony with nature'. In order to reclaim land rights from states, the San are obligated to portray themselves as an essentialized, cohesive indigenous group.

The critical analysis of Namibia's land reform undertaken in this chapter reveals a contradiction: on the one hand, one can observe growing international recognition of the land rights of indigenous peoples; on the other the enclosure of their land continues nationally. Namibia is one of the world's newest nations and, in its focus on creating a unified state, its multilayered German and South African colonial past looms large. For example, colonial tribal chieftaincy rule marginalized San hunter-gatherer bands. Today, the San are Namibia's poorest, most vulnerable group, living as scattered itinerant labourers, often on the outskirts of cities or settlements, and their communities are rife with social and health problems.

The fieldwork described in this chapter indicates that there is little reason for optimism about their sustainability, and a key reason is the long shadow cast by colonialism. It transformed land use from a practice that regulated social organization through property relations into one in which property boundaries affirm political-economic power structures.

Keywords ancestral land rights • indigenous peoples • indigenous rights • Namibia

S. Vermeylen
Lancaster Environment Centre, Lancaster University, Lancaster LA1 4YQ, United Kingdom
e-mail: s.vermeylen@lancaster.ac.uk

8.1 Introduction

It is not uncommon for indigenous peoples to make an explicit link between rights over knowledge, culture, natural resources and land (see e.g. Posey and Dutfield 1996; Simpson 1997; Greene 2002, 2004; Berman 2004; Riley 2004; Solomon 2004; Tucker 2004; Gibson 2005). Indigenous peoples regard knowledge as something that is closely tied to land; knowledge encapsulates spiritual experience and relationships with land (Barsh 1999; McGregor 2004). Agreements[1] drawn up by indigenous peoples themselves highlight the fact that rights to land, traditional institutions, cultural practices and intellectual property rights are inseparable and interrelated, a statement that has also been recognized by some UN institutions.[2]

Not only are indigenous peoples struggling to get their legal rights over land and resources recognized, they also want to have the freedom to make their own decisions about how to use and manage natural and cultural resources (Tucker 2004). One of the major stumbling blocks in indigenous peoples' quest for recognition of their user and ownership rights over land, resources and knowledge is the fact that throughout colonial history their territory and organizational structure have been perceived as, respectively, *terra nullius* and *res nullius*.[3]

The contemporary plight of indigenous peoples can be traced to land enclosures, which began in Western Europe about half a millennium ago, with vast areas of common grazing land enclosed by landlords and made into private property. The enclosure of commons was a motif of European colonialism, which began in the same era. Political thinkers such as John Locke provided intellectual and ideological justification for colonizing indigenous peoples and expropriating their territories (Keal 2003; Scott 1997). Enclosures consistent with this imperial thinking resulted in new land tenure systems, so that land use by indigenous peoples now varies from traditional open-range hunting and gathering, through

---

[1] For example: the Charter of the Indigenous-Tribal Peoples of the Tropical Forests, the Indigenous Peoples' Earth Charter and the Declaration of Principles of the World Council of Indigenous Peoples.

[2] For example: the Declaration on the Rights of Indigenous Peoples, the COICA-UNDP Regional Meeting on Intellectual Property Rights and Biodiversity, the UNDP Consultation on the Protection and Conservation of Indigenous Knowledge and the International Labour Organization's Convention 169 (ILO 1989) on Indigenous and Tribal Peoples.

[3] *Res nullius* is a principle derived from Roman law according to which *res*, which are objects in the legal sense, are not yet the objects or rights of any specific subject. In other words, *res nullius* are considered ownerless property and therefore usually free to be owned. *Res nullius* also has application in public international law, viz. *terra nullius*, referring to unclaimed territory: a nation can assert control of *terra nullius*. Building further on the philosophy of John Locke and Emeric de Vattel, *terra nullius* was the principle used to justify the colonization of Africa: even though there may be people residing on the 'newly discovered' land, it is the right of the 'more civilized' to take the land and put it to 'good use'.

open-range herding of European breeds, to fenced cattle feeding and plough agriculture (Olson 1990), while property holding now ranges from a usufruct system[4] to a European system of free proprietorship. The colonial enclosures continue to bear on the living conditions of indigenous peoples today. After decolonization, new national governments ratified the original land 'grabs' and in some cases extended them. Moreover, some argue that intellectual property rights are the newest version of enclosure (see May 2000).

Focusing on territorial rights, though, indigenous peoples in various parts of the world are increasingly reclaiming ancestral land, while indigenous peoples in Australia, New Zealand and Canada have been able, in some cases, to obtain land and resource rights. However, as confirmed by Hitchcock (2006), indigenous peoples in Africa (such as the San) have encountered more difficulties in their quest to gain land and resource rights. Looking at territorial rights as a proxy for the wider struggle to recover from colonial subordination allows us to see the debate on the protection of knowledge in the broader context of their daily struggle to protect their livelihoods and culture.

The complexity of the San's current socio-economic and political situation vis-à-vis land and resource rights can best be understood as part of a historical framework that describes the colonial history of land enclosure. This chapter looks more closely at the reasons behind the San's struggle to restore their rights. It reflects upon the state and how the institutions of the state deal with the legacy of colonialism and the enclosure of land.

While the San across southern Africa have experienced the loss of their land, this chapter deals in particular with the history of the San's struggle to gain and retain land rights in Namibia.[5] Namibia was the last African country to become independent, and the Namibian government prides itself on having brought a new social order in the country and erased the socio-economic, legal and political remnants of colonialism (Harring and Odendaal 2006). This chapter will examine whether this change has also improved the socio-economic, political and legal position of the San in Namibia. Even though the impact of land enclosure and colonialism affects people differently depending on their history, experience and location, a focus on the colonial legacy of the Namibian state can give valuable insights into the process of subordination and enclosure of the San across the countries in southern Africa where they live.

---

[4] Usufruct is the legal right to use, derive a profit from and benefit from property that belongs to another person.

[5] The formal status of ethnic minorities in Namibia is expressed in the Constitution (Republic of Namibia 1990). The guiding principle is the separation of ethnic and national identity, with the latter given priority. This is intended to assert the primacy of the state without disregarding the reality of ethnic diversity (Suzman 2002). Article 19 stipulates: 'Every person shall be entitled to enjoy, profess, maintain and promote any culture, language, tradition or religion subject to the terms of this Constitution and subject to the condition that the rights contained in this article do not impinge upon the rights of others or the national interest.'

8.2 Land Reform in Namibia

After independence in 1989, Namibia's state formation was rooted in a constitutional framework embedded with liberal democratic thinking that embraced concepts of human rights and the rule of law (Erasmus 2002). It was assumed that the atrocities caused by colonialism and apartheid could best be dealt with by building a unified nation. In Namibia, it was hoped that the Constitution would prevent another example of a failing African state and bring stability and progress instead. In this respect, the Constitution could be judged somewhat successful. However, with regard to land reform, the literature agrees that the Constitution was less successful (Daniels 2003; Harring 1996, 2002; Suzman 2002). The non-recognition of indigenous peoples and of socio-economic rights (the so-called second-generation human rights) hampered much-needed land reform.

One of the key objectives of the Namibian independence struggle was to return the land to the people who had been dispossessed during colonialism. However, according to Daniels (2003), the Constitution perpetuated colonialist policy by explicitly stating that land, water, and natural resources belonged to the state if they were not otherwise lawfully (privately) owned. In other words, communal land became state property. This was emphasized by Prime Minister Hage Geingob's statement that 'people in the communal lands have no acknowledged right, independent of the will of the State, to live and farm in the Communal Areas' (cited in Harring 1996). This means that the vast majority of Namibians have neither ownership nor tenure security of land, even if they have been living on it for many generations.

According to Harring (2002), the Constitution enables the continuation of German and South African racist, colonialist practices. While 70% of blacks live in the communal areas, hardly any whites do. It is impossible for most blacks to acquire land from whites because they lack the means to buy it. The effect is that poor blacks living on communal lands can move only to other communal lands. Because of their vulnerable social structure and poverty, the communal areas used by San communities are under threat of land-grabbing by stronger and better organized groups (Daniels 2003). When San complain to the Ministry of Land Resettlement and Rehabilitation about the fencing of communal land by other groups, the official reply is that Namibia needs to prioritize the productive use of land so that it feeds not only small groups of rural dwellers such as San hunter-gatherers, but also the nation at large (Widlok 2001). Thus the Namibian government is responsive to the stronger ethnic groups, an ironic continuation of colonial rule through powerful tribal chieftains.

Even in areas where San are allocated communal land, they are frequently dispossessed because the state does not give them adequate protection. As a result, the San have argued that land allocated to them should be firmly and exclusively under their control. The San desire for an exclusive use of land is motivated in part by their awareness that others perceive land that they use to be open and 'unproductive' land. This echoes the colonial practices of *terra nullius* and *res nullius* (see e.g. Martin and Vermeylen 2005, for a historical analysis of these practices). The situation for the San is particularly difficult because the return of their ancestral land is refuted by the post-colonial government on moral grounds (Widlok 2002). The Namibian government argues that nation-building is important to counter ethnic segregation and that it has a moral responsibility to cater to all the members of the population without consideration of ethnic identity.

The 1991 National Conference on Land Reform and the Land Question reported unequivocally that the restitution of ancestral land claims by any group or individual would not be entertained in Namibia. This decision was later incorporated in the National Land Policy of 1998. For most San, existing rights to land are de facto rights, not guaranteed by civil customary law. This is most evident on commercial farms, where the rights of San workers to residence are contingent on their employment by a farm owner, or on a farmer granting squatting rights. Whereas the majority of rural Namibians can claim at least partial tenure rights in terms of state or customary law, most San (outside the Tsumkwe district) cannot claim such rights.

Thus there are no specific provisions in the Constitution that protect the rights of indigenous peoples or minorities, and Namibia is not a signatory to International Labour Organization Convention 169 (ILO 1989), the only international convention recognizing the rights of indigenous peoples (Daniels 2003). Both the Constitution and public opinion are biased against conceding group rights on the basis of ethnicity. As a result the San are not recognized and there are no government-led affirmative action plans on their behalf (Widlok 2001).[6] The government defines indigeneity by reference to historical European colonialism. Accordingly, almost everyone born in Africa of an African bloodline is indigenous. Furthermore, the Traditional Authorities Act defines all Namibian traditional communities as indigenous.

The devolution of limited powers through the advisory role assigned to Namibia's traditional authorities is another example of the Constitution continuing German and South African colonialist practices (Daniels 2003). Devolution is replicating the old divide-and-rule patterns used by colonialists, and the creation of tribal reserves has revived – or, in some cases, reinvented – the traditional strong leadership structure among the Herero, Mbukushu and Kwangali tribes. Unlike them, the San were dispossessed of their land during the colonial period and their authority structure was dismantled. As a result, only three of the six established San traditional authorities, East and West Tsumkwe and the Hai//om, are formally recognized by the government. Although the Traditional Authorities Act provides for the recognition of leaders of communities who in the past did not have leaders or whose leadership structure was destroyed during the colonial period, the government has shown on numerous occasions a reluctance to recognize the three other San communities that have applied for their chiefs to receive

---

[6] The National Land Policy identified the San as the principal beneficiaries of any anticipated land reform initiative. When the Ministry of Lands, Resettlement and Rehabilitation came into existence in 1990 with the aim of alleviating poverty and improving access to scarce resources, including land, the San were prioritized as the most needy beneficiary of the Namibian resettlement policy. However, the resettlement policy has failed for a number of reasons, one of the main ones being the lack of participation of the San in the implementation policy (Harring and Odendaal 2002).

official recognition, namely Omaheke North and South and the Khwe. The government's policy of non-recognition has contributed to the further marginalization and poverty of the San and created an opportunity for other groups to further oppress them by grabbing their land.

In summary, the Namibian government's ownership of communal lands represents a continuation of colonial land policy. In claiming ownership rights over communal land previously seized by colonial regimes, the government missed an opportunity to rectify an unlawful land seizure. According to Harring (1996), '[n]o modern authority would cite these seizures of native land as either legal, or justifying modern Namibian land law; the fact is that these land seizures are the modern basis of the idea that the state 'owns' Crown land, and the derivative idea that communally held land is a form of Crown land.'

While the Constitution states in its preamble that it will deal with the injustices of the colonial period, it does not adequately address the legacy of enclosure. Instead, the Namibian government relies heavily on South African colonial law that denied communal land holders secure rights (Harring 2002). After independence, South African state property in Namibia was transferred to the new state; at no point has the Namibian government questioned South Africa's title to this land. Thus it copied South Africa's policy in claiming ownership rights of what in all likelihood was unlawfully seized communal land.

8.3 'Kill the Tribe to Build the Nation'

Namibia's failure to transform the situation of its most vulnerable population is no exception in the context of southern Africa's post-colonial land reform processes. While independence created an opportunity to rehabilitate the inherited institutions of the colonial period – often described in the literature as authoritarian and racist (Seidman and Seidman 2005) – many African governments missed the opportunity. The new leaders believed in the capacity of the rule of law to restructure the colonial legacy of inequality and exploitation. However, the new laws were badly implemented. Thus, even after independence, southern Africa's major political institutions remain imbued with an institutional legacy of colonial rule (Mamdani 2005).

Just as elsewhere in southern Africa, land reform in Namibia calls for a redistribution of land. Land is crucial for making a living in developing economies, for use in either commercial or subsistence farming and grazing. Access to rural land is a major source of affluence. In Namibia one of the major causes of poverty is the continuing unequal access to land and unequal ownership of natural resources (Smit 2002). The appropriation of land in the colonial period remains the basis of this inequity (Werner 1993; Gordon and Douglas 2000; Widlok 2002).

Africa's 'obsession' with its nation-building process is fed by a well-intended drive to homogenize sociocultural differences among ethnic groups (Okafar 2000). The slogan, 'Kill the tribe to build the nation,' exemplifies the policy of banning ethnicity, and was used by FRELIMO, the ruling party in Mozambique in the 1970s. Prior to that Tanzania launched an ambitious 'Ujamaa' policy after independence, described as African socialism, which forced peasants off traditional lands and into allotted communal villages. This policy, which ultimately failed, attempted to reduce traditional land tenure rights in favour of a broad collective national effort (Chachage 1999). The attempts by African leaders to form cohesive nations out of culturally heterogeneous populations can be traced to a popular belief reflected in international law: that the European-style nation state guarantees not only a tighter territorial demarcation, but also a more monocultural nationhood.

In other words, the rules and norms of Western law not only influenced African statecraft, they became fundamental building blocks of African nations. Namibia's failure to adequately address the degraded position of the San is not attributable solely to unwillingness on the part of its government. It represents a continuation of norms that have formed an essential part of international law-making since the eighteenth century. Accordingly, only those people with a level of social organization similar to that of European states, an implication of which is a fixed relationship with a specific land area, are entitled to have rights over land. Other people, such as hunter-gatherers, who are not 'modern,' have their occupation and use of land nullified. Rights over land are granted to those who use the land in the manner described in Locke's *Second Treatise of Government* and Emeric de Vattel's *The Law of Nations, or the Principle of Natural Law*. Following this tradition, people like the San are not using or occupying land as prescribed in international law, and this allows colonial powers to take control of their territory (Dodds 1998). Postcolonial governments, including Namibia's, have continued this practice.

8.4 Customary Law and Communal Areas

Based on the Namibian experience, it is apparent that there is a need for a new land reform process. It is unlikely that simply redistributing existing property (according to the model used in Zimbabwe) will result in an equitable allocation of land. Dividing the land into small parcels of fee simple land[7] is not economically viable and builds upon the existing system of land ownership, i.e. 'using the model of white agriculture as the implicit model for land reform' (Harring 2002).

During the colonial period, many people were dispossessed. Ownership of land and resources by black people was severely restricted; they were allowed access only to the communal areas. Through a system of communal land tenure ('native reserves') every household had access to land, but the land allotted was so small that at least one member of the household had to engage in wage labour to support the household. At the same time employers argued that they could pay a wage below the value of labour because the workers and their dependants lived off the

---

[7]'Fee simple' is an estate in land in common law and represents the most 'absolute' ownership model of real property.

land (Werner 1993). The colonialists regarded the land under the control of Africans as *res nullius* because, so they argued, Africans were incapable of managing its private ownership. Seizing the land of native Namibians not only provided white settlers with the land they needed, it also denied natives access to commercial agricultural production and forced them into wage labour (Werner 1993).

During colonialism, land tenure security could only be achieved through individual ownership rights. The allocation of private rights attached a market value to land, which facilitated its development. In the post-independence period it is still believed that property individualization contributes to the development process (Bruce 2000). However, doubts have been registered about enforced individualization in the African context because it is based on Euro-American economic and technocratic views of land (Smit 2002). The assumption that narrowly defined individual property rights guarantee more secure land rights and more economic development has been criticized (Platteau 1996; Firmin-Sellers and Sellers 1999). The facts appear to support the conclusion that none of Africa's major economic or environmental problems decreased when land tenure was changed from community to individual rights (Bruce 2000). Moreover, economists have argued that a market for land does not exist and anthropologists have criticized individualization for ignoring the complexities of customary tenure, including that it provides for multiple users to hold rights to a single plot.

Taking these criticisms on board could produce a land policy that recognizes existing customary tenure instead of one which copies Western-style private property rights (Firmin-Sellers and Sellers 1999). Namibia has a policy that allows people in rural communal areas to register their customary rights for farming, residential or other purposes. People can exercise these rights for a limited period of time – a human lifespan.[8] It is the chief or the traditional authority of a particular community that allocates or cancels customary land rights, a decision which has to be approved by the communal land board.[9] Apart from these land rights based on customary law, people in communal areas can also apply for grazing rights and leasehold. The former can be part of customary tenure and are allocated by the chief or the traditional authority, while the latter is for agricultural or tourism projects and needs to be approved by the traditional authority and the communal land board.

So far the San have not embraced the registration of customary land rights. For one thing, the application needs to be done in writing, a task which is problematic for the largely illiterate San. Secondly, the application has to go through a chief, which puts the San in an awkward position, because not all their leaders are recognized as chiefs. Finally, some San have argued that the maximum size of land eligible for the rights prescribed in the regulations is far too small for their needs. Unlike other occupants of communal land, the San want to use the land for reintroducing or preserving their traditional lifestyle of hunting and gathering, and 20 ha is clearly insufficient for that purpose.

---

[8] Joint registration (usually by spouses) is allowed. After the title has expired, the land reverts back to the traditional authority.

[9] For the establishment, functions, and composition of communal land boards, see LAC (2003).

Although it seems that a direct and explicit recognition of traditional communal lands in the Constitution would be the best way to change Namibian land policy, it remains to be seen what this would mean for a marginalized community like the San. Whether they would benefit is questionable, since they are not explicitly recognized in the Constitution either as indigenous or as a socio-politically and economically vulnerable group. Furthermore, it would be incorrect to assume that indigenous African land tenure systems are inherently communal. Even in apparently communal tenure systems, individual appropriation of resources and land exists (Bruce 2000).

Chanock (1991a) argues that colonial regimes simplified tenure systems in order to undermine the indigenous use of land. They emphasized their communal elements and ignored their more subtle gradations such as various tenure arrangements for land put to individual uses. Numerous studies have confirmed that prior to the colonial period native land was not solely communal (Harring 1996; Mann and Roberts 1991). Colonial authorities selectively used various rules in support of their position that traditional land tenure consisted solely of communal land tenure, including customary law (Chanock 1991a, b). While customary law was labelled as indigenous law by Europeans, in reality it was manipulated by both Africans (mainly tribal chiefs) and Europeans under colonialism (Mann and Roberts 1991). Local chiefs were, in fact, often the administrative creations of the colonial state (Ribot and Oyono 2005). After independence these customary chiefs were promoted as legitimate local leaders. The government relied on dominant ethnic groups as political allies in the land reform process.

The reality is that in indigenous land tenure systems each category of tenure meets the needs of specific community members. Chanock (1991a) describes a community's territory as a landscape that is divided into areas of land used for various purposes and managed under different tenures. Each area represents a particular 'tenure niche', a space in which access and use are governed by a common set of rules. Different niches can be identified within a single area, ranging from open access (grazing areas), through common property (medicinal field plants), to individual property (small agricultural plots). To phrase it differently,

> each person in a community had rights of access to the land depending on the specific needs of the person at the time; for example, in any given community, a number of persons could each hold a right or bundle of rights expressing a specific range of functions; a village could claim grazing rights over a parcel, subject to the hunting rights of another, the transit rights of a third and cultivation rights of the fourth. (Nzioki 2002)

In order to identify different tenure niches it is crucial to ask who uses the resource and on what terms (Bruce 2000). The legacy of colonialism has made it difficult for the San to answer this question, for several reasons. First, the traditional rules of land allocation have been eroded, making it difficult to identify who uses what part of the land and for what purpose. The San struggle just to understand what kind of rights they have as occupants of communal lands. Second, there is evidence that prejudices exist against the San (Woodburn 1997). Other ethnic groups regard the San's traditional use of land (hunting and gathering) as backward, a view supported by the Namibian government (Suzman 2001). Third, the government is reluctant to recognize alternative forms of social organization and landholding, and it actively supports modes of subsistence that exploit land through 'labour', which is defined as agricultural or pastoral (Widlok 2001). Together with enclosure, Europe's 'enlightened' individualism and independent, self-sufficient farmers were exported to the colonies (Lemert 2002). The ideal has been adopted by Namibia's post-colonial government. Finally, it is common practice in rural Africa that access to land in communal areas is dependent on an applicant's location, culture, social status and use (Nelson 2004), eligibility factors which put the San out of the running. It is therefore unlikely that the San's traditional, precolonial use of land will be recognized under existing national law. Both colonial and post-colonial regimes have ignored the rights of former hunter-gatherer groups because they do not invest (i.e. by tilling or grazing) in the land and because they are politically weak.

The fact that only three out of six San traditional authorities are recognized by the Namibian government hinders their general use of customary law. As unrecognized San chief Sofia Jacobs of the Omaheke region explains: 'The traditional laws as recognized or promoted by the traditional authority laws are different from the laws of the San; these laws fight with each other.'[10] Furthermore, of the three recognized traditional San leaders, two (the Hai//om and !Kung chiefs) are not well regarded by their communities because the elections were hastily held, and the chiefs are perceived to be supported by and in the pocket of the government (for more details, see WIMSA 2005). It is commonly believed that the chiefs of these two groups are not defending the well-being of their communities. In particular, the chief of the !Kung has been accused of allowing powerful ethnic groups (mainly wealthy Herero cattle farmers) to move with their herds into their San territory at the expense of local San.[11] This has caused friction in the community because the San have to compete with cattle for their scarce field food and water.

8.5 Aboriginal Title

Aboriginal title[12] is a *sui generis* proprietary interest in land that is recognized in common law jurisdictions such as Australia, Canada, the US and New Zealand. Australia, in particular, is an interesting example of the principle of aboriginal title. For Australian Aboriginal and Torres Strait Islanders an important step towards rights over land was taken with the Australian High Court's decision in the *Mabo* case of 1992 (Mabo v Queensland), which overturned the *terra nullius* principle. Although in the decision it was declared that the Crown's acquisition of sovereignty could not be challenged in a court, the decision established that native title could be claimed over unappropriated

---

[10] Field notes, 12 August 2005.

[11] Field notes, July 2005. For more details about the tension between the San and Herero farmers see e.g. Harring and Odendaal (2006).

[12] Aboriginal title is a common law property interest in land sometimes also referred to as native title. However, native title is strictly taken as a concept in the law of Australia that recognizes the continued ownership of land by local indigenous Australians.

Crown lands. As a result of the *Mabo* judgment – which, some have argued, was far more conservative than both the debate over its implications and subsequent developments have suggested – native title (or aboriginal title in the general sense) exists only where there is an aboriginal group that has maintained its connection with traditional lands. The group has to be able to prove that it is looking after its land, discharging obligations under traditional law, and enjoying as far as practicable the traditional rights of use and occupation (Brennan 1995 in Keal 2003: 124). In spite of being limited in this way, the recognition of native title in the *Mabo* judgment was, for Australian indigenous peoples, a milestone in the recovery of identity and rights.

The *Richtersveld Community and Others v Alexkor and Another* (2001) was the first case to consider whether or not aboriginal title is part of South African law. In this case the Richtersveld people, comprising the inhabitants of four villages in the Northern Cape province, claimed aboriginal title to land that was the site of diamond mining operations by Alexkor Limited, a public company that owned the land and held surface and subsurface mineral rights. The Richtersveld people alleged dispossession of a portion of the land after 19 June 1913 by a series of racially discriminatory legislative and executive acts. They sought restitution of three alternative rights in land based on the doctrine of native title: ownership, exclusive beneficial occupation and use for specified purposes, and beneficial occupation of the land for a longer period than 10 years prior to dispossession (Ülgen 2002).

The case first went through the Land Claims Court, then to the Supreme Court of Appeal, which overturned the Land Claims Court decision, and then finally to the Constitutional Court. The aboriginal title claim was dismissed at the first hearing on the basis that the Land Claims Court lacked jurisdiction to award restitution of a right to land not recognized under the Restitution of Land Rights Act.[13] This was overturned on appeal, as the Supreme Court of Appeal held that the Richtersveld community was entitled to restitution of the land, based upon the 'customary law interest' in land that predated the annexation of the land. It also held that the disputed land was not *terra nullius*, and that the nature and contents of the rights to land should be determined 'by reference to the law that governed such rights, namely indigenous law' at the time of the annexation (*Richtersveld Community and Others v Alexkor Ltd and Another* 2003 (6) BCLR 583 SCA).

Alexkor took the case to the Constitutional Court, which finally confirmed the Supreme Court of Appeal ruling in a resounding victory for the Richtersveld community that also confirmed the existence of the notion of the primacy of customary or traditional rights to land (*Alexkor Ltd and Another v Richtersveld Community and Others* 2003 (12) BCLR 1301CC). The important international cases of *Mabo* and *Delgamuukw*[14] were referred to with approval, thus incorporating international consensus on issues of aboriginal title into African jurisprudence.

---

[13] Ülgen (2002) has argued that the conceptual framework of the Restitution of Land Rights Act does recognize the principle of native title.

[14] The Canadian *Delgamuukw* case recognized the 'full' meaning of aboriginal title, i.e. full proprietorial rights including ownership of subsurface minerals and the right of aboriginal owners to develop traditional lands in non-traditional ways.

The doctrine of aboriginal title is attractive to the indigenous peoples in southern Africa because it would legitimize their rights over land they occupied prior to colonization. Aboriginal title claims are based on a historical membership of a particular tribe or kingdom in relation to traditional land. Alexkor, expressing the view of the government, argued that in a country still recovering from apartheid, where tribal groups had been uprooted many times, acknowledgement of aboriginal title claims would unnecessarily awaken or worsen destructive ethnic and racial politics. This argument has been generally accepted and adopted by the other southern African governments.[15]

It should be noted that in December 2006, after a historic and long court challenge lasting four years, the Botswana High Court confirmed the rights of the San to reside legally in the Central Kalahari Game Reserve. The San had been evicted in 2001, to which they objected, claiming that they had the right to reside on their traditional lands. The reasoning of the court, which quoted freely from *Mabo* and other international cases, was largely based upon evidence of the San's unbroken occupation, since time immemorial, of their traditional lands. However, the court did not confirm the San right to 'ownership', a distinctly Western concept, but rather that to 'the right to use and occupy the land' (*Roy Sesana and Others v the Attorney General of Botswana*, Misca 52 of 2002).

As was confirmed by the South African Constitutional Court, as well as the Botswana High Court, international law does recognize aboriginal title. The UN Declaration on the Rights of Indigenous Peoples states that there is *opinio iuris* to recognize aboriginal title (i.e. a conviction that the practice is obligatory). Considering the fact that South Africa and Namibia accept in their constitutions the application of international law, it can be argued that they should recognize aboriginal title claims, a vision shared by legal commentators in both countries (Tjombe 2001). Bennett and Powell (1999) have argued that in the context of southern Africa it might be better for indigenous peoples to invoke aboriginal title in international law rather than in national common law, because the latter was responsible for their land dispossession.

8.6 Aboriginal Title and the San: From 'Civilized' to 'Socially Organized'

At first sight it would appear that the San fulfil at least some of the requirements for aboriginal title claims. Although, as mentioned, the validity of indigenous land rights remains an open question in Namibian law, Judge Mahomed indicated in the *Rehoboth Baster* (Cpt Diergaardt of the Rehoboth Baster Community et al v the State of Namibia 1997) appeal that the principles set out in *Mabo* might hold in

---

[15] For example, large parts of South Africa could be subject to overlapping and competing claims where pieces of land have been occupied in succession by San, Khoi, Xhosa, Mfengu, Afrikaner and British people (Ülgen 2002).

Namibia on the basis that the decision did not focus on Australian law, but instead discussed indigenous rights in the context of common law principles. Since common law recognizes aboriginal title (common law is operant in Namibia as the result of the British occupation in 1915) and indigenous land title is inextinguishable by colonial powers (*Western Sahara* case 1975),[16] it is likely that the San could in theory dispute South Africa's original title, which was transferred to the Namibian government (Harring 2002). However, presenting a case of aboriginal title will require extensive anthropological and historical research to prove that neither Germany nor South Africa ever held title to the communal land.

The requirements for aboriginal title which indigenous peoples have to meet have evolved from a 'civilization' requirement to a 'social organization' one (Chan 1997). Until the twentieth century, indigenous peoples had to give evidence of having a civilized legal system in place so courts could establish that the claimants were capable of holding title. Indigenous peoples who failed the civilization test simply did not exist before the law. For example, in *Re Southern Rhodesia* (1919), the Privy Council of the United Kingdom dismissed land claims of the Ndebele as irreconcilable with the legal ideas of a civilized society. In recent common law jurisprudence (the *Mabo* and *Western Sahara* cases), the legally recognized identity of an indigenous group is no longer linked with the civilization requirement. A proof of social organization is now put forward as the decisive factor for aboriginal title claims.

Today, the first step in a native title claim is to show that the indigenous group has its own socio-political structure. In practice this means that it meets the following criteria: community identity, permanence, exclusivity and a pronounced relationship to the land. Some legal scholars have argued that the new requirement, although an improvement, still uses a Western legal yardstick of social structure. However, there has been a shift. For example, it is no longer acceptable for only Western scholars or judges to decide whether or not indigenous peoples conform to the ideas of civilization or social organization. Instead, that decision is now left to indigenous peoples themselves: the group itself must believe that they have a social structure and a relationship with the land, that they adhere to it and, most importantly, that others recognize the group's coherence.

With regard to establishing the test of social organization and aboriginal identity, international law offers some guidelines. In the past, the main guiding principles were language, political affiliations, culture, genetic association and residence. It has been argued that this definition of social organization is too broad. A narrower definition could focus on the purposes of the land claim. In practice this would

---

[16]The sovereignty of the Western Sahara remains the subject of a dispute between the government of Morocco and the Polisario Front, an organization seeking independence for the region. In 1975 the International Court of Justice issued an advisory opinion on the status of the Western Sahara. The court held that while some of the region's tribes had historical ties to Morocco, these were insufficient to establish 'any tie of territorial sovereignty' between the Western Sahara and the Kingdom of Morocco. The court added that it had not found 'legal ties' that might affect the applicable UN General Assembly resolution regarding the decolonization of the territory, and, in particular, the principle of self-determination for its people.

mean that the criteria of social organization are met when an indigenous group shows that it has maintained its lifestyle over time, that as a group it has a pronounced association with the land, and that its group identity is established to the extent that other traditional communities recognize it.

On this basis, it can be argued that in order to secure an aboriginal land claim, the San have to provide evidence of their relationship to a land and their position as an autonomous group with its own identity vis-à-vis other groups. Bishop (1998) has already done this exercise for the San in Botswana and came to the conclusion that the San can provide sufficient evidence of their continued use and occupation of the Kalahari in order to comply with the requirements of aboriginal title. However, she suggests that the common law terminology of 'land tenure' might cause unnecessary further complications for the claim because it cannot be assumed that San use conforms to the common law concept of tenure. Instead, she suggests using evidence of the San's *territoriality*, because it captures the holistic system of land use and occupancy by the San. The term is not frozen, i.e. it does not refer to a specific era of occupation; instead it is inclusive of adaptations to modern circumstances. Furthermore, territoriality captures the identification of various San *n!oresi* (see Vermeylen, Chapter 10) and includes the rules associated with their use and tenancy.

Territoriality opens a new domain of research questions. For instance, on what basis can indigenous peoples claim entitlements to land: moral claims or historical connections and customary practices? These questions are important because they involve mutual exchanges between parties,

> providing all parties with exposure to alternative conceptual frames and qualitatively different environmental relations. Even where groups remain locked in conflicts over land, the mere fact that they are forced to represent their values to each other opens the door to the influence of other ideologies (Strang 2000).

It is important to see if such exchanges will be reflected in laws concerned with land rights, ownership rights and the protection of indigenous environments and cultural heritages.

8.7 Khwe San Land Claims

The Khwe San of the West Caprivi are using a multilayered strategy in their quest for economic and political autonomy. Access to and rights over land in the West Caprivi are linked to settlement history, colonial influences, neighbourly relationships, ethnic identity, the recognition of traditional leaders and economic survival. The main tool the Khwe are using in order to secure access to land is identity. They represent themselves as a cohesive and distinct ethnic group in anticipation of this giving them a legitimate identity in national, regional and global venues.

Besides claiming a special relationship to the land, the Khwe use their language, hunting with bow and arrow, and food gathering as evidence of an authentic tradition. Orth (2003) notes that for the Khwe, 'the importance of tradition was not to be found in their content, but rather in the difference between the Khwe way of doing things and those of other peoples, especially the Mbukushu'. She interprets this expression of difference by the Khwe as a sign of the need to express their own identity in order to survive the threat of subordination posed by the Mbukushu, who perceive them as a subgroup of their tribe. The Khwe authority structure is not recognized by the Namibian government and as a result they fall under the leadership of the Mbukushu king, who denies Khwe access or rights to their ancestral land.

An example of re-traditionalization as a cultural survival strategy is exemplified in the Khwe revival of hunting. Even though hunting is forbidden because of the game park status of the West Caprivi, the Khwe emphasize hunting with bow and arrow as a main feature of their heritage. As one Khwe (based in the Omega camp) testified: 'the wild animals are our cattle, the cattle of Mbukushu are destroying the natural resources, they are killing the wild animals, these Bantus create problems for us Khwe' (31 August 2005). Orth concludes that the Khwe are using a strategy that reinforces their identity as Khwe at the local level, as San at the regional level and as indigenous at the global level.

Khwe claims for land focus on a connection to their use of land. However, this discourse is often based on an unrealistic expectation of recognition of indigeneity. The claiming of aboriginal title is a process that remains embedded in a colonialist frame of reference. For example, non-indigenous peoples define tradition in a narrow and restricted way. Indigenous culture, according to them, is expressed through the use of traditional language, stories, places, ritual practices and kinship ties. Thus aboriginal claims are driven to represent the past as frozen, ignoring that they are part of a complex process of transformation and continuity. As a result, indigenous peoples are required to internalize the non-indigenous (i.e. colonialist) understanding of tradition and authenticity in their strategy to gain land rights (see Altman, Chapter 15).

There are problems with defining indigeneity in this narrow sense, as exemplified by Chennells's reflection on the success of the ≠Khomani San land claim that he represented in 1999.

> The San are now landowners. They'll have to train people to do the tracking and all those things to fill that space. But probably the most major challenge is trying to make *the myth that we've actually created in order to win the land claim now become a reality. It is the myth that there is a community of ≠Khomani San. At the moment there is no such thing* [author's emphasis]. We have to try and find a way of helping the ≠Khomani understand what it means to be ≠Khomani (cited in Robins 2001).

In order to win this land claim, the San in South Africa conformed to the expectations of donors and governments: a strategic narrative of community solidarity, social cohesion and cultural continuity.

8.8 Idealization of Indigeneity

The criteria for indigenous status in aboriginal title claims enforce an engagement with 'primordialist' and 'essentialist' conceptions of culture. In other words, indigeneity is fixed in time and place and is not socio-economically and historically contextualized. This strategy can lead to the exclusion of indigenous peoples who have lost connections with their ancestral lands. For example, Canadian and US courts 'have rendered land claims invalid when plaintiffs do not appear native enough' (Thorpe 2005).

Sylvain (2002) gives the example of the Omaheke San in Namibia, who for generations have been a landless underclass of farm workers and as a result have been incorporated into an ethnically hierarchical class system. Beginning in 1914, large tracts of the Omaheke region were set aside as reserve lands for the Hereros and Tswanas. The reserves became apartheid homelands in the 1970s, and after Namibian independence, they became communal areas. These areas comprise about 35% of the Omaheke land area. The remaining 65% is a 'commercial farming block' dominated by Afrikaner and German cattle ranchers who occupy 900 farms averaging about 7,000 ha. No land in the Omaheke was set aside for the San (Sylvain 2002).

For the non-San in Namibia there are no authentic San in the Omaheke anymore; they no longer hunt and gather and therefore have lost their cultural identity. The Omaheke San, on the other hand, do not consider themselves to be non-authentic. To them, being San means being able to cope with continuing experiences of exploitation. The Omaheke San express a class-shaped conception of territorial identity. However, global discourse expects indigenous peoples to represent themselves as being internally undivided and as untouched by history. While indigenous peoples are expected to represent land struggles in terms of this idealized traditional cultural identity, in reality, as Sylvain (2002) has shown, land rights are tools to obtain contemporary social and economic justice: '[T]he Omaheke San are also seeking land rights, but they are not trying to restore a hunting and gathering lifestyle or regain an evolutionary heritage; rather, they are struggling for access to development, resources for better work conditions, and for political representation.'

Although the San have a strong case to claim aboriginal title over their ancestral land, Suzman (2004) has argued that the fate of the Hai//om San with regard to their claims over Etosha 'ultimately rests on the government's appreciation of their particular predicament of landless underclass and willingness to prioritize them and other San in the land reform process'. Questions can be raised about the continued use of the aboriginal title claim as an 'enforced' strategy. Instead, approaches that are infused with current socio-economic realities might better reflect the needs of present indigenous peoples. For example, a land rights strategy based on compensation for past injustices and discrimination could be a valid alternative.

Normative arguments, such as Locke's concept of *terra nullius*, that justified the colonial acquisition of territory are biased against the political and social organization of indigenous peoples. These biases, according to Tully (1994), influence the current debate about aboriginal title claims and form the basis against which aboriginal title claims are judged. Contemporary property theory does not recognize the sovereignty of indigenous peoples; neither does it approve of indigenous tenure systems. Both Dodds (1998) and Tully (1994) argue that it will be difficult to respond appropriately to compensatory demands for justice with regard to indigenous land claims as long as Western-based property theory is used to judge them. Tully suggests that it is important to assess land claims on the basis of historical unjust practices and whether or not these practices continue. This will require the recognition of alternative property systems, like niche rights.

8.9 Conclusion

This chapter has explored the continuities between the enclosure of the land of indigenous peoples in colonial and post-colonial regimes. Taking the Namibian San as an example, it can be argued that indigenous concepts of land tenure represent centuries of assimilation, subordination and cultural loss, rather than pristine survivals of precolonial eras. Just like other indigenous peoples, the San have started to wield their indigeneity as a basis for claiming the restitution of alienated property. Emphasizing a special relationship to land has become their main weapon.

In their struggle for empowerment, indigenous peoples confront a socio-political climate that drives them to make claims of authenticity, a prerequisite for legitimate status. In practice this inhibits them from developing an inside-out identity, one that flows organically from their contemporary status, a status far removed from an idealized primitive past. Their identity is forged in the context of power asymmetry, so that they have to position themselves between mutually exclusive identities defined by others: as a *backward* people in modernist discourse or as a *natural* people in conservationist discourse.

Those who argue that indigenous peoples should claim property rights over land on the basis of their culture continue to believe – erroneously – that traditional communities are homogeneous and can be represented with one voice. Where society once enforced assimilation upon indigenous peoples, it now encourages re-traditionalization. With regard to aboriginal title claims, it is a continuation of a trend that requires indigenous peoples to link their relationship to land to concepts of identity, culture and personhood. However, as discussed in this chapter, underlying power relations continue to derail land reform processes.

These power asymmetries can be traced back to colonialism and have created an imbalance that still has an impact in the post-colonial period; they have actually been intensified by the forces of globalization. For the San to gain land rights, the government would be required to recognize the historic subordination and domination of the San. This chapter indicates that such a conciliation has not yet been reached in Namibia.

References

Alexkor Limited and Another v Richtersveld Community and Others 2003 (12) BCLR 1202 CC.
Barsh, R. L. (1999). How do you patent a landscape? The perils of dichotomizing cultural and intellectual property. *International Journal of Cultural Property, 8*(1), 14–47.

Bennett, T. W., & Powell, C. H. (1999). Aboriginal title in South Africa revisited. *South African Journal on Human Rights, 15*(4), 449–485.
Berman, T. (2004). 'As long as the grass grows': representing indigenous claims. In M. Riley (Ed.), *Indigenous intellectual property rights: legal obstacles and innovative solutions*. Walnut Creek, CA: AltaMira Press.
Bishop, K. (1998). Squatters on their own land. *Comparative and International Law Journal of South Africa, 31*, 92–121.
Bruce, J. W. (2000). African tenure models at the turn of the century: individual property models and common property models. *Land Reform, Land Settlement and Cooperatives, 1*, 17–27.
Chachage, C. S. L. (1999). Land issues and Tanzania's political economy. In P. G. Forster & S. Maghimbi (Eds.), *Agrarian economy, state and society in contemporary Tanzania*. Aldershot, UK: Ashgate.
Chan, T. (1997). *Land claims and past aboriginal group identity of the Richtersveld Namaqua*, unpublished paper. Ann Arbor, MI: University of Michigan Law School.
Chanock, M. (1991a). Paradigms, policies, and property: a review of the customary law of land tenure. In K. Mann & R. Roberts (Eds.), *Law in colonial Africa*. Portsmouth, UK: Heinemann.
Chanock, M. (1991b). A peculiar sharpness: an essay on property in the history of customary law in colonial Africa. *Journal of African History, 32*, 72–88.
Cpt Diergaardt of the Rehoboth Baster Community et al v the State of Namibia Co.Nr. 760/1997 (Namibia).
Daniels, C. (2003). The struggle for indigenous people's rights. In H. Melber (Ed.), *Re-examining liberation in Namibia: political culture since independence*. Stockholm: Nordic Africa Institute.
Delgamuukw v British Columbia [1997] 3 S.C.R. 1010 (Canada).
Dodds, S. (1998). Justice and indigenous land rights. *Inquiry, 41*(2), 187–205.
Erasmus, M. G. (2002). The impact of the Namibian Constitution on the nature of the state, its politics and society: the record after ten years. In M. O. Hinz, S. K. Amoo, & D. van Wyk (Eds.), *The constitution at work: 10 years of Namibian nationhood*. Pretoria: VerLoren van Themaat Centre, University of South Africa.
Firmin-Sellers, K. P., & Sellers, P. (1999). Expected failures and unexpected successes of land titling in Africa. *World Development, 27*(7), 1115–1128.
Gibson, J. (2005). *Community resources: intellectual property, international trade and protection of traditional knowledge*. Aldershot: Ashgate.
Gordon, R. J., & Douglas, S. S. (2000). *The Bushman myth: the making of a Namibian underclass*. Boulder, CO: Westview Press.
Greene, S. (2002). Intellectual property, resources, or territority? Reframing the debate over indigenous rights, traditional knowledge, and pharmaceutical bioprospection. In M. P. Bradley & P. Petro (Eds.), *Truth claims: representation and human rights*. New Brunswick, NJ: Rutgers University Press.
Greene, S. (2004). Indigenous people incorporated? Culture as politics, culture as property in pharmaceutical bioprospecting. *Current Anthropology, 45*, 211–238.
Harring, S. L. (1996). The Constitution of Namibia and the 'rights and freedoms' guaranteed communal land holders: resolving the inconsistency between article 16, article 100, and schedule 5. *South African Journal on Human Rights, 12*(4), 467–484.
Harring, S. L. (2002). The 'stolen lands' under the Constitution of Namibia: land reform and the rule of law. In M. O. Hinz, S. K. Amoo & D. van Wyk (Eds.), *The constitution at work: 10 years of Namibian nationhood*. Pretoria: VerLoren van Themaat Centre, University of South Africa.
Harring, S. L., & Odendaal, W. (2002). *'One day we will all be equal': a socio-legal perspective on the Namibian land reform and resettlement process*. Windhoek: Legal Assistance Centre.
Harring, S. L., & Odendaal, W. (2006). *'Our land they took': San land rights under threat in Namibia*. Windhoek: Legal Assistance Centre.
Hitchcock, R. (2006). Land, livestock, and leadership among the Ju/'hoansi of north-western Botswana. In J. Solway (Ed.), *The politics of egalitarianism: theory and practice*. New York: Berghahn Books.

ILO (1989). *Convention (No. 169) Concerning Indigenous and Tribal Peoples in Independent Countries*, International Labour Organization, Geneva. www.ilo.org/ilolex/cgi-lex/convde.pl?C169. Accessed 30 July 2008.

Keal, P. (2003). *European conquest and the rights of indigenous peoples*. Cambridge: Cambridge University Press.

LAC. (2003). *Guide to the Communal Land Reform Act: Act No 5 of 2002*. Windhoek: Legal Assistance Centre.

Lemert, C. (2002). Will there be land for community? In J. M. Curry & S. McGuire (Eds.), *Community on land: community, ecology and the public interest*. Lanham, MD: Rowman & Littlefield.

Mabo v Queensland (1992) 175 CLR1; 66 ALJR 408 (Australia).

Mamdani, M. (2005). Identity and national governance. In B. Wisner, C. Toulmin & R. Chitiga (Eds.), *Towards a new map of Africa*. London: Earthscan.

Mann, K., & Roberts, R. (1991). *Law in colonial Africa*. Portsmouth: Heinemann.

Martin, G., & Vermeylen, S. (2005). Intellectual property, indigenous knowledge, and biodiversity. *Capitalism Nature Socialism, 16*, 27–48.

May, C. (2000). *A global political economy of intellectual property rights: the new enclosures?* London: Routledge.

McGregor, D. (2004). Traditional ecological knowledge and sustainable development: towards coexistence. In M. Blaser, H. Feit & G. McRae (Eds.), *In the way of development: indigenous peoples, life projects and globalisation*. London: Zed Books.

Nelson, J. (2004). A survey of indigenous land tenure in sub-Saharan Africa. Land Reform Report 2004/1, Food and Agriculture Organization, Rome.

Nzioki, A. (2002). The effects of land tenure on women's access and control of land in Kenya. In A. A. An-Na'im (Ed.), *Cultural transformation and human rights in Africa*. London: Zed.

Okafar, O. C. (2000). After martyrdom: international law, sub-state groups, and the construction of legitimate statehood in Africa. *Harvard Journal of International Law, 41*, 503–528.

Olson, P. (1990). *The struggle for the land: indigenous insights and industrial empire in the Semiarid World*. Lincoln, NE: University of Nebraska Press.

Orth, I. (2003). Identity as dissociation: the Khwe's struggle for land in West Caprivi. In T. Hohmann (Ed.), *San and the state: contesting land development identity and representation*. Cologne: Rüdiger Köppe Verlag.

Platteau, J.-P. (1996). The evolutionary theory of land rights as applied to sub-Saharan Africa: a critical assessment. *Development and Change, 27*(1), 29–86.

Posey, D. A., & Dutfield, G. (1996). *Beyond intellectual property: toward traditional resource rights for indigenous peoples and local communities*. Ottawa: International Development Research Centre.

Republic of Namibia. (1990). *The Constitution of the Republic of Namibia*. Windhoek: Government Printers.

Ribot, J. C., & Oyono, P. R. (2005). The politics of decentralisation. In B. Wisner, C. Toulmin & R. Chittiga (Eds.), *Towards a new map of Africa*. London: Earthscan.

Richtersveld Community and Others v Alexkor Ltd and Another 2001 (3) SA 1293 (LCC) (South Africa).

Richtersveld Community and Others v Alexkor Ltd and Another 2003 (6) BCLR 583 SCA.

Riley, M. (2004). *Indigenous intellectual property rights: legal obstacles and innovative solutions*. Walnut Creek, CA: AltaMira Press.

Re Southern Rhodesia [1919] A.C. 211 (P.C).

Robins, S. (2001). NGOs, 'bushmen' and double vision: the ≠Khomani San land claim and the cultural politics of 'community' and 'development' in the Kalahari. *Journal of Southern African Studies, 27*, 717–737.

Roy Sesana and others v the Attorney General of Botswana, Misca 52 of 2002 (Botswana).

Scott, C. (1997). Property, practice and aboriginal rights among Quebec Cree hunters. In T. Ingold, D. Riches & J. Woodburn (Eds.), *Hunters and gatherers: property, power and ideology*. Oxford: Berg.

Seidman, A., & Seidman, R. B. (2005). Legal frameworks. In B. Wisner, C. Toulmin & R. Chittiga (Eds.), *Towards a new map of Africa*. London: Earthscan.

Simpson, T. (1997). *Indigenous heritage and self-determination: the cultural and intellectual property rights of indigenous peoples*. Copenhagen: International Work Group for Indigenous Affairs.

Smit, P. (2002). The land issue of Namibia: some environmental, economical and planning perspectives. In M. O. Hinz, S. K. Amoo & D. van Wyk (Eds.), *The constitution at work: 10 years of Namibian nationhood*. Pretoria: VerLoren van Themaat Centre, University of South Africa.

Solomon, M. (2004). Intellectual property rights and indigenous peoples' rights and responsibilities. In M. Riley (Ed.), *Indigenous intellectual property rights: legal obstacles and innovative solutions*. Walnut Creek, CA: AltaMira Press.

Strang, V. (2000). Not so black and white: the effects of Aboriginal law on Australian legislation. In A. Abramson & D. Theodossopoulos (Eds.), *Land, law and environment*. London: Pluto.

Suzman, J. (ed). (2001). *An assessment of the status of the San in Namibia*. Windhoek: Legal Assistance Centre.

Suzman, J. (2002). *Minorities in independent Namibia*. London: Minority Rights Group International.

Suzman, J. (2004). Etosha dreams: an historical account of the Hai//om predicament. *Journal of Modern African Studies, 42*, 221–238.

Sylvain, R. (2002). 'Land, water, and truth': San identity and global indigenism. *American Anthropologist, 104*, 1074–1085.

Thorpe, J. (2005). Indigeneity and transnationality? An interview with Bonita Lawrence. *Women and environments, 68*(69), 6–9.

Tjombe, N. (2001). The applicability of the doctrine of aboriginal title in Namibia: a case for the Kxoe community in West-Caprivi, Namibia. Paper presented at Southern African Land Reform Lawyers Workshop, February, Robben Island.

Tucker, C. (2004). Land, tenure systems, and indigenous intellectual property rights. In M. Riley (Ed.), *Indigenous intellectual property rights: legal obstacles and innovative solutions*. Walnut Creek, CA: AltaMira Press.

Tully, J. (1994). Aboriginal property and Western theory: recovering a middle ground. In E. F. Paul, F. Miller Jr. & J. Paul (Eds.), *Property rights*. Cambridge: Cambridge University Press.

Ülgen, Ö. (2002). Developing the doctrine of aboriginal title in South Africa: source and content. *Journal of African Law, 46*(2), 131–154.

Werner, W. (1993). A brief history of land dispossession in Namibia. *Journal of Southern African Studies, 19*, 135–146.

Western Sahara Case [1975] ICJ Reports 12 (International Court of Justice).

Widlok, T. (2001). Equality, group rights, and corporate ownership of land. Working Paper No. 21, Max Planck Institute for Social Anthropology, Halle/Saale.

Widlok, T. (2002). Towards a theoretical approach to the moral dimension of access. Working Paper No. 37, Max Planck Institute for Social Anthropology, Halle/Saale.

WIMSA. (2005). *Report on activities: April 2004–2005*. Windhoek: Working Group of Indigenous Minorities in Southern Africa.

Woodburn, J. (1997). Indigenous discrimination: the ideological basis for local discrimination against hunter-gatherer minorities in sub-Saharan Africa. *Ethnic and Racial Studies, 20*(2), 345–361.

Some people call me a westernized Bushman. What did those peoples' ancestors wear? Do they still wear that today? Some of the Basters who came here wore skin clothes. Does it change their children into something else if they don't wear that anymore? (Petrus Vaalbooi, Rietfontein, South Africa)

Chapter 9
Speaking for the San: Challenges for Representative Institutions

Roger Chennells, Victoria Haraseb, and Mathambo Ngakaeaja

Abstract This chapter examines the San or Bushmen of southern Africa, with a focus on how their communities have evolved in order to represent themselves in the modern world. The various principles and customs that guided their lives prior to modernity as semi-nomadic hunter-gatherers are described, emphasizing close-knit egalitarian societies devoted to daily survival in often harsh environments.

The acknowledged status of the San as being the most poor and dispossessed peoples in southern Africa raises the question: why have they collectively been unable to compete and succeed in the modern world? Three suggested factors in the San's current marginalization are examined in turn, namely the legacy of a hunter-gatherer world view, pervasive poverty and landlessness, and collective trauma as a source of societal problems.

The history of attempts to assist and guide the San peoples in Namibia, Botswana and South Africa over the decades is briefly described, followed by a focus on the two forms of San institutions that have finally emerged. The first form, described as 'service organizations', includes the earliest attempts to assist the Ju/'hoansi of Nyae Nyae in 1981 and the Kuru Development Trust formed in Botswana in 1986. Each of these organizations has grown and evolved over the years, experiencing internal challenges relating to the particular context of the San, and devising better ways to resonate with and assist their San beneficiaries. The second form is San representative organizations, chiefly the formation of the regional San network

---

R. Chennells (✉)
Chennells Albertyn: Attorneys, Notaries and Conveyancers, 44 Alexander Street, Stellenbosch, South Africa
e-mail: scarlin@iafrica.com

V. Haraseb
Working Group of Indigenous Minorities in Southern Africa (WIMSA),
P.O. Box 80733, Windhoek, Republic of Namibia
e-mail: victoria.haraseb@gmail.com

M. Ngakaeaja
Working Group of Indigenous Minorities in Southern Africa (WIMSA), P.O. Box 934
Ghanzi, Botswana
e-mail: mngakaeaja@yahoo.com

named the Working Group of Indigenous Minorities in Southern Africa (WIMSA). WIMSA coordinates a democratically elected San council in each of the three countries, and has represented the San in the *Hoodia* case.

The chapter concludes with an analysis of the challenges that currently face WIMSA involving concerns about leadership and the continued creation of structures appropriate for the representation of San interests.

Keywords non-governmental organization • indigenous peoples • institutions • representation • San

9.1 Introduction

Who are the San or Bushmen of southern Africa? Are they a 'community'? (Were they ever?) If so, how have they organized themselves in order to represent the views and interests of their scattered peoples? Does this system of organization work, and, in particular, has the collective voice of the San been adequately expressed in the negotiation and signing of the *Hoodia* benefit-sharing agreements?

As discussed by Vermeylen in Chapter 10, there is considerable variation in opinion and practice among San communities and individuals regarding the commercialization of traditional knowledge. The assumption that there is such a thing as 'community' ownership of traditional knowledge already complicates identification of the relevant community that can legitimately make decisions and provide the required 'consent' regarding the commercialization or commodification of traditional knowledge. As demonstrated in Chapter 10, the diversity of attitudes in a community relating to trading in or dealing with traditional knowledge often depends on the socio-economic history of the community and previous exposure to 'trading' in traditional knowledge. In other words, one of the major challenges in negotiating benefit-sharing agreements is identifying the community and who represents it, while remaining aware of the diversity of opinions and potential tensions that exist and can arise within that community.

Comparing the notion of prior informed consent (PIC) in two bioprospecting cases – projects funded by the International Cooperative Biodiversity Groups (ICBG) in Peru and Mexico – Rosenthal (2006) observes that the involvement of an established, credible and politically representative governance system of indigenous communities was one of the most defining and crucial factors that gave the Peruvian case the advantage. Rosenthal argues that in all likelihood the prior notion of (Western-style) governance among the Aguaruna people contributed to the more successful procedural development of PIC in the Peruvian case in comparison to the Mexican case. The Maya communities in Chiapas, Mexico, were not as 'organized' as the Aguaruna people in Peru, in the sense that in Mexico there were hardly any indigenous political organizations authorized to represent and speak on behalf of the communities in relation to local and national natural resources (Rosenthal 2006).

Based on comparative analyses of the Mexican and Peruvian ICBG cases (Tobin 2001; Berlin and Berlin 2003, 2004; Brown 2003; Hayden 2003; Rosenthal 2006; Greene 2004), it can be assumed that the existence of a representative and democratic form of governance is one of the key factors that drives the process of PIC and benefit-sharing negotiations forward. As Chapter 6 explains, the San formed a representative and democratic organization, the South African San Council, to negotiate the *Hoodia* case.

This chapter will analyse some aspects of existing San organizations and institutions, both generally and in the context of the *Hoodia* benefit-sharing negotiations between the CSIR and the San peoples. First, the chapter documents patterns of organization and association that existed prior to modernity. Second, the marginalized state of the San peoples is discussed as a background to an examination of the emergence of organizations. Third, the rapid emergence of non-governmental organizations (NGOs) as well as representative bodies in the San world over 2 decades is described and discussed. Fourth and finally, some of the most telling criticisms and problems associated with the San and their institutions in the past decade, collected from the fieldwork referred to by Vermeylen in Chapter 10, as well as from workshops and wider sources, are discussed. The authors of this chapter were personally involved in many of the processes discussed below, which have not been the subject of published research. The chapter therefore relies heavily on communications from key informants in the field, as well as a decade's worth of private notes and personal observations.

9.2 San Institutions Prior to Modernity

The San have by all accounts lived for millennia without significant alteration in their ways of being. The lifestyle of semi-nomadic hunter-gatherers, now severely compromised by the loss of their *n!oresi* (traditional lands), was founded upon certain basic principles, all influenced by the need to survive as an intrinsic part, rather than conquerors, of nature. In his book *The Other Side of Eden*, Brody (2001) describes how hunter-gatherers simply accept nature and adapt to it exactly as they find it, as opposed to pastoralists and agriculturalists, who are programmed or moved to conquer and transform nature in order to feed their expansionist desires.

The San traditionally used to live in semi-nomadic family groups of 20–30 individuals, who would meet and join other groups from time to time to form larger clans. Their lives revolved largely around the business of survival, which entailed the daily hunting of animals and gathering of bush foods provided by nature, moving from place to place in accordance with the seasonal distribution of game and ripening of food plants. The availability of liquid sustenance in the form of watermelons, sip-wells, springs and natural pans would determine the number of people living in one *n!ore*.

Women contributed a significant proportion of the daily food supply and did a great deal of the household work. The elderly, both male and female, were respected for their knowledge and experience, and older people played important roles, doing numerous domestic tasks, taking care of children and passing knowledge on to younger generations.(Hitchcock et al. 2006). Tanaka and Sugawara (1999), describing the /Gui and the //Gana of Botswana, write that San women enjoyed equal status with men. Decisions would be made collectively, by consensus, with an inordinate amount of time spent discussing matters of importance, including those of a spiritual nature. Conflict of all sorts would be argued about incessantly, through unique interactional devices such as simultaneous discourse, redundant narrative and direct quotation of past speech, fuelled by a belief that to hide anything was a vice which would destroy the community. Force of any form used on others, whether explicit or even coercive, was shunned, and communal decisions were formulated in a most gradual and implicit manner. There was no word in the language for 'chief', and the healing dance was a central ritual performed collectively to draw on the healing powers of the ancestors (Tanaka and Sugawara 1999). Wynberg et al. further examine these intricate societal processes in Chapter 12, and the manner in which decisions were traditionally made by San communities.

Biesele and Kxao Royal /o//oo (1999), writing on the Ju/'hoansi, describe the San egalitarian philosophy as follows.

> Ju/'hoan political ethos abhorred wealth and status differences. No one should stand out from the rest of the group. If someone returned from a successful hunt showing excessive pride, he was put firmly in place, even if the kill was large. Emphasis on sharing and lack of status roles produced a high degree of egalitarianism. These rules, based on living in small groups of kin, worked successfully. Anger and resentment were low as each person's opinion was respected. Conflicts could be terminated by a disputant leaving to join another group. ... Equality and sharing remain important. ... N!ore ownership remains non-exclusive ...

In San society, leadership in all matters, whether of custom and ritual, medicine and healing, hunting, trading or politics, would be by different individuals, and general leadership by one individual was not known. Institutional arrangements were informal, flexible and situationally determined. The San had no use for written language, and all laws, norms, myths and customs were orally preserved and transmitted from one generation to the next. Conflict was abhorred, and withdrawal from the conflict was a commonly accepted and culturally appropriate response. It is not an exaggeration to conclude that the San people, famously termed the 'harmless people' by the author Lorna Marshall, in reference to their gentle and humour-loving natures, were intrinsically ill-equipped to compete in the modern materialistic society.

9.3 Marginalization of San Peoples

Before discussing San modern institutions, we summarize some of the ways in which the San are highly marginalized today and suggest reasons.

Numerous studies have attempted to understand this issue. The *Regional Assessment of the Status of the San in Southern Africa*, an in-depth analysis of all San communities in southern Africa commissioned by the Legal Assistance Centre in Windhoek in 2001, concluded that the San persistently remained by far the poorest, most marginalized and dispossessed of all communities. Various factors, including the hunter-gatherer factor, poverty and poor health, and collective trauma, have been mooted as possible contributors towards this particular vulnerability.

9.3.1 *A Hunter-Gatherer World View?*

Diamond (1999) posed the burning question: why is it that over 12,000 years of development, some societies acquire material wealth through pastoralist or agriculturalist economies, and others, such as the San, show little interest in bettering themselves materially, remaining fundamentally hunters and gatherers? This is a gross oversimplification of a complex subject, but it seems safe to suggest that the San peoples today are essentially a hunter-gatherer society in transition. Many of the traits of hunter-gatherers, such as a focus on the present, a lack of interest in long-term planning, a discomfort with hierarchies and formal structures, and a lack of understanding of material wealth, still imbue their world view in the rural settlements.

Another author's attempt to explain why indigenous peoples of today seem to steadfastly resist 'advancement' (Brody 2001) equates humankind's quest to conquer and control the world with the curse that God placed upon Adam, as recorded in the biblical book of Genesis, and postulates that indigenous peoples who show no inclination to manage or control nature are therefore free of what is characterized as the 'curse' of the drive for material advancement.

San elders often reflect sadly on the quality of the way of life that they remember, and bemoan the loss of land, dignity, culture and order that characterizes their lives in the modern resettlement villages. These elders are vocal in their rejection of the modern aspirations for possessions and pursuit of alcohol, but lack the means to influence the youth or to suggest a course of action that might lead to retention of some of the old values. Newly resettled residents of New Xade in Botswana, recently evicted from ancient traditional lands in the Central Kalahari Game Reserve by their 'development'- focused government, express sadness at the lack of understanding and empathy shown by the authorities. Any acknowledgement of the intangible but potent value of the evicted San's culture – which, linked to their traditional land, was at the core of their sense of identity and self-esteem – is totally absent from the thinking of the Botswanan government (Sugawara 2002).

The vital linkage between land and culture of indigenous peoples is acknowledged in numerous international legal instruments (ILO 1989; UN 2007), but governments continue to ignore them with impunity. How can one place a value on the loss of land and culture? Vermeylen in Chapter 8 examines the state of San land rights in this context. San leaders and organizations today tend to acknowledge their hunter-gatherer roots in a manner that is neither romantic nor primordialist, but accepts the need to adapt to the challenges of modern life.

9.3.2 Poverty and Poor Health?

San NGOs are able to access donor funding readily, largely because the people they serve are the poorest of the poor. Behind the stark reality of this poverty lurks a nagging question associated with other hunter-gatherer societies: why are the San peoples so poor, so subservient to other groups, so lacking in assertiveness or apparent ambition or a desire to 'better' themselves? The objective fact that they earn by far the lowest per capita income in southern Africa, as concluded in Suzman's (2001) regional assessment, masks far deeper problems.

Poverty is the clearest objective index of the status of the San, and throughout southern Africa San communities are characterised by widespread unemployment and an acute reliance on welfare services (where available), casual labour, begging and/or charity.

Especially where they are removed from their traditional lands, the pervading air of lassitude and hopelessness recalls the similar fate of the first peoples of America, Canada and Australasia. Ingstad and Fugelli conclude that 'loss of land results in loss of health via loss of self-esteem' (Ingstad and Fugelli 2006). Renée Sylvain's description of the state of the San farm labour communities in the Omaheke region of Namibia (Sylvain 2006) is especially depressing, in that the social dysfunction seems deeply entrenched. She describes how the San egalitarian values provide space for aggressive drunkenness in response to their new conflicts and the hopelessness of their plight. However, the same values enable ready tolerance and ready forgiveness of the drunken wrongdoer, with the result that there is little incentive to curb the drinking (Sylvain 2006). Poverty thus lies at the root of a range of societal problems which tend to perpetuate and exacerbate the San's meagre circumstances.

9.3.3 Collective Trauma As a Cause of Societal Problems?

The trauma[1] inflicted upon indigenous populations by colonial invasions has been remarkably similar, from the Americas to Australasia to Africa. Superior weaponry devastated entire populations, and the convenient *terra nullius*[2] doctrine provided comfort to governments responsible for atrocities committed in their name.

---

[1] 'Trauma' is a condition of psychological shock or severe distress from experiencing a disastrous event outside the range of usual experience, causing a disturbance in normal behaviour'. A widely accepted definition of trauma is "any injury, whether emotionally or physically inflicted" (http://medterms.com).

[2] This doctrine of colonial empires held that land occupied by indigenous or local peoples, who did not maintain a recognized system of 'ownership' of the land, was in fact empty land and thus open to occupation by the civilizing invaders.

The eradication of the San was rationalized as retaliation against their theft of cattle, imposing law and order on a 'lawless land', and clearing farming land of 'vagrant and treacherous savages'. During the eighteenth century thousands of San were systematically exterminated by hunting parties, and their traditional lands were 'legally' occupied by waves of farmers, supported by colonial law. By the end of the eighteenth century their culture and society had been devastated and those that survived did so in a state of servitude, having lost access to the land and the game that previously had been their livelihood. Much of this shocking history has been submerged, but the exhibition *Miscast* at the South African National Gallery in 1996 (Skotnes 1996) provided a shocking visual reminder of the sustained, merciless carnage wreaked on generations of San in the name of civilization.

The present social conditions of the San resonate with and reflect the traumatic experience of centuries: dispossession, loss of language, loss of traditional lifestyles and values, genocide, slavery and humiliation. Irène Staehelin, writing on San loss of identity, describes how the alcoholism, lethargy and general sense of disorientation are not only the legacy of this dark history, but also contain a dangerous potential for self-perpetuation (Staehelin 2001). Abadian (1999), in a study of the collective trauma of indigenous peoples, makes the critical point that cultural dispossession is connected to alienation, and that unresolved collective trauma is an essential cause of present-day dysfunction. She describes 'unresolved' trauma as being a failure to productively integrate, move through and release traumatic experiences. Communities and individuals are unable to create meaning from the repressed experiences in a manner that enhances rather than debilitates their lives. Abadian argues that the experience of trauma profoundly distorts individual perceptual filters, values and behaviours, and adds that money alone is not sufficient to bring healing. Alcoholism, violence and apathy continue to destroy the social fabric of indigenous societies that receive well-meaning support from NGOs and governments. This leaves donors scratching their heads or blaming these people as being 'doomed to extinction' and 'not wanting to help themselves.' Staehelin (2001) explains addictive behaviour, such a blight on the lives of indigenous peoples, as a by-product of non-validated and suppressed grief, rage and shame.

Abadian (1999) confirms that indigenous children carry and perpetuate the burden of their parents' unresolved pain in a never-ending spiral of trauma and violence. A state of low personal capacity is the norm, linked to and driven by a pervasive low personal and collective self-esteem. On a collective level, these factors lead to dysfunctional communities whose members are not able to interact with their leaders or among each other in constructive ways. In summary, this crowded constellation of self-destructive behaviours in indigenous communities, and particularly the San, is a result of unresolved traumas suffered through past experiences of colonial violence, compounded by the ongoing situation of alienation and hopelessness.

It is suggested that the state of unresolved trauma underpinning the state of the San has received far too little acknowledgement. San NGOs and representative organizasations, as described in the sections below, have the task of moving creatively away from this victimhood and towards empowered San development.

9.4 San Modern Institutions

After millennia of relatively unchanged existence, San institutions have evolved rapidly over recent decades in response to the urgent demands of the prevailing political and social environment. Generally they have faced incomprehension, both well-meaning and otherwise, from governments and individuals alike. In 1936, for example, South Africa's Minister of Native Affairs P.G.W. Grobler was so impressed with a San exhibit at the Empire Exhibition in Johannesburg that he said, 'we must treat these Bushmen as fauna. They must be allowed to continue to live and hunt in the Kalahari Game Reserve.'[3]

On the other hand, the British Resident Commissioner, C.F. Rey, in response to the proposal to cede a portion of the Kalahari to the San, said:

> '[I]n the first place I see no reason whatsoever for preserving Bushmen. I can conceive no useful object to the world in spending money and energy in preserving a decadent and dying race, which is perfectly useless from any point of view, merely to enable a few theorists to carry out anthropological investigations and make money by writing misleading books which lead nowhere.'[4]

It is encouraging to note that in Africa, a continent not known for its attention to concerns emanating from the weaker sections of society, the African Commission on Human and Peoples' Rights has formed a working group on indigenous peoples in order to address their particular human rights plight. In a booklet outlining this working group's mission, entitled *Indigenous Peoples in Africa: The Forgotten Peoples?* they describe indigenous peoples of Africa as follows.

> 'The overall characteristics of groups identifying themselves as indigenous peoples are that their cultures and ways of life differ considerably from the dominant society, and that their cultures are under threat, in some cases to the point of extinction. ... They suffer from discrimination as they are regarded as less developed and less advanced than other more dominant sectors of society. They ... suffer from various forms of marginalization, both politically and socially. They are subjected to domination and exploitation within national political and economic structures that are commonly designed to reflect the interests and activities of the national majority. This discrimination, domination and marginalization violates their human rights as peoples/communities [and] threatens the continuation of their cultures and ways of life ... (ACHPR and IWGIA 2006)'.

Nowadays San communities bear scant resemblance to the informal consensus-based groupings that characterized their living arrangements over past millennia, and the modern organizations that have evolved and are described below bear the onerous responsibility of providing appropriate leadership to the erstwhile hunter-gatherer communities, scattered across national borders

---

[3] *The Cape Argus*, 25 August 1936, quoted in Hitchcock et al. (2006).

[4] Botswana National Archives, BNA file S 469/1/1, quoted in Hitchcock et al. (2006).

and speaking at least seven different languages,[5] that they purport to represent. Their task is to provide substance and content to that most elusive of concepts, namely the modern 'San community'.

9.4.1 San Development and Support NGOs

In Namibia, attempts were made to organize and assist the Ju/'hoansi of Nyae Nyae from as early as 1981, when Claire Ritchie and John Marshall established a cattle fund with a formal constitution and managed by a board including members of government and the private sector as well as Ju/'hoansi representatives. Much discussion took place with the Ju/'hoansi San, and by 1986 the 'Ju/wa Farmers Union' was formally founded in tandem with its support organization, the 'Ju/wa Bushman Development Foundation' (JBDF) (Hitchcock 1992). The JBDF helped raise funds and provided technical support for the Ju/wa Farmers Union, which later became the Nyae Nyae Farmers Cooperative and then, in 1996, the Nyae Nyae Conservancy.

In Botswana, the first formal institution representing the San, namely the Kuru Development Trust[6] was established in August 1986 at D'Kar, near Ghanzi. With guidance and facilitation from founders Braam and Willemien le Roux, ten San individuals from the D'Kar community were chosen to sit on a registered trust body, which initially represented the different income-generation and development programmes focused on poverty alleviation and early childhood education. These San trustees were initially appointed by their groups, but as the programmes expanded and reached further afield, they were appointed from these communities as well as the local church council, the founder, and individuals who had shown a willingness to engage and speak and assume responsibility on the board. Over the decades the work and size of Kuru grew exponentially, and within 10 years the trust was presiding over an annual budget of more than 10 million pula (US$1.57 million) and employing over 120 staff. Some of these employees were skilled expatriates and local Batswana, who were contracted to manage the organization under the formal control of the Kuru Development Trust. In the process of professionalizing the increasingly complex organizational structure, an evaluation recommended internal changes, including forging regional links with other San organizations.

Tensions arose as the original beneficiaries of Kuru in D'Kar felt threatened by the loss of resources they feared would occur if they were to share power with other communities and submit to a more democratic election process.

---

[5] Amongst the San languages still spoken are the G/wi, G//ana, Nharo, !Kung, Ju/'hoansi, Hai//om, Khwe, Xun, ≠Khomani (N/u) , ± X'ao//'aesi and !Xoo.
[6] The word *kuru* means 'to do, to create' in five San languages.

Board members who agreed with the recommendations of the evaluation were accused by these 'dissidents' of receiving benefits in the form of status, salaries and the use of vehicles. Other staff members attempted to secure their positions by forming alliances with the dissident movement, causing divisions and further tensions.

The dissidents turned increasingly against the expatriates in management and attempted to secure support from donors and government in order to 'keep Kuru for themselves'. Thus the acknowledged effectiveness of Kuru as a San development organization alone was not sufficient to protect it against these internal tensions (Le Roux, 2007, personal communication).[7] Matters came to a head in 2000 when resentment grew after rumours were spread by certain aggrieved individuals over perceived inequities such as the high salaries and living conditions enjoyed by the skilled expatriate staff, as well as over the anticipated sharing of Kuru resources with San communities from different language groups. This resentment led to a crisis at the highest level of Kuru, namely the board of trustees. Efforts to negotiate the changes within the board were to no avail, as the San trustees were presented with structural and corporate challenges for which they were ill-prepared. Local politicians and government became involved, placing the dispute in the public domain. Consensus could not be reached. Meetings and discussions aimed at addressing and resolving the perceived problems continued for close to a year, until all of the allegations of the dissidents were formally put to rest. The core work of Kuru continued during this time, but many donors withdrew in discomfort at the crisis, and the subsequent painful restructuring of the entire organization commenced in 2001.

In evaluating the reasons for the breakdown of Kuru, commentators regarded it as unwise to have placed San community leaders in positions of board control over such complex structures and large budgets (Le Roux, 2001, personal communication). San board members, invariably from a rural upbringing, with little formal education and often living in the same state of poverty as the NGOs' intended beneficiaries, had been expected to oversee complex management structures and the employment conditions of people far better educated then themselves, to evaluate proposals written to foreign funders, to manage complex management issues, to understand and approve budgets involving large sums of money, and to apply notions of 'good governance' rooted in distant Europe.

Following the restructuring process, Kuru became effectively 'unbundled' and the important development work previously managed by the organization became reconstituted into a number of far smaller and more manageable organizations based in Botswana and South Africa, each with its own appointed San

---

[7] The description of the complex Kuru restructuring process is extracted from various letters and official reports and primarily based on personal communications with Braam le Roux, the founder and then director of Kuru, in November 2007.

9 Speaking for the San: Challenges for Representative Institutions

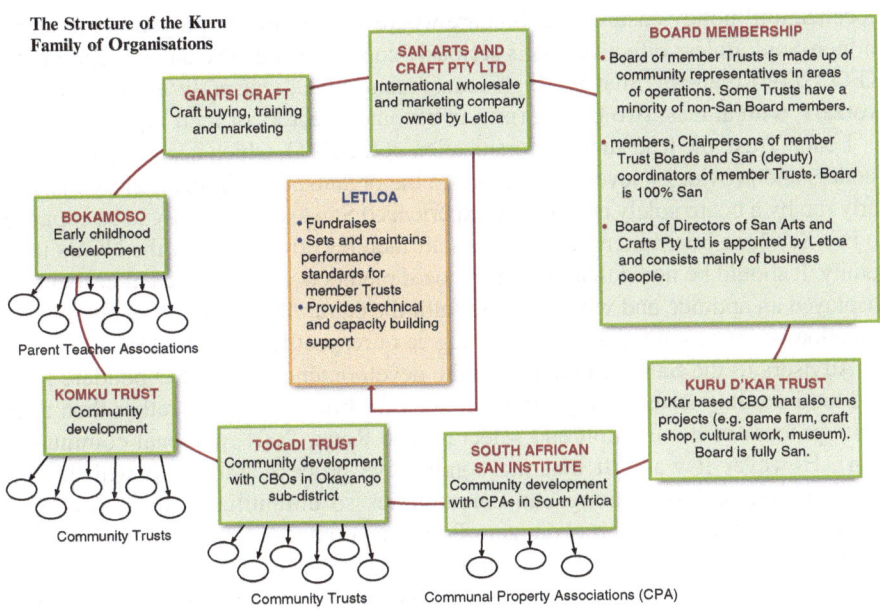

Fig. 9.1 An Organogram of the Kuru Family of Organisations

board. Figure 9.1 shows an organogram of what is now known as the Kuru Family of Organisations (KFO).[8]

Each of these NGOs is committed to a policy of 'holism' which incorporates all aspects of the lives of communities supported, while focusing on a particular geographical area and/or field of development. Another policy is that of San 'embeddedness', which requires that San should own, govern and, if possible, also staff their own development (KFO 2006). The KFO governing boards are thus predominantly San, ensuring San 'ownership' of their own development agenda. In addition a San

---

[8] The Bokamoso Trust focuses on early childhood education and the Komku Trust on livelihoods, youth, health, natural resource management and development. The D'Kar Trust is responsible for development of the D'Kar community, including culture, art, language and youth, whilst Gantsi Crafts is dedicated to the development of craft livelihoods in the region, and San Arts and Crafts (Pty) Ltd helps San producers market their art and crafts. These five NGOs are all based at D'Kar, near Ghanzi. TOCaDI (the Trust for Okavango Cultural and Development Inititatives), which develops community-based organizations in the Okavango subdistrict of Ngamiland, is based in Shakawe, northern Botswana, whilst the Letloa Trust, the KFO's lead support organization, is based in both Shakawe and D'Kar. In addition to providing leadership and guidance to the KFO, Letloa also runs the Land, Livelihood and Heritage Resource Centre, which supports communities with land rights and community-based natural resource management, a culture and education programme that also focuses on heritage in the education context, and a major Tsodilo Hills development programme in partnership with the government. The South African San Institute (SASI), based in Kimberley and Upington, coordinates development work with South African San communities, having joined the Kuru family in 2003. See www.kuru.co.bw and KFO (2006).

'localization' policy ensures that San are wherever possible employed in preference to 'non-San', unless effectiveness is seriously compromised. The annual report of the KFO (2006) confirms that the eight constituent NGOs currently employ 100 San workers, with at least two having risen to become directors of their organizations.[9]

Letloa Trust, the lead support organization of the KFO, which accounts formally to funders and provides a level of guidance to the organizations in the 'family', is similarly run by a board solely comprising experienced San leaders, shortly to be expanded to two San (one board member and one staff member) from each of the NGOs in the family. It should be noted that these San board members are appointed after they have displayed an aptitude and willingness to fulfil the culturally novel task of board representation and have earned a significant degree of respect from their peers.

Advisers to the San who facilitated the development of the KFO pondered long and hard on the merits of requiring elections for board members rather than a less finite process of nomination and selection (Le Roux, 2001, personal communication). However it was felt that the competitive process associated with holding elections, particularly among San communities so unfamiliar with the notion of voting, would introduce a real danger of dividing communities. A further danger of elections was that they might encourage the emergence of ambitious individuals with purely political skills not necessarily appropriate for a board (Le Roux, 2001, personal communication). Board members are therefore currently chosen by a process of nomination and confirmation, rather than election.

While this might seem a rosy picture of the state of the San-owned NGOs, it is far from our intention to claim that these institutions are free of problems. The capacity of San board members and staff to fully understand and assimilate the expectations of the approximately 27 international funder organizations that support their work to a value of approximately P20 million per annum, including adherence to the so-called Western values of productivity and accountability, is among the perennial challenges. An additional concern is the fact that the entire organization remains largely reliant upon the fund-raising skills and credibility of certain key individuals, despite attempts over years to reduce this reliance, which is an acknowledged source of vulnerability.

In 2005 the KFO commissioned The Proteus Initiative, a specialist organizational development firm, to conduct an in-depth participatory evaluation of the entire organization. It was concluded more than a year later, and the recommendations were adopted by the entire KFO in April 2006 (Kaplan and Davidoff 2006). The particular focus of the evaluation was on whether the KFO was adequately serving the San communities it aimed to assist and what changes were needed to improve its performance. The review report listed some of the organization's achievements, which included significant policy shifts in the indigenous arena, the tabling of San issues on national agendas, the building up of San political aspirations and the increasing confidence of KFO individuals. One of the observations of the

---

[9] In November 2007 one of these two San directors, Kabo Motsweu, died tragically in an accident.

evaluation was that pioneer and founder Braam le Roux should increasingly define and modify his role, that core responsibilities should be delegated to others, and that the KFO should simultaneously work to equip young San to eventually take control of the whole organization (Kaplan and Davidoff 2006).

The process of giving effect to the Proteus recommendations continues and will no doubt require adaptation along the way. However, the KFO's proven commitment to brutally honest self-evaluation and questioning of its roles and efficacy provides ample assurance that the San are not being subjected to powerful foreign agendas. While progress towards highly effective San-led NGOs is slow, and while San in far-flung communities remain mired in poverty, it is generally accepted that the KFO provides an essential level of critical and self-reflective development support to all San within its reach. Board members, staff and beneficiary San communities all receive sustained, long-term development support through this network of San-owned, yet essentially 'Western' or modern, institutions. The Proteus strategic review also confirmed KFO's commitment to an ambitious and self-generated 'Vision 2012', which combines a commitment to core San values with a host of specified development targets.

In August 2007 the KFO, with 130 staff members serving over 50 San communities, celebrated 21 years of existence (KFO 2007). The words of the San chairperson of the Letloa Trust, Selina Magu, writing in the foreword of a booklet entitled *The Kuru Story* (2007), deserve to be quoted in full.

> 'People who come to see us today, should not just think what they see now is how it always was. In the past the San people were afraid to speak or appear before other people. Kuru has changed our lives, things are better for us today because of the work of our own organizations. Nowadays a San person can speak for him/herself, can stand up and speak freely, no matter where. The San people's lives are far better than before, because of Kuru. In the past it was difficult for a San person to visit any government office or even to go to Gaborone, we were afraid to meet other people or ask anything, but now I can do that easily. I now know that I am somebody, that I have the right to got to the relevant office and ask what I need.
>
> Because of Kuru San people from other places can now visit each other and their voices are heard worldwide. Our languages are now read, even by other people, because if I go to the clinic I can now talk in my mother's tongue, and the nurse will check a word list to see what I mean, to help me. It was not like this before. Kuru has made me proud of who I am. I can apply for anything that any other human being can, that is why some of us even now have cattle syndicates, we have driver's licences, knowledge and skills like any other kind of person. There are so many more things to say about the way we worked and built up this organization, but the most important for me is to know that I can now plan for my family and my future, and I know what is meant by leadership. This is all because of Kuru.'

9.4.2 San Representative Organizations

At the Regional Conference on Development Programmes for Africa's San Populations held in Windhoek, Namibia, in 1992, the San representatives resolved that 'San peoples should be assisted to represent and articulate their interests at local, regional and international levels'. This was followed a year later by a conference in Botswana and then a needs assessment in 1994 involving San representatives, government officials, NGOs and academics. This needs assessment concluded with support for the San's plea for a forum where they could become part of their own regional development (Brörmann 2002).

A new assessment study covering San communities in Namibia, Botswana, South Africa, Zambia and Zimbabwe was then launched, in which the two appointed consultants were allocated specific countries and priority was given to consultations with San settlements perceived not to have been adequately represented at the earlier two regional San conferences (Thoma and Le Roux 1995). In line with the recommendations of this assessment study, the Working Group of Indigenous Minorities in Southern Africa (WIMSA)[10] was formed in mid-1996 as a San-owned representative organization, with a formal constitution and the core objective of giving the San a network and a strong voice on all regional and international human rights matters. With its head office in Windhoek and a regional branch in Botswana, WIMSA proceeded to develop the capacity of San individuals in its member community-based San organizations.

In cooperation with the San NGOs described above, and with the support of a handful of international donor and support organizations, WIMSA launched an assertive human rights agenda. A board of nine San leaders was elected, three each from Botswana, Namibia and South Africa, which met regularly to decide on broad policy in pursuit of its human rights objectives, including education and organizational support. This entailed engaging with the emerging international indigenous peoples' movement and becoming an active part of the United Nations' first International Decade of the World's Indigenous People, which had commenced in 1994.

Soon the voice of the San became recognized in international forums, particularly on matters of land, natural resources and cultural rights. During a meeting of African indigenous organizations attending the UN Working Group on Indigenous Populations in Geneva in July 1996, WIMSA took the lead by proposing and then leading the formation of the Indigenous Peoples of Africa Coordinating Committee (IPACC),[11] a broader networking body based upon the principles and structure of WIMSA, tasked to coordinate the indigenous struggle on the continent of Africa. WIMSA remains an active member of IPACC, which is currently the only activist organization formally recognized by United Nations agencies as representing indigenous peoples in Africa.

While the international struggle focused on securing a Permanent Forum for Indigenous Issues and on negotiating and adopting a Declaration on the Rights of Indigenous Peoples,[12] WIMSA soon realized that it did not have the resources to

---

[10] The word 'San' was not used in the name as the governments of Namibia and Botswana were at the time opposed to any notion of racial differentiation.

[11] www.ipacc.org.za.

[12] The UN Permanent Forum on Indigenous Issues was formed in 2002 and the Declaration on the Rights of Indigenous Peoples was adopted by the UN General Assembly in August 2007 (UN 2007).

participate fully in the increasingly demanding international arena, as its priorities were perceived to lie with the poor and scattered San communities at home. General meetings were held at least twice per annum, at which issues such as heritage and intellectual property rights were discussed and hotly debated by gatherings of 70–100 San representatives. Matters involving heritage were always the subject of rapt participation, being the one issue that encapsulated the reality of a single San identity across the numerous linguistic and cultural differences between the groups.

At a WIMSA general assembly in 2001, after days of the customary consensus-seeking debate, delegates formally decided that the San culture and heritage in its entirety, including intellectual property, was and should be collectively owned by all San (Brörmann 2002). They decided further that if any financial or other benefits accrued in future, these would be shared fairly among all the San (Brörmann 2002). This decision was to prove of crucial importance some years later when it came to deciding on the practical challenges of benefit sharing in the *Hoodia* case.

For most of the first decade of WIMSA's existence, rightly known as the 'pioneer phase', the WIMSA coordinator and chief executive, reporting to an all-San board of trustees (three from each of the three countries), was a non-San individual, namely Axel Thoma. During the years leading to his resignation in December 2005, he was supported by a San counterpart, who took over as coordinator. Over the ensuing period WIMSA came near to closure, and the board learned some harsh organizational lessons in the process. One of these lessons was that 'San empowerment' was not necessarily advanced by the appointment of San to all key posts, unless they were suitably qualified. After an internal process of review and reflection, the WIMSA board determined that in future important posts would be filled by the most qualified individuals capable of fulfilling the task, whether San or not.

In early 2001, following years of discussion, WIMSA decided that it should evolve constitutionally towards its original vision, namely the creation of a regional democratic umbrella structure made up of elected San councils representing the San of each constituent country.[13] An initial draft constitution proposing an overall regional WIMSA board comprising three of the elected San council leaders from each of the three constituent countries was approved in principle, pending the election and formation of representative San councils in each country. It was decided after some debate that WIMSA should provide equal seats on this regional San board or 'governing body' to each country, rather than have different numbers on the board reflecting the different populations in each country (Thoma 2005). When the issue of the CSIR's patenting of the active ingredients of the *Hoodia* plant erupted in South Africa in June 2001, the WIMSA board met and formally mandated the then unregistered South African San Council to act on its behalf. The council proceeded to represent all of the San in mounting its challenge to the CSIR.

---

[13] Namibia, Botswana and South Africa have been members of WIMSA since its inception. The approximately 5,000 San of Angola are not yet organized, but it is the stated intention to include Angola as soon as its San become organized. Representatives from Angola have been invited to WIMSA general meetings since 2005.

The organogram of the South African San Council, illustrated in Fig. 9.2, warrants brief analysis, as it represents the modern form of San leadership, aiming and professing to democratically represent the San communities within each country. The South African San Council is made up of three representatives from each of the three San communities in South Africa, namely the ≠Khomani, the !Xun and the Khwe. These linguistically and culturally distinct communities each have their own formal constituent structures, as required by South African law.[14] Each is further required to

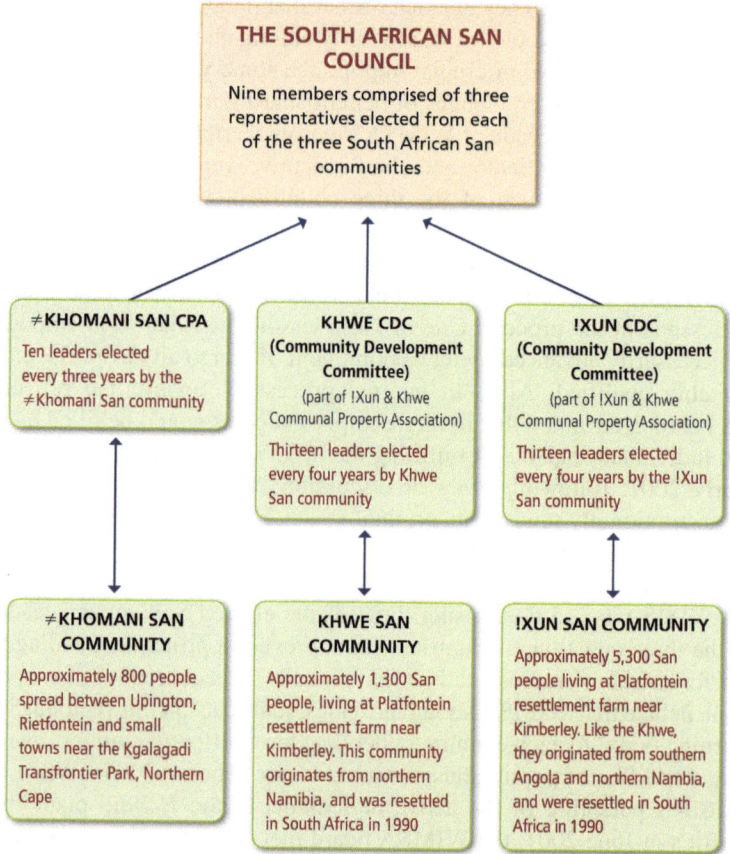

Fig. 9.2 An Organogram of the South African San Council

---

[14] The Communal Property Associations Act 28 of 1996 requires formal elections to be held in each community in accordance with an approved constitution, based upon a communal trust system. The elected representatives then elect a chairperson, vice-chairperson and secretary, who become the formal leaders of the committee.

hold regular elections, based upon formal written constitutions, in which San community members, some of whom are functionally illiterate, are entitled to cast their votes. Elections involving the nomination of candidates and secret voting at ballot boxes take place with the assistance of the Department of Land Affairs and the local KFO service organization, the South African San Institute.

These elected San leaders, with their new mobility, visibility and access to power, immediately become a 'new elite' in their communities and find themselves firmly placed in the modern world of Western representative politics. They are required, without preparation or warning, to endure the inevitable jealousy of their former peers and to deal with the novel responsibilities placed upon them. With varied levels of formal education and far too little training, they are expected to carry out their functions in accordance with principles of democratic accountability and efficiency as stipulated by their modern constitutions, including the injunction to resist the understandable temptation to use their new-found power to feather individual nests. A further challenge is posed by the long-standing traditional leadership structures in South African San communities, which are based upon custom rather than elections. These structures and their incumbents tend to compete more or less openly with the elected bodies, and to espouse conservative values more closely associated with the former egalitarian San hunter-gatherer culture.

In March 2007, after some years during which successive Namibian San councils were elected and received training from WIMSA, the current Namibian San Council was elected, representing six linguistic and geographical San constituencies.[15] In an attempt to avoid power struggles between the two sources of power, it was decided that in each constituency, one member would be elected from the tribal or traditional authority, in which an elected 'tradional leader' is both recognized and paid by the government, and one from the 'civil society' or non-traditional sector. In this manner, WIMSA attempted to ensure that the two sectors of each constituency (namely the 'traditional' and the 'civil society' sectors) were both fully represented on the council. In the previous 3 years, power struggles had erupted over representation on the contested San council, and this compromise, together with a San decision that San traditional leaders or chiefs[16] may not be elected to the San council, represents the ongoing attempt by WIMSA to ensure that the Namibian San Council constitution is responsive to the unique features of Namibia's San communities.

Botswana has by far the most San residents[17] and the choice of an appropriate structure to represent them has proved elusive. The KFO provides practical and development assistance to the San, and their political representation in Botswana

---

[15] The six constituencies are the Hai//om, the !Kung, the Ju /'hoansi, the Khwe, the ±X'ao//'aesi of Omaheke North and the !Xoo of Omaheke South.

[16] In Namibia San chiefs are recognized in three of the San regions under the Traditional Authorities Act of 1994. These three are the !Kung, the Ju/ 'hoansi and the Hai//om. The chiefs of the Khwe, the ±X'ao//'aesi of Omaheke South and the !Xoo of Omaheke North are not yet recognized.

[17] Generally accepted estimates of San populations are Botswana 55,000, Namibia 35,000, South Africa 8,000, Angola 5,000, Zimbabwe 1,500 and Zambia 1,000.

has thus far been pursued not only by the Botswana branch of WIMSA, but also by an NGO called the First People of the Kalahari (FPK), which claims to speak for all San in Botswana. FPK has become well known over the past decade for leading the struggle to protect the rights of the San residents of the Central Kalahari Game Reserve and won a famous court victory against the Botswana government on 13 December 2006. It is generally regarded as a 'one-campaign NGO', despite its aims to achieve wider representation, and has little influence in the country on issues unrelated to the Central Kalahari Game Reserve. An initiative pursued over the past 2 years to form and register a 'Khwedom' council in Botswana is nearly complete. When this San council formally adopts the principles of the WIMSA structure, it will join the other two San councils under the overall umbrella of WIMSA, which will then finally have a complete and democratic regional structure providing equal representation and protection to all San.

The South African San Council duly assumed the role of San negotiator in opposing the CSIR's secret registration of the patent on the properties of the *Hoodia*, as described by Wynberg and Chennells in Chapter 6, and in the process climbed a steep learning curve in the field of negotiation and intellectual property rights. Meeting regularly with their legal team and having a clearly defined objective, the council negotiators acquired a working understanding of the field and finally concluded the two benefit-sharing agreements that are the subject of much of this book. The entire negotiation was conducted at a pace set firmly by their research and pharmaceutical opponents and concluded within 18 months. Because the South African San Council had limited time or money to hold regular, in-depth consultative feedback sessions with its constituencies, it is little wonder that many of the remote San communities whose interests it claimed to represent had scant knowledge at the time of the final agreement being negotiated on their behalf.

9.5 Some Reflections and Concerns from the Field

9.5.1 *Leadership*

One of the most frequently recurring problems observed by an anthropology researcher visiting San communities in Namibia, South Africa and Botswana was the disappointment and lack of trust that community members expressed in their leaders (Vermeylen 2007). Complaints ranged from accusing some San leaders of being aloof, selfish and distant to allegations of bribery, corruption, manipulation and even blackmail.

> Our leaders – what can I say – my heart is paining if only […][18] could be our leader again, our problems would be solved; he would manage the land in an appropriate way; he was a

---

[18] The name of the person referred to has been withheld.

strong and honest man but now he is an old man and he is not feeling very well so he cannot represent us anymore. He knew how to keep the community together so we could all work together to reach the same goal, he always said work with me and not against me'. (Vermeylen 2007)

Modern San leaders, with the high profile accorded to their new status, are subjected to harsh scrutiny and criticism similar to that directed at leaders in the West. In Namibia, a Tribal Authorities Act provides for the election of a chief in three of the San regions, which is a cause for much concern. The Act recognizes the Hei// om, Ju/'hoansi and !Kung communities, which means that they were obliged to hold elections for a tribal 'chief'. Two of these three chiefs are, however, widely regarded by their people as corrupt and corruptible, and receive scant respect around the fireside.[19] One in particular, Chief John Arnold of the !Kung, is not only a cattle owner, but is known to be sympathetic to the interests of the powerful cattle farmers whose expansionary zeal, supported by the ruling party, threaten the traditional lands of the !Kung San. NGOs working with the !Kung and observing an imminent dispute with the government over the intended allocation of San conservancy land to pastoralist farmers anticipate that the !Kung chief will take the side of the farmers rather than his people.[20]

Many San do not like having large San communities presided over by formally acknowledged chiefs, which is not surprising in view of the fact that the San traditionally did not know or acknowledge single leaders. As described above, 'leadership' was almost fluid and could only be attained when an individual had gained sufficient respect in the community on the basis of particular skills. Furthermore, as explained by Guenther (1999), the San's 'traditional' institutions and processes of leadership and decision-making were ad hoc and ambivalent, varying from community to community and between different language groups. Barnard (1993), for example, described this 'loose' structure as an 'anarchy à la Kropotkin'. Where leaders were accepted in the past, explains Guenther (1999), they were usually individuals possessing a charismatic personality or a particular skill who took decisions that were subsequently respected by the other people in the community.

Leadership as currently exercised is thus a modern phenomenon, and one that does not sit easily with the San. WIMSA has attempted at various stages, in acknowledgement of the criticisms being voiced by communities, to create 'terms of reference' for leaders, requiring qualities such as honesty, sobriety, diligence and responsibility. At the consultative workshop held in November 2006, attended by over 30 San leaders, 'leadership' was identified as one of the current problems that prevent the San from achieving their goals. When groups discussed individual problems with leadership, the San leaders themselves listed the following as being the particular issues that stood in their way: a lack of management training and skills, the lack of a work ethic, the lack of a proper support system, the abuse of power,

---

[19]This is the impression gleaned from numerous informants by the authors, Axel Thoma, Cameron Welch, Ben Begbie-Clench and others.
[20]Field notes, Cameron Welch (PhD student, 2006).

and a lack of accountability to their communities (EED 2006). It was noteworthy that this group, representing most of the existing San leadership, were brutally honest in their collective assessment of the difficulties that they experienced in attempting to fulfil their responsibilities.

Another characteristic frequently observed in the field among communities with elected leaders was that other and more charismatic people assumed key roles in the daily management of the community. These 'alternative' and non-official or non-elected leaders seemed, in some cases, to gain more respect in the community than the officially elected leaders. For example, in Shaikarawe, Botswana, it was observed that one person in particular, even though not part of any management committee, was more respected by the community members than any of the elected leaders. He acted almost as if he were the chief, even though he never took any action on his own and would always consult with the other leaders attending a meeting.

It is interesting to observe that in the communities that were visited during the fieldwork, the elected leaders seemed less charismatic than the 'natural' leaders, and were not always respected to the same extent. The question arises again and again among the San: why are the apparently most talented or 'natural' leaders not elected more often?

Further problems with the notion and practical application of elections among the San were explored with key informants. Although community members claimed to be dissatisfied with their leaders, their repeated complaint was that the same leaders got re-elected, indicating a failure of elections to deliver change. Some blamed logistical problems, which prevented all community members from participating in the elections, and others claimed that they stayed away due to a sense of apathy and general disinterest. In some cases, for example in Dobe, Botswana, a widespread criticism was that the elections were prone to fraud. Alcohol had been given to community members prior to the election, with the result that people voted for the 'wrong' sort of leader or just did not show up because they were too drunk.

In short, leaders were accused of a range of wrongs including self-interest, bribery, corruption and nepotism, and of being elected simply because of their high education levels or their membership of certain families. One is tempted to conclude that these criticisms of democracy are shared by most citizens in the developing world.

9.5.2 Organizational Structures

Organizational structures based upon Western notions of governance that are new to rural communities, but required by national governments and donor organizations, have proved vulnerable to breakdown and disfunction. The by now well-documented problems of the ≠Khomani San in South Africa are a good example of the collapse of these imposed formal community structures. This community received over 40,000 ha of land following its land claim, which was settled in 1999, and was obliged to form a modern constitution with elections and governance provisions as set out in South Africa's Communal Property Associations Act 28 of 1996. During 2002 the communal property association's management committee had mismanaged its assets to such a degree that its best farm was up for auction in order to pay the debts of creditors, and the government was obliged to step in and secure a court order placing the elected management committee under the administration of the Department of Land Affairs (Chennells 2006).

The South African Human Rights Commission (SAHRC) launched an inquiry process in 2004 to investigate complaints of human rights violations in the ≠Khomani community. After its investigation, the South African Human Rights Commission (2004) reported that reinventing a community from dispersed San descendants was one of the major challenges this community had to face. The task was, according to the SAHRC, particularly demanding since there was no unifying system of leadership. This led to community divisions and malfunctioning of the communal property association when it came to taking decisions over the management and utilization of land. A perennial complaint reported to researchers during field visits in 2004, repeated in 2007, was that San leaders simply did not communicate with their constituencies (Vermeylen 2007).

The San-*Hoodia* Benefit-Sharing Trust, referred to in the overview chapter, also suffers from the general complaint by San communities of too little communication. Expectations were raised by optimistic media reports predicting millions of rands for the San, and many individuals took such reports of imminent wealth as correct. The majority of San individuals interviewed during 2007 were unhappy about the lack of information, and in the absence of reliable facts were prone to imagine all sorts of suspicious goings-on. An additional challenge facing the trust and other institutions needing to disseminate information to the San is the fact that San reside in far-flung communities, well outside the modern communications network and barely served by postal or telephone facilities.

The most common response from San respondents asked to suggest improvements was to propose forming smaller organizations in which individuals would be better able to understand decision-making processes and communicate with one another. Such smaller-scale organizations would most closely resemble the band or extended family structure that characterized San prehistory and provide a pertinent reminder of the mode that the San are generally most comfortable with. The KFO's decision to 'go smaller', referred to above, reflects this key understanding.

A practical example of the effectiveness of clan level organization was experienced in Dobe, where a particular extended family were regarded as as being the most successful and admired in the community for getting themselves organized, taking decisions and being self-sufficient in the provision of water (Vermeylen 2007).

The structure and functioning of San councils are likely to remain hotly debated issues, as there appears to be no viable way of representing the San peoples within a country other than through an elected body or council. During a workshop at the

Molopo Lodge in the Kalahari[21] in December 2006, held to discuss benefit sharing among the San, it became clear that communities in Botswana had a strong view of how their San council should be organized that was implicitly critical of the structures of the San councils in Namibia and South Africa. Many of the Botswana representatives objected to the proposed San councils, which they believed had been imposed by WIMSA.

The fact is that the San in each country are free to choose the representative system that they prefer, and are not obliged to join the overarching international structure provided by WIMSA. While San communities in Namibia and South Africa are largely language-based, those in Botswana are linguistically mixed, and much work has gone into devising a participatory governance model that takes into account the far-flung and differentiated San communities in that country.

The inescapable fact exacerbating the problems referred to above is that the San communities are still relatively unorganized and have not yet created effective mechanisms for controlling their leaders and holding them to account.

9.6 Discussion and Conclusion

Robins's (2002) analysis of the ambiguities and contradictions of donor and NGO development discourses relating to local constructions of community, cultural authenticity and San identity with regard to the 1999 ≠Khomani land claim contains some valuable lessons for organizational structures and leadership. He questions the current practices of NGOs and donors, arguing that some of these organizations (especially in the context of the land claim) have promoted contradictory objectives. On the one hand, the NGOs and donors emphasized the importance of the traditional values, culture and identity of the San, but simultaneously they encouraged the San to adapt to the 'modern' ideas of accountability and democratic decision-making.

In a sense this dichotomy in development discourses has the effect of exacerbating a tension between 'traditionalist' and 'Western' San values. Robins argues that development agencies struggle to understand the hybrid identity of the San and seem to get stuck in a binary typology, a framework that continues to dichotomize tradition and modernity. However, as Robins has argued and as observed during Vermeylen's fieldwork, this rather neat dichotomy is not part of the San's everyday practices. The dual mandate of the NGOs to promote the cultural survival of indig-

---

[21] As part of the project on best practice in benefit sharing sponsored by the Wellcome Trust (of which this book is a product), a workshop was organized in September 2006 to discuss issues and disseminate information regarding the *Hoodia* benefit-sharing agreement. San representatives from Namibia, Botswana and South Africa attended the two-day meeting in Andriesvale, South Africa. As part of this members of the community drafted the Molopo Declaration, which set out the principles that should bind all communities in giving effect to benefit sharing. (See Chapter 12 by Wynberg et al which describes the process followed and includes a copy of the Declaration).

enous peoples and simultaneously 'shape' them into modern citizens within a global civil society potentially diminishes the possibility of the San recreating and reproducing their own cultural ideas and practices while engaging with the challenges of 'modernity'.

After more than 2 decades of experience in managing their own development organizations, as described above, the San have learned valuable lessons and are beginning to demonstrate the ability to understand the rapidly changing socio-economic and political environment that they inhabit. There is no doubt that as a marginalized community they face challenges of unique complexity. What is clearly desirable is that they should increasingly find their own authentic collective voice and make that voice heard through the organizations that claim to represent them, independently of funders, governments, lawyers, anthropologists and other external influences on their thinking. It is essential that they become free from the temptation to emphasize their 'otherness' as San in order to secure funding, or from blind devotion to development trajectories and imperatives rooted in Western paradigms. San organizations should become increasingly free and able to set their own priorities and chart a course of development with the most appropriate balance between their unique history and culture as San peoples, on the one hand, and their desire to exercise their normal human rights as citizens, on the other.

San leaders will, it is hoped, determine how to blend traditionalism and modernity – a constructed divide, as argued by Vermeylen (2007), which merely serves to perpetuate the myth of San 'otherness'. Finally, as they distribute the financial benefits flowing from the *Hoodia* agreement, the San must determine how best to advance and empower both the traditional and the modern interests of their own, self-created communities.

References

Abadian, S. (1999). From wasteland to homeland: trauma and the renewal of indigenous peoples and their communities, PhD thesis. Harvard University, Cambridge, MA.

ACHPR and IWGIA (2006). Indigenous peoples in Africa: the forgotten peoples? Banjul, African Commission on Human and Peoples' Rights, and Copenhagen, International Work Group for Indigenous Affairs.

Barnard, A. (1993). Primitive communism and mutual aid Kropotkin visits the Bushmen. In C. Hann (Ed.), *Socialism*. London: Routledge.

Berlin, B., & Berlin, E. A. (2003). NGOs and the process of prior informed consent in bioprospecting research: the Maya ICBG Project in Chiapas, Mexico. *International Social Science Journal, 55*, 629–638.

Berlin, E. A., & Berlin, B. (2004). Prior informed consent and bioprospecting in Chiapas. In M. Riley (Ed.), *Indigenous intellectual property rights: legal obstacles and innovative solutions*. Walnut Creek, CA: AltaMira Press.

Biesele, M., & Kxao Royal /o//oo (1999). The Ju/'hoansi of Botswana and Namibia. In R. Lee and R. Daly (Eds.), *The Cambridge encyclopedia of hunter gatherers*. Cambridge: Cambridge University Press Syndicate.

Brody, H. (2001). *The other side of Eden: hunters, farmers, and the shaping of the world*. New York: North Point Press.

Brörmann, M. (2002). WIMSA. *Cultural Survival Quarterly, 26*(1, Spring), 45–47.
Brown, M. F. (2003). *Who owns native culture?* Cambridge: Harvard University Press.
Chennells, R. (2006). Legal Report on land rights, commissioned by the Free State and Northern Cape Regional Land Claims Commission, Bloemfontein.
Diamond, J. M. (1999). *Guns, germs and steel: the fates of human societies*. London: W.W. Norton & Company.
EED (2006). Proceeds of consultative workshop held at Maun, by Evangelische Entwicklungsdienst e.V (EED), 28–30 November 2006.
Greene, S. (2004). Indigenous people incorporated? Culture as politics, culture as property in pharmaceutical bioprospecting. *Current Anthropology, 45*(2), 211–238.
Guenther, M. (1999). *Tricksters and dancers. Bushman religion and society*. Bloomington, IN: Indiana University Press.
Hayden, C. (2003). *When nature goes public. The making and unmaking of bioprospecting in Mexico*. Princeton, NJ: Princeton University Press.
Hitchcock, R. (1992). *Communities and consensus: an evaluation of the activities of the Nyae Nyae Farmers Cooperative and the Nyae Nyae Development Foundation in Northeastern Namibia*. New York/Windhoek: Ford Foundation/Nyae Nyae Development Foundation.
Hitchcock, R., Ikeya, K., Biesele, M., & Lee, R. B. (Eds.) (2006). *Updating the San: image and reality of an African People in the 21st century, Senri Ethnological Studies, vol 70*. Osaka: National Museum of Ethnology.
ILO (1989). Convention (No. 169) concerning indigenous and tribal peoples in independent countries. International Labour Organisation, Geneva. www.unhchr.ch/html/menu3/b/62.htm. Accessed 10 May 2008.
Ingstad, B., & Fugelli, P. (2006). "Our health was better in the time of Queen Elizabeth": the importance of land to the health perception of the Botswana San. In R. Hitchcock, K. Ikeya, M. Biesele & R. B. Lee (Eds.), *Updating the San: Image and Reality of an African People in the 21st Century, Senri Ethnological Studies, vol 70*. Osaka: National Museum of Ethnology.
Kaplan, A., & Davidoff, S. (2006). Participatory evaluation and review of the Kuru Family of Organisations, conducted by the Proteus Initiative, Cape Town, April.
KFO (2006). KFO Annual Report on 2005. Kuru Family of Organisations, Ghanzi, Botswana. www.kuru.co.bw.
KFO (2007) *The Kuru story* (2007). Naro Language Project, Letloa Trust, D'kar.
Le Roux, B. (2001). Personal communication with Braam le Roux, founder and director of Kuru, August.
Le Roux, B. (2007). Personal communication with Braam le Roux, founder and director of Kuru, November.
Robins, S. (2002). 'NGOs, "bushmen", and double vision: the ≠Khomani San land claim and the cultural politics of "community" and "development" in the Kalahari'. In T.A. Benjaminsen, B. Cousins, & L. Thompson (Eds.), *Contested resources: challenges to the governance of natural resources in South Africa*, Programme for Land and Agrarian Studies, School of Government, University of the Western Cape, Cape Town.
Rosenthal, J. P. (2006). Politics, culture, and governance in the development of prior informed consent in indigenous communities. *Current Anthropology, 47*(1), 119–142.
South African Human Rights Commission (2004). Report on the Inquiry into Human Rights Violations in the ≠Khomani San Community. Andriesvale-Askham area, November 2004. www.sahrc.org.za.
Skotnes, P. (1996). *Miscast: negotiating the presence of the Bushmen*. Cape Town: UCT Press.
Staehelin, I. (2001). Lost, contested and found: the recovery of Bushman identities through access to history and cultural heritage, Masters thesis. Boston University, Boston, MA.
Sugawara, K. (2002). Voices of the dispossessed. *Cultural Survival Quarterly, 26*(1), 28–29.
Suzman, J. (2001). *The regional assessment of the status of the San in Southern Africa*. Windhoek, Namibia: Legal Assistance Centre.
Sylvain, R. (2006). Drinking, fighting and healing: San struggles for survival and solidarity in the Omaheke region, Namibia. In R. Hitchcock, K. Ikeya, M. Biesele & R. B. Lee (Eds.), *Updating the San: image and reality of an African people in the 21st century, Senri Ethnological Studies, vol 70*. Osaka, Japan: National Museum of Ethnology.

Tanaka, J., & Sugawara, K. (1999). The /Gui and //Gana of Botswana. In R. Lee & R. Daly (Eds.), *The Cambridge Encyclopedia of Hunter Gatherers*. Cambridge: Cambridge University Press Syndicate.
Thoma, A. (2005). Personal communication with Axel Thoma, coordinator and chief executive of WIMSA, October.
Thoma, A., & Le Roux, B. (1995). Project proposal for the establishment of the Working Group of Indigenous Minorities in Southern Africa, a Regional Network in Windhoek, Namibia, and D'Kar, Botswana, Working Group of Indigenous Minorities in Southern Africa.
Tobin, B. (2001). Redefining perspectives in the search for protection of traditional knowledge: a case study from Peru. *Review of European Community and International Environmental Law*, *10*(1), 47–64.
UN (2007). United Nations Declaration on the Rights of Indigenous Peoples, Adopted by United Nations General Assembly Resolution 61/295 on 13 September. www.un.org/esa/socdev/unpfii/en/drip.html. Accessed 10 May 2008.
Vermeylen, S. (2007). Between law and lore, PhD thesis. University of Surrey at Guildford, Guildford.

When other groups in the region became aware of the land's richness in resources, they occupied the richest land. Our ancestors were dispossessed, killed, forced into slavery on their own land. But the San continued to pass on to the younger generations the traditional knowledge of fauna and flora

(Kxao Moses ‡Oma , 24 March 2003, Chairperson of WIMSA, signing of Hoodia benefit-sharing agreement, Molopo, South Africa)

Chapter 10
Trading Traditional Knowledge: San Perspectives from South Africa, Namibia and Botswana

Saskia Vermeylen

Abstract One of the most controversial aspects of the access and benefit-sharing debate is the way in which traditional knowledge is used and commercialized. Many critics have pointed out the inherent contradictions between traditional knowledge systems, which are typically collective, based on sharing and of a non-barter nature, and Western approaches to knowledge protection such as patenting, which by contrast are monopolistic and individualistic. Few, if any, empirical studies have documented the relationship between these systems and community perceptions of the so-called commodification of traditional knowledge. Based on fieldwork conducted in South Africa, Namibia and Botswana, this chapter examines how these issues are perceived by San communities.

While indigenous peoples are often portrayed in the literature as homogeneous groups voicing uniform opinions, the scenario surveys used in the fieldwork clearly indicate that within the communities studied, there were many different opinions on whether or not to commodify traditional knowledge. This diversity of voices is not surprising when one takes into account the local context or the current and historical socio-economic and political circumstances of individuals and communities.

Although there was widespread acceptance of commodification in principle, it is important to be aware of its cultural, symbolic, and economic value. At the same time, the scenario surveys showed that many respondents wanted to keep control of their knowledge rather than part with it for economic benefit (royalties) only. Notably, a gender divide could be observed, with women more likely to settle for royalties – to finance their children's education, for instance – and men more likely to either reject all commodification or opt to be co-holders of patents.

S. Vermeylen
Lancaster Environment Centre, Lancaster University, Lancaster, LA1 4YQ,
United Kingdom
e-mail: s.vermeylen@lancaster.ac.uk

Keywords benefit sharing • commodification • Convention on Biological Diversity • indigenous communities • traditional knowledge

10.1 Introduction

Western consumers are being swamped with advertisements for *Hoodia* products: 'Slim without effort!' and 'Take appetite-suppressing pills instead of dieting!' are the messages. One could almost say that *Hoodia* has developed into a symbol for the commodification of traditional knowledge. At the same time, commodification has been subjected to strong criticism (e.g. Dove 1996; Nijar 1996; Shiva 1997, 2001; Takeshita 2001; Heath and Weidlich 2003; Halbert 2005). Commodification is seen as characteristic of a market-based economy and therefore something that should not be incorporated into the so-called indigenous economies of gifts and reciprocity (e.g. Gudeman 1996; Zerda-Sarmiento and Forero-Pineda 2002; Posey 2002). Indigenous peoples have rejected the commodification of their natural and intellectual resources, as have academics, on the basis of conflicting values between industrialized economies and local, indigenous practices. The (mis)use and commercial exploitation of indigenous heritage by non-indigenous parties has been called 'sacrilege' and 'defamation' (Greene 2004). On the other hand, it has been argued that commodification can reduce serious poverty (Ertman and Williams 2005).

In other words, the debate about the commodification and commercialization of traditional knowledge tends to be highly polarized, with opinions ranging from categorical rejection of the process to accepting it as a liberating act (Vermeylen 2007). A number of authors have questioned this dichotomous thinking and have noted that some of the depicted controversies with regard to the commodification of traditional knowledge project indigenous communities as bounded and discrete, ignoring their changing environment and circumstances (e.g. Strathern 2000; Tobin 2000; Moran et al. 2001; Castree 2003; Heath and Weidlich 2003; Greene 2004; Riley 2004). The importance of their critique is underlined by the fact that even some of the most recent literature about traditional knowledge and intellectual property rights (e.g. Gibson 2005) pays little or no attention to the variety of ideas and perceptions encountered on the ground. The debate about traditional knowledge is often still muddied by an implicit assumption that indigenous peoples speak with one coherent, authentic voice and see the defence of their traditional knowledge as their sole and maybe last stand against the advance of Westernization.

A similar concern must also be raised with regard to the *Hoodia* benefit-sharing agreement. Although it has been applauded for the opportunity it created for the San to take increased control over their knowledge and for stimulating capacity-building among them, to date little is known about how the San perceive the commodification and commercialization of their traditional knowledge. This chapter seeks to analyse some of the key concepts that govern the debate about the *Hoodia* benefit-sharing agreement by exploring the views of the San themselves.

Positioned within a wider remit of developing more 'emic' or culturally specific insights into the commodification of traditional knowledge, this chapter aims to document the responses of San individuals to the commodification of *Hoodia*. First, it explores the San's generic views about the commercialization of their traditional knowledge by using scenario surveys that reflect the ways in which Western-style commodification may take place. Then follows a more culturally embedded approach, which records 'life stories' about *Hoodia* to reveal how some of the San feel and experience its commercialization. The concept of recording life stories of *Hoodia* has been inspired by Kopytoff's (2005) biographical approach to commodities. The final part of this chapter reflects upon the issues raised by the San and revisits some of the most prominent concepts in the ongoing debate about the commodification and protection of traditional knowledge.

10.2 Methodology

10.2.1 Scenario Survey

A scenario survey was conducted to assess and clarify attitudes to the commodification of traditional knowledge in view of demands from the outside world. It consists of a hypothetical story about a businessman coming to a community because he has heard about a medicinal plant that he would like to sell outside the community. He meets three fictitious San individuals who respond in different ways.

Scenario 1: The first individual refuses to share knowledge of the plant.

Scenario 2: The second agrees to share knowledge in exchange for a one-off payment. By accepting money, the San lose control and ownership of the knowledge. It is clear that commercial products might be developed from the knowledge without further consultation with the San.

Scenario 3: The third is willing to share knowledge on condition that the San keep legal rights over their knowledge and can therefore control its use by others. Any decisions about commercialization must be taken in consultation with the San. Money might be available, but only from case-by-case negotiations and without guarantees. For instance, if the bioprospecting company's research is not successful, no funds might be forthcoming, in contrast to scenario 2, which always includes monetary benefits.

The participants were asked to choose the response they liked best and comment on it or say what their own response would be. The survey was based on Soleri and

Cleveland's (1994) scenario survey,[1] which was adapted after testing in Omatako[2] and subsequently replicated in the settlements of Vergenoeg, Blouberg, Andriesvale, Shaikarawe and the Dobe area (mainly Qangwa, G!oshe and //aari/nxo).

The communities in Vergenoeg and Blouberg are small, each consisting of a handful of families, so the sample included the majority of adults present during the fieldwork. In Andriesvale, Shaikarawe and the Dobe area, the key informant, who was also the translator, suggested who should participate. With the exception of Vergenoeg, the key informants, who also acted as translators, were community members proposed by the Working Group of Indigenous Minorities in Southern Africa (WIMSA) in Namibia, the South African San Institute (SASI) in South Africa and the Letloa Trust in Botswana. In Vergenoeg the translators and key informants did not belong to the community and were proposed by the Centre for Research Information Action in Africa – Southern African Development and Consulting (CRIAA SA-DC) and the Omaheke San Trust (development agencies that work in the region). The data for the communities in Namibia and South Africa was collected between July and October 2004, but the interpretation of the data includes observations made during subsequent fieldwork in July–September 2005, September 2006 and June 2007. The data for the communities in Botswana was collected in July 2007.

A total of 114 people participated in the scenario surveys, of whom 73 did the exercise as part of a longer in-depth interview. Of the total sample,[3] 64 were women and 50 men of various ages, with a minority younger than 20 and older than 59.[4] Almost 80% of the participants were 'ordinary' community members in the sense that they did not belong to any of the community committees that either 'governed' the community or represented it at the local or national level. The remaining 20 per cent were what could be called 'elite' San – community members fulfilling leadership or representative functions in the community.

Although this survey instrument has some limitations, in the sense that it invites San participants to respond to the issue of commodification according to Western options, the scenarios represent potential ways to understand commercialization.

---

[1] Soleri and Cleveland (1994) argue that the scenarios they have developed are 'an instructive example of the sort of questionnaire that can be effective for assessing and clarifying attitudes toward the proper use of traditional cultural knowledge'.

[2] This relatively large and diverse San community is located in West Tsumkwe district, Namibia (part of an area that during apartheid was known as 'Bushmanland').

[3] Because of very high mobility, poor communication facilities and questionable census data, statistically representative sampling of the San was not a realistic or useful approach. People typically relate themselves to places through their extended family, but many family members are 'away' at any given time – for instance, working (or looking for work) on farms or staying with relatives. However, scenario interviews were carried out with a diverse range of community members (across age groups, genders and socio-economic positions) until saturation was achieved (i.e. when no new or additional insights were gained). The sample was thus considered large enough to reflect fairly the range of views held by community members present at the time of the survey.

[4] Not all the participants knew their ages. The government officials who issued their identity cards often simply made up the birthdates.

They are therefore a useful tool of analysis that can be used to introduce the San to the dominant (i.e. 'Western') concept of commodification and commercialization of traditional knowledge and then encourage them to think about and comment on these issues.

At this point, it is useful to highlight that the scenarios were designed in the context of examining the San's perception of the commodification of their knowledge. Even though scenario 3 differs from scenario 2 in the sense that it recognizes the San's prior rights over their knowledge, the scenarios are not intended to reflect upon the current international intellectual property regime. In order to keep the scenarios understandable and applicable in the field, the options had to be clarified and simplified in such a way that it was relatively easy to explain the different options to the San. But as a result the scenarios could not illustrate the legal complexities of intellectual property rights.

While acknowledging that the scenario surveys are useful in examining the extent to which San individuals accept or reject the concept of commodification in a hypothetical case, it is equally important to gain culturally embedded insights into this issue. This method is part of the cultural school of thought on commodities, first introduced by Appadurai ([1986] 2005). The cultural study of commodities allows one to examine the changing meaning of things (including knowledge) when they pass through various local and global circuits and cultural meanings (Radin and Sunder 2005). The recorded 'life stories' of *Hoodia* in particular provide a deeper insight into how some San individuals actually feel about a real case study where their medicinal knowledge has been used and commercialized on a large scale.

10.2.2 San Communities

Blouberg and Vergenoeg (Namibia). These formerly white-owned farms in the Omaheke region were resettled and are now community-based farms where the San live together with other communities (mostly Hereros) that are typically more oriented towards farming than the San. The San here do not have *de jure* land rights and seem to be more marginalized (both culturally and from a socio-economic and political perspective) than, for example, the San in East and West Tsumkwe,[5] who are the only San communities with *de jure* land rights. The San in the Omaheke region own little livestock and the opportunities for the collection of veld food[6] and for growing crops are much more limited than in East and West Tsumkwe, both owing to landownership issues and because the physical environment is drier. Some of the San in Vergenoeg harvest devil's claw (a root exported for use in rheumatism

---

[5]The scenarios were first tested in West Tsumkwe. The results of that pilot study are not incorporated in this study.

[6]A substantial caloric intake consists of 'wild' food that is collected in the bush, such as nuts, tubers and watermelons.

and arthritis products) under a scheme run by CRIAA SA-DC (2003)). While devil's claw is harvested across rural Namibia, this particular scheme is characterized by price guarantees for contract harvesters and the promotion of sustainable harvesting methods. CRIAA SA-DC is not active in Blouberg.

Andriesvale-Witdraai (South Africa). The San were labelled 'coloured' under apartheid and widely scattered. The ≠Khomani San were thought to be linguistically and culturally extinct in the early 1990s, when the ≠Khomani land claim was being researched. In the course of tracing claimants to the land, the sociolinguist working for SASI carried out a search across the Northern Cape province and discovered a handful of elderly individuals who could still speak the ancient N/u language. A community of their descendants has been more or less 'reconstructed' to claim back their land rights in and around the South African part of the Kgalagadi Transfrontier Park. Known as the ≠Khomani San, they are still adapting to their new status as collective landowners and struggling to develop viable livelihood strategies. Whilst this is regarded as one of the more privileged San communities, thanks to its successful land claim and proximity to services, it is clear that it will take many years before social and leadership structures adapt fully to the changed circumstances.

Shaikarawe and Dobe area (Botswana). Both are in the Ngamiland district. While Shaikarawe is situated in the Okavango Delta, one of Botswana's most resource-rich areas, the Dobe area is located in the dry and sandy soils of Ngamiland.

Shaikarawe is 10 km west of the Okavango river and just south of the Namibian border. This area is close to other areas in the Delta inhabited mostly by the agriculturalist Hambukushus, but only Khwe reside in the village itself. Most people older than 35 were born into a hunter-gatherer existence, whilst most of the older men served in the South African army during the war against SWAPO, the South West Africa People's Organization. In doing so they crossed the border that divided the traditional territories of their people and joined their Namibian Khwe relatives. Financial support from the Letloa Trust (mainly for transport) has recently enabled many Khwe to resettle in this community – coming from a range of cattleposts and other small settlements in the area where they worked for large cattle farmers, often for a small token salary and without any formal rights to stay on the land. The resettled San all have relatives in Shaikarawe and many claim ancestral links to the location. The primary rationale for these relocations appears to be the creation of a Khwe village large enough to gain governmental village status, which enables access to more government-led projects and services such as schooling and health care).

The Dobe area is one of the most remote parts of Botswana. It lies north of the Ghanzi block of commercial (mostly white-owned) farms, west of the Okavango panhandle and east of the Namibian border, just opposite Tsumkwe on the Namibian side. As with the Khwe further north, here the Namibia-Botswana border cuts across the traditional territory of the Ju'/hoansi, separating family groups and relatives. Originally this area was exclusively inhabited by the Ju'/hoansi, who lived around a number of permanent waterholes over 100 km from any permanent settlements. Over the past century a few extended families of Hereros and Tswanas have settled in the area, and they now dominate the Ju'/hoansi economically, politically and culturally (Lee 2003). The Ju'/hoansi still constitute the majority of the population in this area and are strongly organized along family lines. The wish to return to their *n!oresi*[7] is severely hampered by the problematic hydrology of the area: many drilled wells are drying up or are becoming too salty, and this results in people moving to different waterholes or settlements. The area's extreme isolation limits its scope for development.

10.3 Perceptions of Commodification

10.3.1 Scenario Surveys: San Responses to the Commodification of Medicinal Knowledge

Table 10.1 summarizes the results of the scenario survey, dividing the participants into categories with distinct responses. Clear differences were found between men and women. Overall, men had a very strong preference for option 3 (legal protection), which was three times more popular than option 1 (refusal to share knowledge) and almost five times more popular than option 2 (once-off payment). The opinions of women were more evenly distributed.

Table 10.1 Responses to Commodification of Medicinal Knowledge: Breakdown by Gender, Community, Country and Income

Breakdown by		Option 1[a]	Option 2[b]	Option 3[c]	No idea	Total
Gender	Men	11	7	31	1	50
	Women	17	22	19	6	64
Community	Vergenoeg (Nam)	0	3	15	6	24
	Blouberg (Nam)	9	19	1	0	29
	Andriesvale (SA)	13	3	19	1	36
	Dobe area (Bot)	2	3	8	0	13
	Shaikarawe (Bot)	4	1	7	0	12
Country	Namibia	9	22	16	6	53
	SA	13	3	19	1	36
	Botswana	6	4	15	0	25
Income	Pension	7	3	4	N/A	14
	Nothing	2	6	2	N/A	10
	Child care (only Andriesvale)	3	0	5	N/A	8
	Occasional	1	2	14	N/A	17

Income only recorded for the communities in Namibia and South Africa.
[a] Refusal to share knowledge
[b] Agreement to share knowledge in exchange for money
[c] Willingness to share knowledge in exchange for legal protection

---

[7] *N!oresi* (plural for *n!ore*) are named territories without fixed boundaries. Usually important resources can be found on *n!oresi*, such as permanent and semi-permanent waterholes or highly valued food or medicines.

Even though women's preferences were almost equally dispersed among the three scenarios, option 2 was narrowly the most preferred option. When asked why they opted for scenario 2, their view was very utilitarian. Generating money was important to feed children, pay for school fees and buy clothes. It was thought that giving children a decent education might enable them to climb the social ladder and become teachers, civil servants or even members of parliament; this would help them shake off their stigmatized identity and become full and equal citizens.

Another reason for women choosing this option was that money could give them the chance to start their own development projects so they would not continue to depend on government handouts. Starting small farming and agricultural projects topped the list of what could be done with the money. Often they also mentioned that in order to start small cultivating and herding projects, they first had to have access to land, and they hoped that the money would allow them to buy land. They also thought that if they had money, other people would treat them with more respect. Buying land, farming and empowerment were all expressed as community-based achievements. The women repeatedly mentioned that people had to work together as a community to achieve something. Even when they chose money as the preferred option, they made it clear that their motivation was not the accumulation of personal wealth, but support for community-based development projects.

The difference between the preferences of men and those of women may relate to gender inequality. Exposure to other cultures has undermined the traditional gender equality of the San (Becker 2003). San women have lost influence and autonomy as a result of sedentarization, the wide-ranging impact of land loss, the shift to pastoralism and wage-labour, and the influence of male-dominated neighbouring communities (Kent 1993; Felton and Becker 2001; Becker 2003). Furthermore, the labour market in which the San have been employed (agriculture) favours men over women. This has pushed San women further into the margins of the cash economy, which may explain why more women chose option 2. Since more men than women have access to money, men also tend to have more control over the financial resources within families. Interestingly, subsistence gathering for family sustenance remains predominantly a woman's activity, but the harvesting of natural resources for cash is typically done by men. It is expected of San women to take care of their families, while it is the men who are more widely involved in the cash economy.

On the other hand, most of the men who participated in the survey strongly supported legal rights and protection (option 3). Gaining rights was not limited to property rights over knowledge. When discussing what sorts of problems the community faced, men often mentioned the lack of access to land and the lack of rights over natural resources as the two most important causes of their poverty. Some men (especially respondents struggling to get access to land) also argued that gaining rights over knowledge, natural resources and land was crucial for restoring their human dignity. Men who mentioned that they wanted to keep the knowledge to themselves felt this way because they were worried that something might go wrong if they started to share the knowledge on a large scale: the medicinal plants could stop working or become poisonous. (Women also used this as an argument for keeping the knowledge to themselves.) Furthermore, respondents lacked confidence in the benefit-sharing option (option 2) and, with their experience of marginalization in mind, did not believe that legal rights (option 3) would be granted. Keeping the knowledge for themselves seemed, then, the safest option. Some, especially women, thought that keeping knowledge to themselves would give them a chance to restore the traditional way of life. The protective behaviour of women could be explained on the basis that, traditionally, women were in charge of collecting plants.

There were also significant differences in opinion between the three communities. Vergenoeg was characterized by a strong preference for option 3, which was five times more popular than option 2. In Blouberg, two-thirds of the respondents chose option 2 and one third chose option 1. Opinions were most divided in Andriesvale, where just over half the respondents chose option 3, just over a third chose option 1 and 1 in 12 preferred option 2.

The outspoken preference for option 3 in Vergenoeg is likely to be related to the Sustainably Harvested Devil's Claw Project (SHDC), which started there[8] as a pilot scheme in 1997. The project has made the San in Vergenoeg aware that their natural resources are valued in the marketplace and that they need control over both harvesting and selling in order to demand a fair price. While the SHDC project has alerted the San in Vergenoeg to the commercial value of their natural resources, it has also demonstrated to them that not only the natural *resources* but also the *knowledge* related to the resources has commercial value and therefore needs to be protected.

The situation in Blouberg illustrated that extreme poverty and exclusion from the market or cash economy could translate into a more pragmatic and utilitarian response: opting for the benefit-sharing agreement. People in Blouberg complained that they went without food for days and had to live on handouts. Unlike those in Vergenoeg, the people in Blouberg – at the time of the fieldwork in 2004 – had not participated in the SHDC or any other project related to the use and commercialization of natural resources and related knowledge. Sharing knowledge in return for money was seen as a means to end poverty. Comparing the options chosen against income reveals similar results: that is, respondents with an income were more likely to choose option 3 and, to a lesser extent, option 1. However, interviewees without a source of income opted for the benefit-sharing agreement (option 2) in the hope that this could generate an income. Also, the respondents who received a pension (65 years of age and older) seemed more protective of their knowledge than the younger generation: they were more likely to pick option 1, not sharing their knowledge. The respondents in Blouberg who chose this option were mainly women. As explained previously, women seemed in general to be more protective than men about sharing medicinal knowledge.

---

[8] The NGO CRIAA SA-DC started to organise groups of registered harvesters in order to set up networks of knowledge exchange about sustainable resource use and management. Harvesters became increasingly involved in ecological surveys to determine sustainable harvesting quotas and to monitor compliance with the surveys and quotas. As a result of this pilot scheme, the harvesters deal directly with the exporters and are getting a much better price for harvested devil's claw.

All the respondents in Andriesvale knew about the *Hoodia* benefit-sharing agreements, yet most of them rejected the benefit-sharing option as their preferred solution. In all likelihood, the fact that most of the interviewees, as they complained, had not been involved in the process or been kept informed might have influenced their responses. The people who were interviewed in Andriesvale expressed feelings of exclusion and neglect. Therefore the strong rejection of option 2 in that area must be interpreted in the context of their experiences with the *Hoodia* benefit-sharing agreement. The majority of interviewees in this community complained that the community leaders (also outside the remit of the *Hoodia* agreement) were not adequately reporting back to the community about land and community issues in general and the *Hoodia* benefit-sharing agreement specifically. When probed about this finding, Gert Bok,[9] the former chairperson of the Community Property Association (CPA), argued that the supporters of the traditionalist group amongst the ≠Khomani San in particular harboured 'bad' feelings about the *Hoodia* benefit-sharing agreement because in their opinion this agreement was an example of how the San's tradition could be 'misused' by being shared with non-San. Gert Bok's observation could indicate that the historical schism between the 'modernist' group and the 'traditionalist' group[10] has had repercussions on the ≠Khomani San's attitudes to the commodification of their traditional knowledge. This is demonstrated by the following excerpt from his comments:

> In the past when we still had rainy periods, the *Hoodia* was growing, strong, big and juicy. Now that we are experiencing droughts, succulents like the *Hoodia* have died out. The moment the *Hoodia* was shown to other [non-San] people it disappeared; we showed it to too many different people such as the white people. We did not know that our plants, our knowledge would be turned into pills. This has caused friction in the community; we blame each other for showing it to other people.[11]

Lack of communication between members of the CPA, the South African San Council, the San-*Hoodia* Trust and 'ordinary' community members remains a recurring problem.

Probably as a result of the *Hoodia* and CRIAA SA-DC experiences, the interviewees in Andriesvale and Vergenoeg were more aware of the value of their knowledge and natural resources and keen to gain greater control of the dissemination and commodification of that knowledge. The respondents in Andriesvale were less concerned about their poverty than those in Vergenoeg and Blouberg, highlighting the fact that control over their natural resources and knowledge would empower them and bring recognition of their human rights and identity. But then the people in Andriesvale were visibly better off: pensions were double those in Namibia and young mothers received money for childcare. Most of the people interviewed also confirmed that winning the land claim[12] had improved their social situation and made them feel proud to be San.

---

[9] Interview with Andriesvale informant, 17 October 2004. Interview translated from Afrikaans to English.

[10] See, for example, Robins (2001) for more details on the intracommunity tensions between the self-assigned 'traditionalists' and the 'western' or 'modern' ≠Khomani San.

[11] Interview with Andriesvale informant, 21 June 2007. Interview translated from Afrikaans to English.

[12] For more information on the land claim see Chennells (2002).

Although the sample of interviewees in Shaikarawe and the Dobe area was too small to allow any major conclusions to be drawn, it is worth mentioning that they expressed a preference for option 3. Most of the people who chose this option argued that it was important for the 'outside' world to know about the San and their rich cultural heritage and knowledge of botany and medicine. Furthermore, the 'outside' world would acknowledge the San as the owners or custodians of the knowledge and cultural heritage. The participants who chose option 3 argued that they preferred it over option 2 because it allowed them to set up a continuous 'social relationship' with all the parties involved in the commercialization of their knowledge. This network of social relationships would let the San keep control over what happened to the knowledge and could potentially generate a continuous and guaranteed stream of income. Option 2 was perceived to be a once-off deal and as such more likely to generate less money over time than option 3, because option 3 would stimulate long-term cooperation between the San and the fictitious company.

The participants who chose option 1 were led by previous experiences. In the past both the Khwe and the Ju'/hoansi in Botswana 'traded' their medicinal knowledge with other ethnic groups. This relationship was described as problematic because the more dominant (non-San) ethnic groups considered the San inferior and treated them accordingly. Consequently, it can be suggested that those respondents who had had a bad encounter with other ethnic groups when 'trading' with their traditional knowledge were showing a preconceived distrust of the concept of commercializing and commodifying traditional knowledge.

The results of this survey represent generic snapshots of individual San views of the process of commodification and cannot be said to be statistically representative. However, the scenarios indicate that the most prominent factors in determining attitudes towards the commercialization and commodification of traditional knowledge are, first, whether or not the individual or community has previously engaged in selling medicinal knowledge and, second, the current socio-economic status of the individual or the community. As has been mentioned, however, a major shortcoming of this scenario approach is the fact that it is embedded in a Western framework and therefore gives little understanding of how the San feel about the commercialization of *Hoodia* from a cultural perspective. The next section looks at this question in more detail.

10.3.2 *Perceptions of the Commercialization of the Hoodia from a Cultural Perspective*

Because of the limitations of the scenario survey method, as described, other data was collected to reflect the San's 'emic' perceptions about the commodification of their medicinal knowledge. Twenty-eight San inhabitants of Andriesvale were invited to tell 'life stories' about *Hoodia* and to reflect upon the following questions:

- What does *Hoodia* mean to them?

- Has that meaning changed since *Hoodia* was commercialized and made available to the outside world?
- How do they feel about cultivating *Hoodia* and processing it into a commercial product for consumption by non-San people?

Inviting people to express their feelings about *Hoodia* triggered stories about the San's traditional life when they were still nomadic people who roamed freely in the Kalahari in search of food and water. *Hoodia* was described as one of the most important plants for the San; it was their 'life force', giving food, water and energy. The meaning of *Hoodia* for the San in Andriesvale appeared to be symbolic of their former identity and representative of a certain way of life which was nostalgically described as 'the old days when we could still go on hunting trips and collect food in the veld'.[13] The following quote captures some of these poignant feelings:

> When you eat *Hoodia* you can feel the supernatural powers coming from above. When you smell *Hoodia* and taste it on your tongue you will feel how it stimulates you, how it controls your hunger, how it gives you power and energy. ...You cannot experience these powers and energies of *Hoodia* in pills; we gave the power away for money. Everything that we had here is gone because we traded the supernatural powers for money, for simple things. ...You cannot enjoy *Hoodia* when it grows in containers. You will walk past it with all your diseases and you will contaminate the plant. *Hoodia* must grow in its natural environment. ... When diseases were introduced on our land [through contact with other people] we had to look for medicines to cure ourselves; that is how we got the knowledge. All the knowledge that Unilever and CSIR have comes from the 'Bushmen' but they [Unilever and CSIR] have nothing; the knowledge stays ours.[14]

This quote indicates that some San challenge the very process of commodifying *Hoodia* (i.e. cultivating it outside its natural habitat and capturing its medicinal properties in a commercial product) as one that results in *Hoodia* losing its life force and power to heal. The commodification of *Hoodia* is seen as another step in the historical process of marginalizing the San's culture and way of life. While the scenario survey reveals that many San recognize that their medicinal knowledge may have potential use and exchange value (i.e. economic value) in the wider world and are prepared to commodify that knowledge, the life stories show that the San also continue to value their medicinal plant knowledge for symbolic, supernatural and ritual reasons. When participants in Andriesvale (i.e. the San who had been closely involved in the development of the *Hoodia*) were asked to describe how the meaning of *Hoodia* had changed since it became a commodity, this triggered stories of the symbolic, ritual and cultural meaning of *Hoodia* and of the San's marginalized socio-economic and political position in society.

It seems that the San recognize the potential economic value of their medicinal knowledge and are, to a certain extent, willing to commercialize this knowledge. At the same time, however, they find it hard to come to terms with the process of commodification in the sense that it changes cultural meanings. This raises the

---

[13] Field notes, June 2007.

[14] Interview with Andriesvale informant, 21 June 2007. Interview translated from Afrikaans to English.

question: to what extent is the pro-commodification 'choice' informed by the experience of economic hardship, rather than by culturally embedded convictions, whether traditional or of recent origin? In other words, is this simply a desperate choice, in the sense that it can provide a sorely needed source of income? Or could it be that the role and value of traditional knowledge have been culturally, economically and socially transformed as a result of changing market ideologies, both locally or globally? Or is it a utilitarian acceptance by the San of a market-based economic system? Or is it a combination of all of these?

It is difficult to answer this question, but some insights are provided by unrelated anthropological research into the changing meaning of the San's trance dance (e.g. Katz et al. 1997; Guenther 1999, 2002). These studies conclude that both internal and external factors can explain the transformation of the trance dance from a ritual performance carried out in a traditional and cultural spirit to a practical service provided for a fee by a professional dancer for a client. As a result of the changing socio-economic and cultural landscape – many of the San now live in a cash economy – the San started to commercialize the trance dance. However, this process has been reinforced by internal changes in the socio-cultural organization of the San, such as the eroding social position of the trance dancer in the community. In other words, the acceptance of the commodification of the trance dance has been brought about through both changing symbolic values (internal factors) and changes in the socio-economic situation (external factors).

10.4 Conclusion

One of the greatest challenges that has to be dealt with when negotiating benefit-sharing agreements is getting prior informed consent from the knowledge custodians. While indigenous peoples are often portrayed in the literature as homogenous groups voicing uniform opinions, the scenario surveys clearly indicate that within this particular sample of communities, there were many different opinions on whether or not to commodify traditional knowledge. This diversity of voices is not surprising when the local context is taken into account, as well as current and historical socio-economic and political circumstances at the individual and community level. It must also be acknowledged that although some indigenous peoples may recognize the economic value of their knowledge and accept its commodification, at the same time – as the *Hoodia* life stories show – they may also continue to value their medicinal knowledge for symbolic and ritual reasons.

This finding chimes with those of anthropologists (e.g. Malinowski [1935] (1978); Davenport 2005) that indigenous communities have always made a distinction between ordinary commodities and valuable ones, and that the same thing can simultaneously have an economic or material value and a mythical, supernatural or symbolic value. The problem is that in the current processes of commodifying traditional knowledge – 'regulated' through benefit-sharing agreements – only the economic value is recognized and compensated for. This can lead to situations in which indigenous peoples develop feelings of mistrust, inequality and betrayal because the cultural and symbolic value of the commodity has been ignored and even damaged by the very act of recognizing only the commercial value of traditional knowledge.

Based on the findings of this study it can be argued that indigenous peoples' acceptance of commodification in principle does not equate to the acceptance of a commodification practice that is driven only by economic compensation or exclusively financial benefits. One must therefore ask whether the remit of benefit-sharing agreements should not go beyond compensatory justice and whether policymakers should not open up the debate and discuss whether benefit-sharing agreements can also deal with questions of redistributive justice, procedural justice and even proprietary justice.

At first sight it seems that the debate is wide open. Certainly, from the perspective of the San, the debate about how to compensate for the commercial use of traditional knowledge by other parties seems to be connected to the wider context of their eroding socio-economic, political and cultural position in society. This could imply that for the San, benefit-sharing agreements need to include some of the major components of the justice debate:

- Compensatory justice: Are the San fairly compensated for the wrongs done in the past?
- Redistributive justice: Does the benefit-sharing agreement redress the socio-economic inequality of the San?
- Procedural justice: Has the process of negotiating the benefit-sharing agreement been based on the principle of fairness?
- Proprietary justice: Are the San's property rights recognized in the agreement?

It remains to be seen whether policymakers will act upon this challenge.

References

Appadurai, A. (Ed.) (2005). *The social life of things: commodities in cultural perpsective.* Cambridge (first published in 1986): Cambridge University Press.
Becker, H. (2003). The least sexist society? Perspectives on gender, change and violence among southern African San. *Journal of Southern African Studies, 29*(1), 3–23.
Castree, N. (2003). Bioprospecting: from theory to practice (and back again). *Transactions of the Institute British Geographers, 28*(1), 35–55.
Chennells, R. (2002). The ≠Khomani San of South Africa. In J. Nelson & L. Hossack (Eds.), *Indigenous peoples and protected areas in Africa: from principles to practice.* Moreton-in-Marsh, UK: Forest Peoples Programme.
CRIAA SA-DC (2003). The Namibian national devil's claw situation analysis: socio-economic analysis of devil's claw harvesting and trade issues in Namibia. Centre for Research Information Action in Africa and Southern African Development and Consulting, Windhoek.
Davenport, W. H. (2005). Two kinds of value in the Eastern Solomon Islands. In A. Appadurai (Ed.), *The social life of things: commodities in cultural perspective.* Cambridge: Cambridge University Press.

Dove, M. R. (1996). Center, periphery, and biodiversity: a paradox of governance and a development challenge. In S. B. Brush & D. Stabinsky (Eds.), *Valuing local knowledge: indigenous people and intellectual property rights*. Washington, DC: Island Press.

Ertman, M. M., & Williams, J. C. (Eds.) (2005). *Rethinking commodification: cases and readings in law and culture*. New York: New York University Press.

Felton, S., & Becker, H. (2001). *A gender perspective on the status of the San in Southern Africa*. Windhoek: Legal Assistance Centre.

Gibson, J. (2005). *Community resources: intellectual property, international trade and protection of traditional knowledge*. Aldershot: Ashgate Publishing.

Greene, S. (2004). Indigenous people incorporated? culture as politics, culture as property in pharmaceutical bioprospecting. *Current Anthropology, 45*(2), 211–238.

Gudeman, S. (1996). Sketches, qualms, and other thoughts on intellectual property rights. In S. B. Brush & D. Stabinsky (Eds.), *Valuing local knowledge: indigenous people and intellectual property rights*. Washington, DC: Island Press.

Guenther, M. (1999). *Tricksters and dancers: Bushman religion and society*. Bloomington, IN: Indiana University Press.

Guenther, M. (2002). Independence, resistance, accommodation, persistence: hunter-gatherers and agropastoralists in the Ghanzi veld, early 1800s to mid-1900s. In S. Kent (Ed.), *Ethnicity, hunter-gatherers, and the 'other': association or assimilation in Africa*. Washington, DC: Smithsonian Institution Press.

Halbert, D. J. (2005). *Resisting intellectual property*. Abingdon: Routledge.

Heath, C., & Weidlich, S. (2003). Intellectual property: suitable for protecting traditional medicine? *Intellectual Property Quarterly, 1*, 79–96.

Katz, R., Biesele, M., & St. Denis, V. (1997). *Healing makes our heart happy: spirituality and cultural transformation among the Kalahari Ju/'hoansi*. Rochester, VT: Inner Traditions International.

Kent, S. (1993). Sharing in an egalitarian Kalahari community. *Man, 28*, 479–514.

Kopytoff, I. (2005). The cultural biography of things: commoditisation as process. In A. Appadurai (Ed.), *The social life of things: commodities in cultural perspective*. Cambridge: Cambidge University Press.

Lee, R. (2003). cf. Solway, J. *The politics of egalitarianism*. Oxford: Berghahn Books, 151.

Malinowski, B. ([1935] 1978). *Coral gardens and their magic: a study of the methods of tilling the soil and of agricultural rites in the Trobriand Islands*. London: Allen and Urwin.

Moran, K., King, S. R., & Carlson, T. J. (2001). Biodiversity prospecting: lessons and prospects. *Annual Review of Anthropology, 30*, 505.

Nijar, G. S. (1996). *In defence of local community, knowledge and biodiversity*. Penang: Third World Network.

Posey, D. A. (2002). Commodification of the sacred through intellectual property rights. *Journal of Ethnopharmacology, 83*(1), 3–12.

Radin, M. J., & Sunder, M. (2005). The subject and object of commodification. In M. M. Ertman, & J. C. Williams (Eds.), *Rethinking commodification: cases and readings in law and culture*. New York: New York University Press.

Riley, M. (Ed.) (2004). *Indigenous intellectual property rights: legal obstacles and innovative solutions*. Walnut Creek, CA: AltaMira Press.

Robins, S. (2001). NGOs, 'bushmen' and the double vision: the ≠Khomani San land claim and the cultural politics of 'communty' and 'development' in the Kalahari. *Journal of Southern African Studies, 24*, 17–24.

Shiva, V. (1997). *Biopiracy: the plunder of nature and knowledge*. MA: South End Press Boston.

Shiva, V. (2001). *Protect or plunder? Understanding intellectual property rights*. London: Zed Books Ltd.

Soleri, D., & Cleveland, D. (1994). Gifts from the creator: intellectual property rights and folk crop varieties. In T. Greaves (Ed.), *Intellectual property rights for indigenous peoples: a sourcebook*. Oklahoma City, OK: Society for Applied Anthropology.

Strathern, M. (2000). Multiple perspectives on intellectual property. In K. Whimp & M. Busse (Eds.), *Protection of intellectual, biological and cultural property in Papua New Guinea*. Canberra: Asia Pacific Press.

Takeshita, C. (2001). Bioprospecting and its discontents: indigenous resistances as legitimate politics. *Alternatives, 26*, 259–286.

Tobin, B. (2000). The search for an interim solution. In K. Whimp & M. Busse (Eds.), *Protection of intellectual, biological and cultural property in Papua New Guinea*. Canberra: Asia Pacific Press.

Vermeylen, S. (2007). Between law and lore, unpublished PhD thesis. University of Surrey, Guildford.

Zerda-Sarmiento, A., & Forero-Pineda, C. (2002). Intellectual property rights over ethnic communities' knowledge. *International Social Science Journal, 54*(171), 99–114.

In the old days we had animal folk tales, and these were our lessons in those times. These tales were narrated by our grandparents, parents and elder people, and included animal songs, bird songs etc. and it was a lesson of life, and our leisure times. It was done in the evening when it got dark and the lesson was to teach us the way of life for the future generations.
(Peter Goro, Tobere, Botswana)

Chapter 11
Putting Intellectual Property Rights into Practice: Experiences from the San

Roger Chennells

Abstract This chapter outlines the basics of the intellectual property rights system before proceeding to describe its challenges and advantages for the San. The theft of music, folklore, traditional art and innovations shows that the current system is inadequate to secure the full protection of indigenous rights. Yet there is room in that system for flexible, local initiatives driven by indigenous peoples to remedy the situation.

One example is the 'research and media contract' drafted by a San NGO and now used widely, which requires prospective researchers not only to provide full details of the applicant and of the nature, content and purpose of the research, but also to negotiate terms with an appointed San leader. This chapter shows that there are practical methods for regaining control over traditional knowledge and heritage, but indigenous peoples need to be proactive in asserting their own rights and using existing laws and tools.

Keywords indigenous communities • intellectual property rights • research • San • traditional knowledge

11.1 Introduction

Intellectual property rights (IPRS) are an important cog in the intricate machine of international laws and policies underpinning the economic realities of the modern world. As international bodies attempt to resolve the inequities between rich and

---

R. Chennells
Chennells Albertyn: Attorneys, Notaries and Conveyancers, 44 Alexander Street, Stellenbosch, South Africa
e-mail: scarlin@iafrica.com

poor nations, they subject many of these laws and policies to intense scrutiny. The Millennium Development Goals represent tangible commitments to reducing poverty and hunger, and have inspired a range of collaborative initiatives to examine ways of reviewing and reforming international laws and policies in order to advance the achievement of the goals.

The IPR regime was recognized as an area that needs to be thoroughly examined, playing as it does a central role in the flow and distribution of information, technology and wealth. In the words of Sir Hugh Laddie, a UK high court patent judge: 'For too long IPRs have been regarded as food for the rich countries and poison for poor countries' (CIPR 2002). While developed countries regard IPRs as crucial to stimulating economic growth and innovation, developing countries argue that they increase the costs of essential medicines and inputs. Developing countries, though technologically poor, are often custodians of rich treasures in the form of genetic resources and traditional knowledge that are of potential value to the world at large, but are not adequately protected by the IPR system.

One of the initiatives launched to analyse the world IPR regime in the light of the objective of reducing poverty was the Commission on Intellectual Property Rights (CIPR), among whose tasks was to consider:

- How national IPR regimes could best be designed to benefit developing countries within the context of international agreements, including TRIPS (the World Trade Organization's Agreement on Trade Related Aspects of Intellectual Property Rights)
- How the international framework of rules and agreements might be improved and developed – for instance in the area of traditional knowledge – and the relationship between IPR rules and regimes covering access to genetic resources

The recommendations of this commission were far-reaching (CIPR 2002) and have been fed through to the working bodies under the auspices of the World Intellectual Property Organization (WIPO) tasked with reaching consensus and negotiating improvements to IPR laws and policies.

Traditional knowledge, as well as 'access and benefit sharing', has been discussed broadly by Dutfield in Chapter 4 and Wynberg and Laird in Chapter 5. The protection of traditional knowledge, usually held by traditional or indigenous peoples, is not adequate under the current IPR regime. Various new methods of protecting traditional knowledge have been recommended and are being debated.

This chapter will broadly sketch the international IPR regime, including the overlap of this field with intangible heritage rights, as the context within which the San have begun to articulate and protect aspects of their heritage and traditional knowledge. While the most important types of IPR, such as patents, copyright and trademarks, have evolved over centuries, they have significantly failed to protect the tradition-based knowledge systems held by indigenous peoples such as the San. Sustained criticism of the international intellectual property regime by indigenous peoples has contributed to initiatives by WIPO and the CIPR to examine the less than adequate protection enjoyed by holders of traditional knowledge. While these initiatives promise to bear fruit in the future, the way of life of indigenous peoples remains extremely vulnerable to domination by more assertive and dominant cultures, which partly explains the rapid loss of culture and traditional knowledge common to these peoples worldwide.

Noting that the plight of indigenous peoples is seldom alleviated by the IPR regime, the San have taken steps to identify the many ways in which their knowledge and IPRs are wrongly appropriated and to actively protect their knowledge. The chapter concludes with a description of the practical steps the San have taken to protect their IPRs.

11.2 International and National Intellectual Property Rights

Notions of the value and ownership of property which is not physical or tangible, but is located in the field of the intellect, have their origins in the emergence of recorded thought. Over successive millennia, mankind as the innovator developed norms and laws to recognize and protect the rights of inventors and originators of all forms of creative and artistic endeavour. More than 3,000 years ago, Indian craftsmen engraved their signatures on their artistic creations before sending them to Iran. More than 2,000 years ago, Chinese manufacturers sold goods bearing their marks to Mediterranean countries. It is recorded that at one time about a thousand Roman marks were in circulation, some of which were copied and counterfeited in an early example of intellectual property theft (WIPO 1998).

Following the industrial revolution, the field evolved rapidly, closely linked with the need of governments and corporations expanding their interests worldwide to trade in and exploit not only physical goods, but also the associated ideas. IPRs became known and defined as the legal rights which result from intellectual activity in the industrial, scientific, literary and artistic fields. Generally, intellectual property law aims at safeguarding creators and other producers of intellectual products by granting them certain rights, limited by time, to control the use of those products. IPRs are 'rights of exploitation in information' (Drahos 1999a).

The United Nations Universal Declaration of Human Rights, adopted in 1948 by the international community as a common standard of human rights, included in article 27.2 what is regarded as the foundation of intellectual property as a human right.

> Everyone has the right to the protection of the moral and material interests resulting from any scientific, literary or artistic production of which he is the author.

Many subsequent international legal instruments have fleshed out the ambit of IPRs, which reach into fields as diverse as trade, health, scientific progress, culture, heritage and the environment.[1] Article 15 of the International Covenant on

---

[1] They include the International Covenant on Economic, Social and Cultural Rights (UN Economic and Social Council), 1966; the Convention for the Safeguarding of the Intangible Cultural Heritage (UN Educational, Scientific and Cultural Organization – UNESCO) 2003; and the Convention on the Protection and Promotion of the Diversity of Cultural Expressions (UNESCO), 2005.

Economic, Social and Cultural Rights repeats the human right referred to above, whereas article 19 of the International Covenant on Civil and Political Rights prescribes the freedom of every individual to 'seek ... and impart information and ideas of all kinds, regardless of frontiers, either orally, in writing or in print, in the form of art, or through any other media of his choice'. The globalization and commodification of knowledge have thus been recognized in terms of human rights.

In 1967 the United Nations formed WIPO as a specialist agency to administer the treaties that establish and facilitate the international protection of intellectual property. The original Paris Convention of 1883 and the Berne Convention of 1886 formed the basis for the numerous subsequent conventions that have been negotiated and updated as the field of intellectual property has exploded in importance (WIPO 1994).

The most important intellectual property rights are considered to be patents, copyright, registered designs, trademarks, plant breeders' rights, geographical indications and confidential information or know-how. States enact laws to regulate these rights domestically, but remain subject to the international conventions, managed by WIPO, which they have adopted and ratified. WIPO not only provides the forum where the nations of the world meet to organize their common intellectual property issues, but has become central to the recent rapid evolution of the rights of indigenous peoples and holders of traditional knowledge.

Patents, copyright and trademarks are some of the most commonly traded intellectual property rights, and all provide monopoly rights to the owner. Patents are instruments issued by governments and used to protect an invention. The state grants a monopoly to an inventor for a limited period, in return for full disclosure of the invention, so that others may gain the benefit of the invention (WIPO 1997a). A patent must meet several criteria, including that it must be industrially applicable (useful), it must be new (novel, thus not 'prior art' and not in the 'public domain'), and it must disclose an inventive step (non-obviousness). These legal criteria are complicated in practice, and have come under the spotlight not only in the case of the patent relating to *Hoodia*, but in many challenges to patent applications where traditional knowledge relating to the patent was claimed to destroy the novelty of the application (Dutfield 2004) The trend over the past 2 decades to issue patents for discoveries of information that already exist in nature, such as genetic sequences of organisms, is perceived as an unjust commodification of biological resources, serving the technology-rich developed countries of the North (Drahos 1999b).

Copyright law deals with the rights of intellectual creators and protects the form of expression of the ideas only. It protects the arrangement of words, shapes and colours, only once they are recorded, from those who 'copy' or imitate the work (WIPO 1997b). The author is always the owner of copyright, unless he or she was employed in producing it, and can license or assign rights to others. Copyright protection usually lasts for the lifetime of the owner, plus a period (usually 50 years) after the author's death. Copyright to ancient art such as San rock art has thus long expired, but certain 'moral rights' to artistic works, which always belong to the author of a work, exist indefinitely and are also acknowledged by law (WIPO 1998). These 'moral rights,' unlike the exclusive economic rights associated with copyright, remain vested in the author and cannot be assigned or licensed to another. They can thus be the basis of a common law legal action against a third party who claims the right to such an artistic work.

A trademark is a sign or mark, registered in a state's trademark office, that individualizes the goods of a given enterprise and distinguishes them from the goods of its competitors. In order to be registered, a mark must fulfil the criteria of distinctiveness and lack of deceptiveness.[2] In addition, related marks can be registered with similar purpose, including 'service marks', 'collective marks', 'authentication marks' and 'certification marks', all of which are variations and extensions of the trademark concept (WIPO 1993).

Advances in information technology and biotechnology have required the intellectual property system to evolve and adapt rapidly, which it has attempted to do. However, indigenous peoples and traditional knowledge holders, empowered by the United Nations International Decade of the World's Indigenous People[3] and strengthened by the growing international consensus on the relevance of their concerns, have objected increasingly over the years to the very foundation of the international intellectual property system, based as it is upon the private 'ownership' of knowledge, a fundamental contradiction of their collective ethos. Article 8(j) of the Convention on Biological Diversity (CBD) articulates the need for states to 'respect, preserve and maintain knowledge, innovations and practices of indigenous and local communities embodying traditional lifestyles' as part of the broad drive to promote the conservation and sustainable use of biodiversity and the equitable sharing of benefits. However, if this knowledge is patented and thus privatized in the hands of a third party, without the formal acknowledgement of the source of such knowledge or the conclusion of a benefit-sharing agreement in accordance with the access and benefit-sharing provisions set out under the auspices of the CBD, the rights to such knowledge are lost to the community whose knowledge provided the research lead. The Bonn Guidelines on Access to Genetic Resources and Fair and Equitable Sharing of the Benefits Arising out of their Utilization (CBD 2002) provide a broadly accepted framework for ensuring that states respect the sources of traditional knowledge, and that the requirement for appropriate benefit-sharing agreements compensates for the imparting of

---

[2] Trademarks that might deceive the public as to the nature, quality or other characteristics of the goods or their geographic origin do not, in the interests of the public, qualify for registration.

[3] 1993 was the International Year of the World's Indigenous People. The International Decade of the World's Indigenous People commenced in 1995. In 2004 the UN declared a second such decade, as the duties of the Working Group on Indigenous Populations of the Sub-Commission on Prevention of Discrimination and Protection of Minorities (mainly the negotiation of an international convention on the rights of indigenous peoples) were not yet complete.

such knowledge. Wynberg and Laird in Chapter 6 trace the progress over the past decade of the rapidly evolving international law in this regard.

The well-documented (Shiva 1997; Dutfield 2004) failure of the intellectual property system to protect tradition-based knowledge, technologies and creations in particular from exploitation by third parties produced a growing clamour of discontent from ever more vocal indigenous peoples. A range of assertive declarations, statements and demands claimed equity in research involving indigenous cultures (Laird and Wynberg 2002). Allegations of biopiracy[4] targeted pharmaceutical companies that used traditional knowledge to lead them to patentable and thus private 'inventions' (Dutfield 2004). Indigenous peoples claimed that the intellectual property system encouraged and legitimized the misappropriation of their knowledge and innovations[5] (Dutfield 2004; see also Dutfield, Chapter 4).

Other forms of misappropriation of unprotected culture took place, in the form of the theft of music, folk law, traditional art and innovations similarly unprotected by the prevailing intellectual property system. Indigenous peoples called for nothing less than a fundamental reappraisal of the entire system. In the United Nations Declaration on the Rights of Indigenous Peoples, negotiated since 1992 and finally adopted by the General Assembly in August 2007, article 31 proclaims:

> Indigenous peoples have the right to maintain, control, protect and develop their cultural heritage, traditional knowledge and traditional cultural expressions, as well as the manifestations of their sciences, technologies and cultures, including human and genetic resources, seeds, medicines, knowledge of the properties of fauna and flora, oral traditions, literatures, designs, sports and traditional games and visual and performing arts. They also have the right to maintain, control, protect and develop their intellectual property over such cultural heritage, traditional knowledge, and traditional cultural expressions.

It is fair to state that the response by indigenous peoples in challenging IPR systems has been linked to a political struggle, not merely to change the existing intellectual property regime, but to pursue the self-determination and even sovereignty of indigenous peoples. In the most extreme interpretation of this struggle, the advancement of weaker peoples' rights to develop politically and economically is regarded as the lifeblood of emancipation, while Western intellectual property regimes, which tend to consolidate the power of the wealthy nations, are criticized as the very epitome of repression (Drahos 1999b). While awaiting possible reforms to the system as they are slowly negotiated between states, indigenous peoples wishing to prevent the commodification of their traditional knowledge by others and the exploitation of aspects of their culture and heritage have little alternative but to use the existing IPR system.

---

[4] 'Biopiracy' normally refers to the unauthorized extraction of biological resources and/or associated traditional knowledge from developing countries, or to the patenting of spurious 'inventions' based on such knowledge or resources, without compensation.

[5] Often such patents make no reference to the relevant traditional knowledge (e.g. the *Hoodia* patent) or merely mention it in a cursory manner as if it is of little importance (e.g. the turmeric patent).

WIPO responded to the challenge to the established intellectual property regime by launching a new programme on global intellectual property issues in 1997, in order to examine, inter alia, the less than adequate protection provided to owners of traditional knowledge. International fact-finding missions were launched to assess the intellectual property expectations and needs of traditional knowledge holders (WIPO 2001), and intergovernmental working committees[6] were created to debate and propose changes. The recommendations made by the CIPR in this regard are practical and intended to be implementable in domestic jurisdictions without too much difficulty (CIPR 2002). The following is one which would, if implemented, immediately curb biopiracy.

> The principle of equity dictates that a person should not be able to benefit from an IP right based on genetic resources or associated knowledge acquired in contravention of any legislation governing access to that material (CIPR 2002).

While sincere efforts are being made to address the concerns of indigenous peoples, the complex international IPR system remains firmly rooted in the market economy foundations of free trade and private ownership. Indigenous peoples have thus had to find creative ways of engaging with researchers and other external agents in order to ensure equity and fairness in the exchange of traditional knowledge (Laird 2002).

11.3 Vulnerability of Traditional or Indigenous Knowledge

Heritage, culture and traditional knowledge are closely interlinked. Several international conventions and documents, largely coordinated by the United Nations Educational, Scientific and Cultural Organization (UNESCO), have attempted to create binding obligations on states to respect, protect and preserve the varied components that constitute the cultural heritage of humankind. But history is written by the victorious, and often the cultures of weaker sections of society, which include entire bodies of traditional knowledge and practice, are extinguished in the remorseless march towards modernization. Over the past decade, the indigenous peoples' movement has become a vocal advocate of the more assertive legal recognition and protection of their heritage, with the associated traditional knowledge.[7]

All aspects of traditional cultures, from cultural manifestations such as art, songs, rituals, stories, dance and symbols to knowledge-based aspects such as plants and traditional medicines, are easy prey to commoditization by outsiders (Dutfield 2004). The term 'traditional knowledge' has been defined in many ways,

---

[6]The WIPO Intergovernmental Committee on Intellectual Property and Genetic Resources, Traditional Knowledge and Folklore held its 11th session in July 2007. This committee engages with representatives of indigenous peoples and makes recommendations to the WIPO General Assembly on its findings (www.wipo.int/tk/en/igc).

[7]For a discussion of the indigenous peoples movement (see Heintze 1993; IWGIA 2007).

all of which attempt to describe knowledge that is built over generations by people living in close contact with nature (Dutfield 2004). The social process by which this knowledge is acquired, used and shared, which is unique to the indigenous culture, lies at the heart of the 'traditionality' of such knowledge (Barsh 1999). 'Indigenous knowledge' is a closely related concept,[8] referring to bodies of knowledge held and perpetuated by people regarded as 'indigenous', a word with an equally complex etymology in international law. 'Indigenous peoples' are defined in the International Labour Organization's Convention 169 concerning Indigenous and Tribal Peoples in Independent Countries as follows:

> Peoples who are regarded as indigenous on account of their descent from the populations which inhabited the country, or a geographical region ... at the time of conquest or colonisation ... and who ... retain some or all of their own social, economic, cultural and political institutions. (ILO 1989)

It is unwise to be dogmatic about precise distinctions between indigenous knowledge and traditional knowledge, as the terms overlap considerably and are subject to continued academic debate. What is of more practical importance is the question: who owns the knowledge in traditional or indigenous societies? The fact that traditional communities have a strong sharing ethos among themselves cannot be used to insist, for example, that all of their knowledge may be appropriated by outsiders – even their own governments – eager to commodify and exploit it. Biopiracy and the patenting of 'inventions' based upon traditional knowledge are among the key challenges facing traditional or indigenous peoples.

The case of the patent on the active ingredients of the *Hoodia* plant is a clear example: the patent made no mention of the fact that the original information came from the San. Many other patents based upon traditional knowledge have been challenged, some successfully (Dutfield 2004). The defence proffered is usually that the traditional knowledge in question was already in the public domain and was therefore no longer protected by intellectual property laws. Some authors have drawn the analogy between this form of theft of traditional knowledge and the colonial appropriation of countries under the *terra nullius* doctrine (literally, 'land belonging to nobody'), which asserted that the lands were 'nobody's property' before their 'discovery' (by explorers, scientists or governments) and therefore could be legitimately appropriated (Dutfield 1999).

Similarly, traditional medicines used by indigenous peoples that are not protected by IPR are in the public domain and therefore free for all to use and exploit.

When the San peoples became collectively organized and their institutions began to connect with the world indigenous peoples' movement in 1997 (see Chapter 9), they recognized that their culture, heritage and traditional knowledge systems were similarly under threat. Not only were the old people dying without having passed on their knowledge to their communities, but the youth, disillusioned with the disintegrating San culture and way of life, were avidly striving for modernity.

---

[8] 'Indigenous knowledge' is understood in at least two ways: first as 'the traditional knowledge of indigenous peoples' and second as knowledge that is itself 'indigenous' (WIPO 2001).

11 Putting Intellectual Property Rights into Practice: Experiences from the San

This is part of a complex and worldwide phenomenon, but San leaders realized that their rights to heritage could be a rallying point around which they could regroup as a people and reaffirm their very identity. In addition, San leaders understood that the right to land was central to the maintenance of culture and heritage, as was demonstrated in the famous Australian case of *Mabo v Queensland*.[9]

11.4 Breach of Intellectual Property Rights and the Use of Law

As described in Chapter 9, the San leadership decided in 2000 to take active steps to protect their heritage and cultural rights, including their traditional knowledge. It was understood, upon analysis of past experiences, that in every single transaction involving traditional knowledge or practices, the need for full prior informed consent was perhaps the most important requirement. Problems with engagements could always be traced back to a lack of information from those responsible for providing authority to proceed.

The San thus recognized that in every instance of assertion of rights to intellectual property, they needed to ensure that they were fully informed and aware of all the possible implications in a matter before making any decisions. The following were recorded as the most common intellectual property issues and potential abuses of their legal rights.

11.4.1 Intellectual Property and Research

Social and natural scientists had for centuries researched the San without ever informing them in advance or even requesting permission from San leaders. Books and doctoral theses were published in the West based upon the research information, which was often obtained from naive elders or community members in exchange for tobacco or even alcohol.[10] Books containing myths, stories and traditions would be written for adult and child audiences. Often elders would report that they had been asked countless questions about their customs, dances, or medicinal uses of plants

---

[9] *Mabo and Others v Queensland* (1992) 175 CLR 1.
[10] Mathambo Ngakaeaja, a delegate from the Working Group of Indigenous Minorities in Southern Africa (WIMSA), stated that 'the San have been treated as objects of research, and more often were not even involved in the research agenda'. He went on to make it clear that the San would in future manage and participate in any research on their people (Khoisan Identities and Cultural Heritage Conference, 12–16 July 1997, University of the Western Cape). Ngakaeaja, Mathambo et al. (1998) A San position: Research, the San and San organisations. In A. Bank (Ed.) *Proceedings of the Khoisan Identities and Cultural Heritage Conference*. Cape Town: University of the Western Cape, Institute for Historical Research.

by friendly researchers who left no details and paid them with cigarettes. During discussions with San leaders on intellectual property issues, the experiences of aboriginal peoples in Australia and the Americas were often referred to in order to elucidate their intrinsic legal rights. A popular and instructive example of the use of law to protect intellectual property rights was the case of the Pitjantjatjara people of Australia,[11] who successfully sued an anthropologist for breach of confidence after he publicly disclosed information given to him without permission (WIPO 2001).

A simple 'media and research contract' (Useb and Chennells 2004) was drafted for the Working Group of Indigenous Minorities in Southern Africa (WIMSA).[12] It required an aspirant researcher not only to provide full details of the applicant, and of the nature, content and purpose of the research, but also to negotiate terms with an appointed San leader. WIMSA then arranged for all San leaders to be trained in the use of this contract, so that they could tell whether a project would be useful for the community, distinguish a commercial from a non-commercial project and negotiate to ensure that the community received fair benefits, which could be copies of the book or thesis or money or both.[13] The level of compensation of trackers, translators and other assistants would also be negotiated by the leaders, in order to avoid exploitation of community members, which was all too common.

There was an immediate outcry from some overseas universities that had begun to regard particular San communities as almost their exclusive sources of research information, claiming that 'special relationships' would be ruined by such a contract. The San leadership stood firm. 'If you want to research us, you have to complete the contract,' was the simple message. Within a few years the tables had turned, and the contract is now routinely completed by all researchers of San communities, including those behind this book.

For a detailed analysis of the complex task of building research relationships with indigenous peoples, including research agreements and prior informed consent, see Chapter 7 of Laird (2002).

11.4.2 Intellectual Property and the Media

There were countless ways in which images of and information about the San were 'captured' by outsiders, to emerge later in films, articles or books. Small cameras were used to unobtrusively film culturally sensitive dancing or singing, afterwards reproduced as a 'scoop' in the Western media. A famous South African photographer[14] brought out a successful range of postcards based upon 'private' pictures that he

---

[11] *Foster v Mountford* (1976) 29 FLR 233.

[12] A San representative organization described comprehensively in Chapter 9.

[13] WIMSA annual reports from 2000 to 2004 report on the IPR training provided to leaders around the media and research contract.

[14] Identity withheld by agreement with research subject.

had taken of San, without acknowledgement or any warning of his intentions. He was firmly challenged and quickly agreed to withdraw the range. However, the matter was complex. San leaders recognized the value of tourism, and did not want to become unfriendly and demand excessive fees or reject requests from individuals wanting to take snapshots for private albums. Similarly, journalists who wished to publish positive articles that would serve the interests of the San needed to be encouraged, rather than discouraged by excessive formality.

WIMSA held workshops with San leaders and those who made their living selling crafts by the roadside, in order to ensure that the San knew and understood their commercial, legal, and private rights. If a commercial film or other media product was to be made, a trained San leader would be required to negotiate the completion of a contract. If it was non-commercial, but nevertheless a product such as an article or video, the applicant was required to record his or her details and undertake to provide the San with a copy of the product. If, however, the entire endeavour was clearly non-commercial, such as a tourist taking snaps or asking questions out of curiosity, then a nominal amount at the most should be requested.

11.4.3 *Intellectual Property, Music and Dance*

Many informal and private sound recordings of traditional San music have been made, particularly of the hypnotic clapping and singing that accompanies the iconic San trance dance. When some of this recorded music started appearing as unaccredited backing to the newly popular 'trance' music, WIMSA determined to put a stop to any further unauthorized San recordings.

In intellectual property workshops dealing with music as cultural 'property', the San fully accepted and understood that traditional music, like traditional knowledge, is in its nature 'collective' and cannot be 'owned' or claimed by the artist. Traditional songs were invariably passed down over generations from musicians to aspirant musicians. It was also accepted that where a musician had added important improvisations or new elements to a traditional song, that person would be entitled to recognition in addition to a negotiated share of the value of the rights.

In 2002 the first registered musical collaboration was done using the WIMSA contract between artist Pops Mohamed and the Gcubi family of musicians from Omaheke South in Namibia. A successful CD, *Sanscapes*,[15] was produced, and the commercial media contract recorded and regulated the payment of royalties to the community. Royalties are still being earned from *Sanscapes*, and it serves as a good example of how collective cultural knowledge can be harnessed and used by a community for its general benefit.

Dance is as elemental a manifestation of a distinctive cultural heritage as music, and, as described by Vermeylen in Chapter 10, is being increasingly utilized by the San

---

[15] Produced by MELT records (www.melt.co.za), search for *Sanscapes*.

as a valued economic commodity. The trance dance was originally a private event held primarily to evoke the healing force of *num*, to connect with the ancestral spirits, and to heal those in the community that needed to be healed (Katz et al. 1997). With the increasing demands of tourism and guided by NGOs, the San have learned to regard the dance as a 'performance' that can be commodified and 'sold' to tourists, but can simultaneously be a joyful way of building community and making core aspects of their culture central to their lives. Dance is an 'expression of folklore', which is a subset of and included in the intellectual property term 'traditional knowledge' under the definition provided by the WIPO 'model provisions' (WIPO 2001).[16]

The Kuru Dance Festival, which is held in August every year at D'Kar in Botswana, hosted over 20 different dance teams from all over southern Africa in 2007 for 3 days of exuberant performance.[17] Far from its private and spiritual origins, it has become a vibrant modern celebration of the distinctions and similarities between the cultures of different San communities, and a manifestation of San-ness in the contemporary world. Naturally filming of this event is now strictly controlled, in keeping with the San understanding of the value of performance in the modern world.

11.4.4 Rock Art and Cultural Symbols

The San are the modern custodians of a treasure-house of art that was painted on cave walls and rock faces by their forebears over many millennia. The sheer abundance and beauty of this artistic heritage has made South Africa a premier rock art tourist destination, and in KwaZulu-Natal the Ukahlamba-Drakensberg Park was proclaimed a World Heritage Site in recognition of its priceless San rock art. San leaders realized with dismay that they had not been consulted in the process, and digested the precept that rights need to be claimed in order to be recognized.

Another difficulty facing the San in the modern field of rock art studies is the fact that San communities have long been driven out of the areas that hold the richest repositories of San rock art (primarily the Drakensberg and Cedarberg mountain ranges). In addition, modern rock art researchers inexplicably do not deem it necessary to consult with modern San leaders. A matter of some amusement to San leaders is the fact that the most famous rock art authors invariably pontificate about their glib and often esoteric theories on the meaning of rock art without having deemed it useful to consult with living San.

When, in July 2002, the KwaZulu-Natal provincial government announced the opening of the Didima Rock Art Centre, dedicated to San rock art, without having consulted with San leaders, WIMSA was ready for assertive action. Preparations were made for a court injunction to prevent the event, on the grounds that the San,

---

[16]In the 1980s, 'model provisions' for the protection of folklore 'against illicit exploitation' were adopted under the auspices of WIPO and UNESCO.

[17]www.kuru.co.bw/dancefestival.htm.

as rightful custodians of the rock art, had not been consulted. The government backed down dramatically in the face of this challenge, apologized profusely, and extended an invitation to the entire WIMSA board to attend the opening as guests of honour. The keynote address was delivered by Petrus Vaalbooi, then chairperson of the South Africa San Council, who expressed the deep feeling of ownership and 'custodianship' felt by modern San leaders over the rock art. Undertakings were made by Ezemvelo KZN Wildlife that the San would forthwith be given a seat on the governing board of the national heritage site, that San youth would be given free admission to visit the museum, and that the San would be fully consulted in future as the 'heritage custodians' for the rock art museum. Sadly, and largely through the San's lack of capacity to follow up this opportunity, the San do not yet play an active role in the management of their rock art heritage.

As the rock art of the San has gained international recognition, so the rock art images have fallen prey to all forms of exploitation. Copyright does not protect rock art from being copied, and many artists sell replicas or copies of this art without explanation or apology. In addition, the powerful and characteristic images, once they have been placed in the public domain through popular publications, make striking trademarks and designs on letterheads and documents. Nothing but conscience prevents a businessperson from copying rock art images and selling them as part of a product range. The San have successfully prevented businesspeople from illegally using the name San (on products as diverse as table salt, playing cards and barbeque sauce) and have found that where their legal rights are lacking, their indisputable moral rights are equally effective. In most cases a letter to the owner of the business explaining that the name or image of the San should not be used without their permission is effective. The prospect of the negative publicity that would follow a court challenge is usually sufficient to uphold these general rights to name and images.

11.4.5 Art and Crafts

Art produced by the San is rapidly acquiring the international status and recognition already accorded to Australian Aboriginal art. The Kuru Art Project was the first formal attempt to encourage San artists and assist them in marketing their art (KFO 2006). Normal copyright in the name of the artist attaches to artworks, but the 19 artists currently involved in the project are assisted in negotiating commercial aspects of copyright, as the artworks are purchased for private collections, publication on calendars and other uses.

Numerous cases in Australia have been taken to court, where designs of Aboriginal paintings have been reproduced on commercial fabrics without permission. The famous Australian case of *Bulun Bulun v R and T Textiles* in 1998 was another encouraging example to the San of the power of the IPR legal regime. In this case the Aboriginal artist successfully sued a clothing manufacture for infringement of copyright, in that it used his painting without permission. The artist was successful, the outcome sending a stern warning to all similar unlawful users of art.[18]

The !Xun and Khwe San Art and Culture Project near Kimberley nurtures similar aims and objectives, and has found a ready demand for the unique artworks produced by the San artists. The previous sale and licensing for publication of these artworks by leaders who had not first acquired a clear understanding of copyright contributed towards the difficulties experienced by this project, which is currently being revived.[19]

Crafts are not protected by copyright, nor by patent law. Where the manufacture of crafts can be copied by others, the need arises for some form of protection. For example, bows and arrows are easily reproduced by non-San, in the same way as the didgeridoo has been manufactured and sold by non-Aborigines. The Tiwi artists of Australia registered an 'authentication label' for their didgeridoos as proof of authenticity (WIPO 2001). Discussions are taking place about designing a registered brand to serve as an authentication label for crafts made by the San. If deemed commercially viable, a trademark will be registered in order to secure the fullest form of IPR protection for an important San brand. These initiatives show that a domestic IPR legal system, creatively used, can provide market protection relatively cheaply.

11.4.6 Traditional Knowledge of Biodiversity

The term 'traditional knowledge' is interchangeable with several other terms, such as 'indigenous cultural and intellectual property', 'indigenous knowledge', and 'customary heritage rights'. WIPO uses the term to refer to 'tradition-based literary, artistic or scientific works; performances; inventions; scientific discoveries; designs … undisclosed information …' (WIPO 2001), and there are numerous categories of traditional knowledge.

Notwithstanding all this terminology, the central meaning is perfectly clear to the layperson and does not depend upon a single unifying definition. San peoples have over countless millennia developed extensive knowledge of the natural world, including knowledge of the medicinal uses of the plants and animals surrounding them. Research shows that San people today still regard the free sharing of this knowledge with neighbours as normal, and not a commercial transaction. It is only recently, because valuable traditional knowledge related to *Hoodia* was patented and commercialized without the San being aware of it, that the notion of knowledge as a 'commodity' with intrinsic financial value has emerged. San healers are now being encouraged not to share plant and medicinal knowledge freely with strangers without being assured, through a completed contract, of the nature and purpose of the enquiry.

---

[18] *Bulun Bulun and Milpurrurru v R and T Textiles Pty Ltd*, Queensland Law Reports 1998, cited in WIPO (2001).

[19] The South African San Institute was requested by the !Xun and Khwe communities to revive the art and craft ventures, under the name //Naoa Djao (KFO 2006).

After the negotiation of the *Hoodia* benefit-sharing agreement with the Council for Scientific and Industrial Research (CSIR), the San leadership was moved to deal with the question of the further appropriate management and protection of their traditional knowledge. Should it be kept secret? Should it be recorded for posterity and possible future commercial use? Much of the San's traditional knowledge is in the public domain, meaning it was freely shared in the past, is now available to the public and can no longer be protected by the IPR legal system. Having gained exposure to the debate around the failings of the IPR system, and aware of suggestions by some indigenous activists that other methods of *sui generis* ('custom-designed' or 'specific to this case') legal protection of IPR should be adopted, the San were required to formulate a practical decision on this question.

Some indigenous peoples called for the documentation of traditional knowledge on public databases, in order to identify and preserve the knowledge and make it available for future generations (Correia 2001). In addition, according to the proponents of this policy, the knowledge, once documented, falls into the public domain, thus placing it in the category of 'prior art'. This prevents any future application for a patent, on the basis that the knowledge is no longer 'novel'. This latter policy is referred to as 'defensive publication'.

However, the decision on whether to document or not was found to be even more complex, for a number of reasons. Publishing the knowledge also prevents the originating indigenous community from applying for any IPRs, as the essential (for patenting) component of 'novelty' is lost to them, too. In addition, the issue of the 'ownership' of the published information, including copyright and related IPR rights, can be problematic in some communal cases. (Who is 'the community'?) Other opponents of defensive publication suggest that publication simply facilitates the unauthorized exploitation of the traditional knowledge by outsiders. In South Africa, an indigenous knowledge Bill under the Department of Trade and Industry has been many years in the making and is not yet promulgated. Significantly, the indigenous knowledge policy that has been adopted supports defensive publication, as it proposes the creation and maintenance of a register of indigenous knowledge.

The San adopted a middle approach, which was to authorize the compilation of a private and protected database of traditional (medicinal) knowledge, in conjunction with a selected bioprospecting partner. An agreement was negotiated in 2003 with the CSIR, which undertook to gather medicinal plant information from San healers and record it on a private database that could only be accessed by the two parties under the agreement. Any information that might lead to a commercial product would be freely shared between the parties, and the San were assured of co-ownership of any patent or other IPR attached to the fruits of the agreement. Negotiations on the precise form of profit-sharing would commence once a product was identified, and a dispute resolution clause in the event of deadlock ensured that an equitable agreement would be reached. A clear provision with checks and balances ensured that the information on the database would not be available to any outsiders, which meant that it was not in the public domain and could not be stolen.

This interesting partnership is currently inactive, awaiting operating funds.

11.4.7 Contract Law

The San have recognized the fact that the law of contract, which is international in its ambit and open to any consenting parties, is a powerful tool for IPR protection. The law of contract is virtually identical in all countries, providing rules based upon common sense and natural justice that guide and bind parties that enter into agreements. The San are aware that they entered into a long-term and binding IPR contract with the CSIR for benefit sharing as a result of their IPR rights long before the laws that were intended to regulate the issue came into existence.[20]

The San have also recognized that communities and individuals have significant legal rights under the constitutions and laws of their countries. Even the media and research contracts referred to above are in essence 'benefit-sharing agreements' based upon the San's IPRs. The contracts clearly set out the manner in which the various rights (to research, film, record or publish) are to be exercised, with binding provisions regarding the provision of benefits (copies of films, books or videos, payment of royalties or lump sum payments).

For this form of protection to succeed, the 'rights holders' must educate and empower themselves as to their legal rights, and as to the ways in which these rights can be commercially exploited. This can require advice and information relating to strategy and commercial realities far outside the normal knowledge of San leaders. Armed with this information, and after the requirements of prior informed consent regarding all aspects of the transaction have been satisfied, the task of the leaders is to ensure that all agreements made – for example, for the selling of natural botanical resources – are governed by explicit and fair contracts.

11.5 Conclusion

IPRs have become a significant reality in the lives of San leaders. Their recent experience in identifying the range of engagements and transactions that exploit their rights, and deciding on practical methods of regaining control, as described above, has confirmed the intrinsic social and commercial value of their unique heritage. In addition, this process has consolidated a collective mindset that it is their own responsibility to acquire a good overall understanding of the IPR and commercial world, rather than to await assistance or salvation from external sources. International conventions and a more accessible international IPR regime are to be welcomed, but indigenous peoples need to assert their own rights, using the existing laws and tools at their disposal.

---

[20] Negotiations on the *Hoodia* patent began in June 2001, and the final agreement was signed in March 2003. The National Environmental Management: Biodiversity Act was only promulgated in 2004.

Not only is the distinctive San heritage a proud source of identity, but it has been shown to be a powerful collectively owned asset in the San's struggle to develop soundly as an emerging nation. Loss of knowledge and culture has been recognized by San leaders and institutions as a matter of concern, which it is their own responsibility to redress. The lessons that they continue to learn in the *Hoodia* case, together with the initially faltering steps taken to protect their intellectual property rights, have provided the San with an understanding that the commodification and exploitation of culture are not necessarily bad, and can be harnessed in a manner that unifies and simultaneously empowers them as indigenous peoples.

References

Barsh, R. L. (1999). Indigenous knowledge and biodiversity, indigenous peoples, their environments, territories. In D. A. Posey (Ed.), *Cultural and spiritual values of biodiversity*. London: UNEP and Intermediate Technology Publications.

CBD (2002). Bonn guidelines on access to genetic resources and fair and equitable sharing of the benefits arising out of their utilization. Decision VI/24, 2002, Secretariat of the Convention on Biological Diversity, Quebec. www.cbd.int/doc/publications/cbd-bonn-gdls-en.pdf. Accessed 22 March 2008.

CIPR (2002). *Executive summary. Integrating intellectual property rights and development policy, report of the commission on intellectual property rights*, September, London. http://www.iprcommission.org/papers/pdfs/final_report/CIPR_Exec_Sumfinal.pdf. Accessed 16 May 2008.

Correia, C. (2001). *Traditional knowledge and intellectual property: issues and options surrounding the protection of traditional knowledge*. Geneva: Quaker United Nations Office.

Drahos, P. (1999a). *The universality of intellectual property rights, origins and development. In Intellectual property and human rights*. WIPO Publication No. 762 (13), Geneva.

Drahos, P. (1999b). Biotechnology, patents, markets and morality. *European Intellectual Property Review, 21*(9), 441–449.

Dutfield, G. (1999). Introduction: rights, resources and responses. In D. A. Posey (Ed.), *Cultural and spiritual values of biodiversity*. London: UNEP and Intermediate Technology Publications.

Dutfield, G. (2004). *Intellectual property, biogenetic resources, and traditional knowledge*. London: Earthscan.

Heintze, H.-J. (1993). The protection of indigenous peoples under the ILO convention. In M. Bothe, T. Kurzidem & C. Schmidt (Eds.), *Amazonia and Siberia: legal aspects of the preservation of the environment and development in the last open spaces*. London: Graham & Trotman.

ILO (1989). *Convention (No. 169) Concerning Indigenous and Tribal Peoples in Independent Countries*. International Labour Organisation, Geneva. www.unhchr.ch/html/menu3/b/62.htm. Accessed 10 May 2008.

IWGIA. (2007). *The indigenous world: yearbook*. Copenhagen: International Workgroup for Indigenous Affairs.

Katz, R., Biesele, M., & St Denis, V. (1997). *Healing makes our hearts happy: spirituality and cultural transformation among the Kalahari Ju/'hoansi*. Rochester, VT: Inner Traditions.

KFO (2006). KFO Annual Report for 2005. Kuru Family of Organisations, Ghanzi, Botswana. www.kuru.co.bw.

Laird, S. A. (Ed.) (2002). *Biodiversity and traditional knowledge: equitable partnerships in practice*. London: Earthscan.

Laird, S. A., & Wynberg, R. P. (2002). Institutional policies for biodiversity research: setting standards for conduct, prior informed consent, and benefit sharing. In S. A. Laird (Ed.), *Biodiversity and traditional knowledge: equitable partnerships in practice*. London: Earthscan.

Shiva, V. (1997). *Biopiracy: the plunder of nature and knowledge*. Cambridge, MA: South End Press.
Useb, J., & Chennells, R. (2004). Indigenous knowledge systems and protection of San intellectual property: media and research contracts. *Before Farming, 2004*(2), article 2, pages 1–11.
WIPO (1993). Introduction to trademark law and practice: the basic concepts: a WIPO training manual Publication No. 653(E), Geneva, World Intellectual Property Organization.
WIPO (1994). Introduction to WIPO: objectives, organisation, structure and activities. Development Cooperation Program, Geneva, World Intellectual Property Organization.
WIPO (1997a). Introduction to basic notions of copyright and neighbouring rights. Document code: WIPO/CNR/KTM/97/1, Geneva, World Intellectual Property Organization. www.wipo.int/meetings/en/doc_details.jsp?doc_id=4734.
WIPO (1997b). Basic notions of industrial property. Document code: WIPO/IPM/POS/97/1 A, Geneva, World Intellectual Property Organization. www.wipo.int/meetings/en/doc_details.jsp?doc_id=5123.
WIPO (1998). Intellectual property reading material. Publication No. 476(E), Geneva, World Intellectual Property Organization.
WIPO (2001). Intellectual property needs and expectations of traditional knowledge holders: WIPO Report on Fact-Finding Missions on Intellectual Property and Traditional Knowledge (1998–1999). Publication No. 768(E), Geneva, World Intellectual Property Organization, April.

*I eat the Xhoba and then I no longer feel hungry or thirsty.
I eat it when I am feeling weak and then I feel strong and virile.
I eat it when I have a bad stomach and then I feel better.*

(Piet Rooi, Andriesvale, South Africa, January 2003)

Chapter 12
Sharing Benefits Fairly: Decision-Making and Governance

Rachel Wynberg, Doris Schroeder, Samantha Williams, and Saskia Vermeylen

Abstract Understanding how decisions were made by the San in the *Hoodia* case and how decision-making and governance structures vary between bioprospectors and indigenous communities is essential for the implementation of effective benefit sharing.

Drawing on academic literature and on interviews undertaken in South Africa, this chapter shows that decision-making processes in benefit-sharing negotiations vary significantly from party to party. In corporate hierarchies, decision-making usually centres on a small number of individuals and does not involve the wider consultation of stakeholders. Decisions are routinely made by highly educated personnel in positions of power who are well versed in the legalities and implications of their decisions. By contrast, decision-making in traditional indigenous communities such as the San often involves a large number of community members, typically with little knowledge of the technicalities and legal implications of their decisions. Discussions are seldom limited to a single event, but rather emerge over time during conversations among friends, relatives and neighbours. In the case of the San, decisions are taken by consensus, which is reached when significant opposition no longer exists.

R. Wynberg (✉)
Environmental Evaluation Unit, University of Cape Town, Private Bag X3, Rondebosch 7701, Cape Town, South Africa
e-mail: rachel@iafrica.com

D. Schroeder
UCLAN, Centre for Professional Ethics, Brook 317, Preston PR1 2HE, United Kingdom
e-mail: dschroeder@uclan.ac.uk

S. Williams
Environmental Evaluation Unit, University of Cape Town, Private Bag X3, Rondebosch 7701, Cape Town, South Africa
e-mail: Samantha.Williams@uct.ac.za

S. Vermeylen
Lancaster Environment Centre, Lancaster University, Lancaster, LA1 4YQ, United Kingdom
e-mail: s.vermeylen@lancaster.ac.uk

These differences in decision-making practice place an obvious burden on negotiations, with one party requiring fast decisions to satisfy shareholders while the other needs significant time to allow meaningful community consultation and digest the implications of different options. This clash over decision-making procedures and speed often turns out to be detrimental to traditional knowledge holders, whose decision-making abilities are compromised by the commercial partners' need for urgent resolution.

One possible solution is embraced by South Africa's National Environmental Management: Biodiversity Act, which now locates support for consultation firmly with the government to ensure that negotiations are on an equal footing when benefit-sharing agreements are negotiated. However, the practical implementation of this requirement remains hampered by constraints of capacity, resources and knowledge.

Keywords benefit sharing • consultation • decision-making • governance • indigenous communities • traditional knowledge

12.1 Introduction

We constantly make decisions. Will we have muesli for breakfast or toast? Will we take the bicycle to work or the tram? Will we make a start on this overdue book chapter or not? Many of us also make decisions for others. Does grandmother need to see a doctor today or not? Will the family have spinach for supper or beans? Which performance objectives will be set for this employee for next year?

Decision-making involves a complicated cognitive process, which ends in a final choice between alternatives. The choice can be rational or irrational, impulsive or well thought through, understandable or obscure, welcome or unwelcome, etc. It can lead to an action ('I'll have muesli today') or an opinion ('Well, since you are asking me, my favourite author is Iris Murdoch'). Things get complicated when those who will be affected by a decision are not directly involved in making it.

Communities engaged in bioprospecting and benefit-sharing agreements face a bewildering array of decisions. First, there needs to be agreement from the community for use of their knowledge and/or biodiversity. This has to be done in accordance with CBD obligations and, often, national laws requiring prior informed consent. As Schroeder (Chapter 3) and Dutfield (Chapter 4) elucidate, prior informed consent implies a complex process requiring relationship-building, capacity development and some level of organization within the community. Such factors often play a central role in shaping the decision about whether or not to provide consent.

Second, once a negotiating relationship has commenced, traditional knowledge holders and/or local custodians of biodiversity need to negotiate a benefit-sharing agreement with the user, typically a research institution or private company, in which fair benefits are provided in exchange for permission to access the resource.

This is often an iterative and lengthy process, involving multiple decisions at different points by all parties, not all of whom would fully accept, or even understand, all the implications of these decisions.

Third, beneficiary communities need to agree on mechanisms for the receipt, use and distribution of benefits. This may involve, for example, the establishment of an institutional structure to administer the implementation of the agreement, the selection of members to represent community interests within this governance structure, or the building up of local capacity to distribute funds or other benefits.

While this sequence of events may be ideal, it is not always followed, as evidenced by the San-*Hoodia* case, where informed consent was not obtained from the San prior to the successful patent application by the Council for Scientific and Industrial Research (CSIR). This chapter therefore concentrates on the second and third steps of decision-making, in which the objective is to enhance understanding about decision-making in benefit sharing and extract lessons and policy recommendations from the approach taken. While perspectives of all parties are included in this analysis, a particular focus is placed on the way in which the San came to decisions.

The chapter begins by introducing the dominant Western approach to group decision-making, namely voting representatives into positions of power according to democratic principles. The second part describes aspects of traditional decision-making in indigenous communities, using the San community as an example. The third part analyses decision-making processes in the *Hoodia* benefit-sharing case. A conclusion summarizes lessons from the *Hoodia* case and the implications of these for other communities engaged in negotiating and implementing benefit-sharing agreements.

12.2 Decision-Making and the Democratic Model

Imagine a lawless region with 20 settlements of 200 adults each. After a time of brute struggles and diminished resources for all, due to the constant war effort, some people from across the villages start to talk. They talk about a new, exciting future – a future in which war will end and all settlements be united in peace. 'How can we achieve this?' they ponder, until one of them says: 'How about this? There is one time in the year when we can meet everybody, the night before the first of May with its open-air festivities. Let us explain to all that we want to bring peace, but we can only do so if all 20 villages agree on a peace accord. It is not possible to gather 4,000 people in one place at one time. Hence, to facilitate decision-making for the accord, each village needs to send five people to a gathering of decision-makers. Everybody can select the person whom they trust the most and who they believe will represent their interests best. The five people who are selected will be given decision-making powers on behalf of their village. And after a long absence from the village, they will return with a result: the peace accord.' Those around the speaker nod approvingly until one person stands up and says: 'But how will we ensure fairness?' The first speaker replies: 'We shall rely on the principles of the autonomy of individuals, equality and popular sovereignty.'

Autonomy of individuals – Every human being has the right to freedom, a right to plan and determine their own future and to act from their own will. The only restriction on this right is the proviso that freely chosen acts must not foreseeably and avoidably harm others (Mill 1910). Under this principle, every villager has a right to influence the peace accord, as the accord will determine each individual's future and therefore their ability to act from their own will.

Equality – All human beings are equal (Locke 1960). There is no difference between one villager who has suffered grievous harm during the war and another who was hardly touched, or between one villager who owns a large herd of cattle and another who does not even own a chicken. They will all take part in the process as potential representatives or as those choosing whom to send.

Popular sovereignty – A sovereign is a supreme authority within a certain area of governance, such as 20 villages with 4,000 adult inhabitants. Popular sovereignty or 'sovereignty by the people' means that the supreme authority lies with the inhabitants. No law can be passed or peace accord signed for the region unless it has been agreed to by the sovereign – that is, the 4,000 inhabitants (Rousseau 1973). If the 4,000 cannot assemble for popular decision-making, representatives are elected in a system called 'democracy'. Those thus elected have to speak on behalf of their people when it comes to agreeing on the accord; they cannot represent their own interests only. This is how it is possible for 4,000 inhabitants to make decisions indirectly through 100 (20 × 5) representatives.

On the first of June, a month after the election of the representatives, a peace accord is signed and the 100 villagers return to their homes to bring the good news. And under the peace accord, they live happily ever after.

That is the ideal scenario of democratic decision-making. However, democratic decision-making has been heavily criticized. Here are the main criticisms as they relate to our village scenario.

- What about the elderly or sick who are absent from the festivities and therefore unable to choose a representative? And what about those, usually women, looking after them at the time? They are able neither to choose a representative nor to be selected themselves. And who knows how the men will represent their interests! Today, say feminists, democracy is not succeeding in achieving one of its highest aspirations, namely the equality of citizens (Mitchell 1987).
- What about the interests of a minority who may not agree with the views of the majority who elected the representatives? How are their dissenting views to be accommodated? In a democracy, minority views are often disregarded.
- How will the representatives be held accountable? As soon as they are away in their Hilton hotels, they will forget about their fellow villagers and arrogantly make inappropriate, irrelevant and self-serving decisions (Craig 2004).
- Wouldn't it be better if the 'mob' didn't make important decisions, such as who should agree the peace accord? Wouldn't it be preferable for the elite in each community either to go themselves or choose their delegates? Democracy is a flawed, unstable system (Plato 1935).

- The system of democratic decision-making is uniquely Western, emphasizing values of individualism and human rights that are not universal (Huntington 1996), and therefore cannot be used across the globe.

On the one hand, democracy is a contentious form of group or community decision-making, as the above points indicate. On the other hand, such highly respected thinkers as Nobel laureate Amartya Sen have emphatically declared it the only game in town (Sen 1999). The next section introduces a competing model in the context of traditional decision-making among the San.

12.3 Traditional Decision-Making Among the San

The village scenario described above depicts a lawless conflict zone where people settle their disputes according to the 'law of the jungle' until they realize the destructive effects of their barbaric behaviour and therefore decide to act in a more civilized fashion. Thus a peace accord is agreed, requiring the negotiation of a fair agreement based on the principles described: autonomy of individuals, equality and popular sovereignty.

In hunter-gatherer societies such as the San, a high priority is accorded to the avoidance of conflicts, and their approach to dispute resolution is believed to be superior to that used in Western societies (Ury 1990, 1995). The resolution of disputes is essential for small semi-nomadic groups that need to cooperate in order to survive. In Box 12.1 we analyse the way in which the San have traditionally resolved conflict, an examination that provides valuable insights into how they make decisions.

Silberbauer (1982) and Lee (2003) have identified and distilled several characteristics of the San's decision-making process. Decisions that affect the band as a whole always come into being through a process in which everyone in the community participates. Discussions seldom take place at a single event, but rather emerge over days during ordinary conversations among friends, relatives, and neighbours. If more serious decisions have to be taken and factions emerge, the San will involve a wider audience and include those members of the community who did not take part in the initial discussion. In this way each faction can find out how the wider community responds to the issues and, possibly, influence members' opinions. Direct confrontation with the opposition is avoided and their inclusion is frowned upon. However, the opposing faction can use the same technique. In other words, when there is a problem in a community that manifests itself as two quarrelling factions, community members discuss the problem in small groups among themselves. If the problem persists, both factions ask those who are not directly affected for their opinion. However, the factions do not talk to each other. In this way, the onlookers or listeners are treated as independent, neutral parties helping to resolve the issue without direct confrontation.

Decisions in band societies[1] such as the San are taken by consensus. This does not mean that the decisions or opinions are unanimous, just as 'egalitarian' does not

---

[1] A 'band society' consists of a small kin group which is usually not larger than an extended family. Two main characteristics of band societies are egalitarianism and consensus-based decision-making.

Box 12.1 Conflict Resolution among the San

Analysing the way in which the San have traditionally resolved conflict provides valuable insights into how they make decisions. Much of this work has been done by Ury (1995), who notes that an effective conflict management system (regardless of the particular cultural setting) must perform six functions: (a) **prevent** disputes from arising and, when this initial buffer fails to operate, (b) **heal** emotional wounds, (c) **reconcile** divergent interests, (d) **determine rights**, (e) **test the powers** of the parties and (f) **contain unresolved disputes** that threaten to escalate into violence. The application of these functions by the San is examined below.

Prevent disputes where possible – One of the most important lessons San parents teach their children is the avoidance of conflict. Sharing resources and creating social relationships through gift exchange are learned practices that can prevent conflict from arising. Gift-giving and mutual exchange foster long-lasting relationships between individuals and groups. When groups have to leave their *n!ore*[2] in times of distress and settle temporarily elsewhere, they can fall back on those relatives with whom they have fostered such reciprocal generosity. Teaching respect for community norms and values also prevents disputes. The people are alert to early signs of conflict, and friends or relatives of the quarrelling parties encourage them to settle their problems before the situation escalates.

Heal emotional wounds – Those disputes that cannot be prevented must be resolved. The first step is to air the negative emotions in public. If the conflict persists, a meeting is called to discuss the problem with the elders and possibly other community members. An important rule is that everyone must have the opportunity to talk and give his/her opinion and that everyone should respect the other's opinion. This process can be very time-consuming and may last for days. If the problem is still not resolved or the tension is widespread in the community, a trance dance (see Vermeylen, Chapter 10) may be organized to 'heal' the community. Even when the dispute has been settled at the meeting, the community often organizes a trance dance to bring back unity and restore social relations. If a dispute spirals out of control, the elders might request one of the parties to spend some time away visiting relatives, allowing calm to return to the community.

Reconcile divergent interests – Conflict resolution among the San is consensual, aiming to meet the needs of all parties and find a solution that everyone supports. Those involved in the conflict and their relatives

(continued)

---

[2] As Lee (2003) describes it, each waterhole in the region inhabited by the San is surrounded by an area of land that provides the food and other natural resources that a group depends on. This territory, or what is called *n!ore*, is owned by a group of related people who are called the *k"ausi* (owners).

Box 12.1 (continued)

are contacted prior to the meeting to discuss the problems and to give everyone an early chance to air their opinions and look for a possible solution. At the meeting itself, members of the extended community can express their opinions and try to persuade others of the best solution. Everyone has the right to ask questions of the quarrelling parties. Gradually the community will come to a solution that everyone finds acceptable. There is no jury or group of people who make this decision, and there is no vote or verdict, as the solution is based on consensus. Biesele (1978) describes this process as 'centrifugal', as opposed to the Western 'centripetal' process. While a centripetal system is characterized by someone taking centre stage, the San's system is characterized by movement away from the centre. As part of the egalitarian ethos of the San, no one assumes authority or wants to draw too much attention to themselves. Instead, decisions are deferred to others and the involvement of the wider community is encouraged.

Determine rights – The community has its own norms and values specifying what is right and what is wrong. Important steps in the San's conflict management system are to gain a better understanding of the facts (i.e. what has caused the conflict) and then to apply the appropriate norms and values to solve the conflict, or determine what norms will prevail. The San organize this process through a community meeting and the practice of using witnesses, who educate and admonish the offending party about the abuse of norms and values. In Botswana, anthropologists (e.g. Lee 2003) record that in more recent years, serious conflicts (usually those that could turn violent) have been taken to the court of the Tswana. This is usually done as a last resort in order to prevent the community from breaking down completely.

Test relative power – Most conflicts take place in the wider context of power and power differentials between parties. In San traditional communities, power is mostly dispersed equally and evenly. They use a strategy that involves one of the parties to a conflict moving away, rather than entering into a negotiated agreement. This makes it difficult for the San to negotiate bilateral agreements, as negotiations are constrained by the unilateral freedom of any party to walk away from the conflict. In other words, it is very difficult to exercise coercion (an important component of bilateral arrangements) over a person or a group of people. The community cannot actively impose a resolution, but they can show their disagreement with and disapproval of the offender.

Contain potential and actual violence – Because many members of San communities used to keep highly poisonous arrow tips in their possession in times gone by, there was a huge fear of violence among the San. Therefore, when tension did mount between parties, relatives or friends would hide the poison and the poisonous arrow tips and ask one of the parties to move to another area.

mean 'equal'. As Silberbauer (1982) argues, it is important to understand that 'consensus' is not a synonym for 'democracy'. Democracy, he continues (1982), 'is about equality of opportunity of access to positions of legitimate authority and the limitations this imposes on the exercise of power'. As such it is an organizational framework 'ruled' by the majority (in a representative democracy) that makes and executes decisions. Band societies, on the other hand, make decisions on the basis of a series of judgements that can be formed because everyone had access to a common pool of information. According to Silberbauer, consensus arrives when 'people consent to judgment and decision' (1982). It is reached when there is no longer significant opposition to the decision, which is different from unanimous acceptance.

The role of the community is a defining element of the San's conflict resolution management system. The community teaches children from a very young age that they must avoid conflict, and, when conflict arises, it tries to pacify opposing parties. The community also organizes meetings and performs healing dances every night in the hope that they will resolve the conflict and ease tensions in the community.

Over the past 20–30 years the life of the San has changed considerably and they have transformed from relatively isolated foragers to peasants (Vierich 1982; Barnard and Taylor 2002). Lee (2003) and others (e.g. Guenther 2002) observe that some of the main guiding principles in the San's social organization, such as sharing, have eroded over time as a result of increased exposure to new social norms and socio-economic and political environments. However, elements of their traditional decision-making process can still be observed.

Table 12.1 presents a simplified comparison of democratic decision-making with traditional indigenous decision-making.

Recently the San were introduced to one of the most highly contested concepts in Western law – intellectual property rights. The next part of this chapter examines how decisions were made by the San in their negotiations with the CSIR about the

Table 12.1 Democratic and Traditional Decision-Making: A Comparison

	Decision-making	
	Modern democratic	Traditional indigenous
Participation	Indirect through representatives	Direct through full involvement in all community decisions
Power	With representatives	Dispersed among community
Accountability	Through re-election and bureaucratic controls	Through diverse systems that vary across communities
Minorities	Exclusion through majority voting	Inclusion through sustained attempt at consensus forming
Values	Focus on individual person and human rights	Focus on community well-being
Time frames	Time-sensitive: following previously agreed schedules	Time-insensitive: as long as it takes to reach consensus

patent rights to *Hoodia*, benefit sharing and the implementation of the benefit-sharing agreement.

12.4 Decision-Making in the Negotiation and Implementation of San-*Hoodia* Benefit-Sharing Agreements

The coming together of Western and traditional systems of decision-making in contemporary San institutions has, unsurprisingly, created a hybrid of these governance models. As Chennells et al. contend (Chapter 9), San institutions today are characterized by a sometimes uneasy blend of features. Individual leadership, not recognized in traditional systems, is often contested and fraught, and hampered by capacity and legitimacy constraints; accountability is perceived to be weak; and the structure and functioning of San organizations is fragile in most cases and strongly debated in all. San institutions are also very new and do not have well-established norms and standards. Just how this influences decision-making is open to question, and the challenges of benefit sharing add a further layer of complexity.

This section describes and analyses the way in which decisions were made in the negotiation and implementation of the benefit-sharing agreement between the CSIR and the San. We arrived at this description and analysis through interviews with key actors engaged in this process, as explained in the appendix, and an exploration of the following three important decision-making phases:

1. The initial negotiation between the San and the CSIR and the later development of a benefit-sharing agreement
2. The establishment of a trust to distribute benefits arising from the commercial use of *Hoodia*
3. The receipt and disbursement of funds by the trust

12.4.1 Negotiations Between the San and the CSIR

As other chapters describe, the negative publicity generated about the 'stolen knowledge' of the San led them and the CSIR to the negotiating table in 2001. The South African San were fortunate in having a leadership structure at the time, developed over six years of participation in meetings of the Working Group of Indigenous Minorities in Southern Africa (WIMSA). However, the *Hoodia* controversy accelerated the registration of a legally constituted South African San Council to ensure, in the words of the San legal adviser at the time, 'that the CSIR took us seriously' (Roger Chennells, September 2006, personal communication, Upington).

Two questions were uppermost in the minds of the San team: first, should the patent be challenged; and second, should they enter into negotiations with the CSIR, and, if so, under what terms? Andries Steenkamp, current chair of the South African San Council and a key negotiator, explained the decision-making challenges:

> It was a difficult process because everyone that was on the San Council then had a very low education level. We did our best to figure out whether or not it was a good thing to negotiate. We had to consider whether or not we were going to resist the patent. And we came to the conclusion that negotiation was the best option, to enter into conversation. It started by them [the CSIR] writing a letter of apology to us and inviting us to talk to them (Andries Steenkamp, September 2006, personal communication, Upington).

The strategy was to insist that the CSIR admit, through a legal memorandum of understanding, that their lead for *Hoodia* had come from the indigenous knowledge of the San, and that the San had intellectual property rights to this knowledge. Once this was achieved, the more difficult process of negotiating an agreement would commence. One negotiator commented:

> The implied threat was, if you don't give this to us, we will get it and we will embarrass you and threaten the entire pack of cards. We needed to get the prize – an admission from them [the CSIR] that the entire intellectual property that they had licensed out was to a significant degree based upon the research lead they got from the San. Because once they had signed that then they had to go through to the second phase (Roger Chennells, September 2006, personal communication, Upington).

A great deal of urgency accompanied the decision to engage with the CSIR. This, combined with the complexities of the CSIR patent on *Hoodia*, resulted in the San legal adviser, Roger Chennells, playing a major role in decision-making at this stage. Just how this influenced the final outcome is open to question. Chennells comments:

> I got buy-in from the San Council ... but I think it would be fair to say that they were completely in the dark about what was right and what was wrong with regards to intellectual property, so at that stage I had a stronger role than had I modern clients, whose leaders are fully aware of everything. Frankly, they [the San] did listen to what I said, but I carried that as a burden and not as a licence to decide in a very quick way what to do. I am intensely conscious of the fact that a lawyer can easily say that my client has decided when you've actually forced them to make that decision (Roger Chennells, September 2006, personal communication, Upington).

The interpretation of this situation by San negotiators reflects this dilemma. 'We were sitting there and relying mostly on our lawyer as we had no knowledge (San-*Hoodia* Trustee 1, personal communication, September 2006, Upington),' remarked one San representative, while another affirmed that final decisions resided with the San: 'We have a legal person that often gives us advice: how he views things, what those people say, how they feel and what they want to do. We take our own initiative to say what we want to after he has given his advice to make our decisions (San-*Hoodia* Trustee 2, personal communication, September 2006, Upington).'

The speed and complexity of negotiations had unintended consequences for the decisions made, limiting opportunities among the San to learn and explore the implications of different options, in particular with regard to the patent. These are the observations of one of the San negotiators:

> It would have been different if we knew before what we found out later. You see, we were not as well educated as other people. Why couldn't we get at least half of the rights to the patent? We actually believed that the patent right ought to be awarded to us ... all the San were very unhappy ...To be honest, I am not sure if we were assisted properly (San-*Hoodia* Trustee 3, September 2006, personal communication, Upington).

To some extent these divergent opinions about the fairness of the patent, and whether or not it should be challenged, mirrored the broader debates taking place at the time. These were primarily concerned with the design of appropriate systems to protect traditional knowledge (e.g. WIPO 2001) and the difficulties of reconciling Western systems of privately held and monopolistic intellectual property rights, based on 'innovations' or 'discoveries', with traditional knowledge systems, which are typically collective and based on prior art (WIPO 2001; CIPR 2002). Ethical concerns were expressed about the patenting of life forms (e.g. Crucible II Group 2001; Ekpere 2001), as were widespread fears about the potential abuse of the patent system to create monopoly ownership of biodiversity (e.g. ETC Group 2001; GRAIN 2001; Then 2004). Roger Chennells explains:

> We had a lot of people, NGOs, who were quite keen on us taking the principled position, which was to challenge the patent. I explained how the San people had chosen to take benefits rather than make a principled point on behalf of mankind against patenting... and that position wasn't at all popular. And I said, well you explain to the San why they should hand away potentially millions when they are at the bottom of the pecking order and probably are the poorest people in southern Africa. When it came down to this practical assessment the principle point [of not accepting the patent] wasn't held very strongly by the San. The other side appeared to have all the attraction, which was that this is something which we, the San, will benefit from (Roger Chennells, September 2006, personal communication, Upington).

With the decision not to challenge the patent and the memorandum of understanding in hand, negotiations on the benefit-sharing agreement moved swiftly forward. To a large extent these were characterized by the conflict avoidance strategy described above in relation to traditional San institutions. 'We saw negotiation was a better option than fighting and arguing with one another,' remarked one San negotiator. 'We saw that we could negotiate to the benefit of the community (San-*Hoodia* Trustee 3, personal communication, September 2006, Upington).' Considerable effort was also put into building the capacity of the team to negotiate and take effective decisions together. All meetings with the CSIR, for example, were attended by the entire San negotiating team, and were preceded by preparatory meetings, led by the legal adviser, to discuss the aims and objectives of each negotiating session and to divide up responsibilities. Two major workshops and a number of educational meetings were also held in the broader southern African San community to help build awareness and knowledge of the case and what it meant.

Whether this was sufficient is debatable. At a workshop of 40 San representatives from South Africa, Namibia and Botswana in 2006, 58% of participants did not believe that there had been adequate consultation and information flow about the negotiating process and benefit-sharing agreement ('those in the offices knew but they never informed the communities'). Access to information was a concern even among those negotiating the agreement. One San negotiator commented:

> [T]he most difficult thing was that the negotiations went very fast and we were not fully informed about what was going on; the process wasn't explained ... During the negotiations it was very hard for us to grasp what it was really about. I was still trying to figure out what was going on and then the negotiation was already done (San-*Hoodia* Trustee 3, September 2006, personal communication, Upington).

Although efforts were continuously made to slow down negotiations in order to improve information flow and increase awareness, there were also constraints in that the CSIR was paying for the process and WIMSA had limited funding and technical capacity. Roger Chennells observed:

> If I could have built up better NGO links and better funding arrangements and all of that it would have been made easier ... we did not consult enough, we didn't have enough time to talk ... because it was all was done in a rush. I was forced into a Western system that required very quick answers for the shareholders – Pfizer [a large pharmaceutical multinational interested in developing and marketing *Hoodia*] was not prepared to go ahead until there was clarity. So if this thing was going to be a success for us, we had to reach an agreement quite soon. Yet ... there was this whole world of people who had not had the opportunity to understand this collective thing. So more time and money and more support would have been good (Roger Chennells, September 2006, personal communication, Upington).

One of the critical decisions required during this period hinged upon the royalty percentages due to the San. Although the CSIR had shared much financial information relating to the case with the San, the San did not have full disclosure of all the financial information between the CSIR and its licensees. The considerable financial investment in the project by the CSIR further complicated matters, along with the fact that an additional two licensing agreements existed between the CSIR and the pharmaceutical companies Phytopharm and Pfizer (see Wynberg and Chennells, Chapter 6).

Based on a 'sense of what was achievable', the San set an initial proportion of 15% prior to the commencement of negotiations. When negotiations began, however, the San tabled a proposal of 10%, in response to the 4% offered by the CSIR. Anything beyond this, according to the CSIR negotiators, was a matter for their board. Noted one, 'We had a mandate from the Vice-President to negotiate up to a maximum of 10% milestones and 8% royalties but we still needed to subtract all our substantial research and development costs [so the actual percentages were less than this]' (Helena Heystek, October 2006, CSIR, personal communication, Pretoria).

The CSIR developed three different models to present these figures, and negotiators on the San team observed that 'when they [the CSIR] got up to 6% there was a very strong feeling that if we went for any more their directors were going to pull the plug on the entire negotiations'. A San representative remarked that 'the CSIR was a little too cautious, they said we were asking too much ... we eventually reached 5% and at the end of the day we got 6%, and then we also worked out the schedules and settled on an 8% milestone payment'(San-*Hoodia* Trustee 2, personal communication, September 2006, Upington).

Much of this toing and froing took place in internal meetings of the San, with the CSIR asked to leave the room to allow the San to discuss their positions. There was also heated debate among the San about the correct approach to adopt. One San negotiator remarked:

They [the CSIR] led and I said no, it is the wrong way to do things. They had to come up and we had to come down from 15% to 10%. [Our lawyer] did not agree and said don't push them because they put in a lot of money. I said no, they took a lot of things away from us, because knowledge is not cheap. I was not satisfied but it was my first time and we did not have a lot of negotiation skills (San-*Hoodia* Trustee 1, personal communication, September 2006, Upington).

Ultimately, negotiators understood the result to be finely balanced, with both parties 'equally unhappy', as one negotiator put it.

12.4.2 Establishment of the San-Hoodia Benefit-Sharing Trust

Separate discussions took place parallel to the negotiations. Over three successive annual general meetings of WIMSA, and at many other meetings in between, the San debated the way in which the proceeds of the benefit-sharing agreement would be shared. To a large extent this was informed by experiences of the San's land claim under South Africa's post-apartheid land restitution programme, which had led to conflict between San communities (Robins 2002), and Central American access and benefit-sharing agreements that had fallen apart because of disagreements as to participating communities and countries (Hayden 2003). In a bid to avoid similar controversies, the San decided in principle that their heritage was collectively owned, and that any benefits arising from its use were to be divided equally among countries with San populations.

This principle provided the foundation for the San-*Hoodia* Benefit-Sharing Trust (San-*Hoodia* Trust for short), established in 2004 as a mechanism to distribute benefits from the commercial development of *Hoodia* (see also Wynberg and Chennells, Chapter 6 and Chennells et al., Chapter 9). A San trustee commented:

[W]e discussed the money cake and how we would go about dividing it. We decided there would be exact division among the various San Councils from the different countries ... and the little that is left we would keep in the Trust for administration (San-*Hoodia* Trustee 4, September 2006, personal communication, Upington).

However, integrating the decision for collective ownership into the trust presented a problem. The CSIR were adamant that the body be exclusively South African and saw no place in it for Namibia and Botswana. They also insisted that the trust agreement be appended to the benefit-sharing agreement and, in what was perceived as a patronizing though understandable demand, that the terms of the trust be agreed to by the government. In the words of a CSIR negotiator, 'It was a precondition of the benefit-sharing agreement that the trust be formed in a way that was acceptable to the CSIR (Helena Heystek, October 2006, CSIR, personal communication, Pretoria). This caused immediate tensions, ultimately resolved through a democratic process. As a San negotiator put it:

It was our knowledge and therefore our money. We wanted to use the money the way we wanted to, but according to the trust deed we couldn't. We didn't want the document to state that the money needed to be used for a certain purpose. We just felt that it wasn't fair. It is our shoes but we cannot wear them where we want to. But we were only four [who opposed the decision] and the majority accepted it. Democracy can either build you up or break you down (San-*Hoodia* Trustee 3, September 2006, personal communication, Upington).

The next major decision concerned the make-up of the trust and its representation, a process that generated substantial debate. The CSIR and San negotiating teams decided to include nine trustees: one representative from the CSIR (required to have financial expertise and be Afrikaans-speaking because the San were more comfortable in that language than in English), one legal professional elected by WIMSA and seven San representatives: three from South Africa (representing the ≠Khomani, !Xun and Khwe communities respectively) and one each from Namibia, Botswana, Angola and WIMSA. The South African government, through the Department of Arts and Culture, was allocated a non-voting seat on the trust, although this was not taken up. No requirements for gender equity were imposed. An executive committee was established comprising a chair, vice-chair, secretary and treasurer, with the explicit brief to act only on decisions already mandated by the whole trust. Because of a law relating to the use of public money (the royalties), the CSIR insisted on an exclusively South African structure, with Namibia, Botswana and Angola nominally represented through South African San. The San considered this process far from ideal and accepted it only after substantial argument and resistance.

The selection process involved sending a letter to San communities, who then nominated people to serve on the trust. A workshop held in Upington, South Africa, brought San together from South Africa, Namibia and Botswana to discuss the process of nomination and election. The apparent simplicity of this approach belies its highly politicized nature, however. *Hoodia* promised to deliver significant financial benefits, and participation in the trust, although unpaid, offered numerous opportunities to develop networks, build capacity, secure political capital, and gain greater knowledge of *Hoodia*.

Not surprisingly, therefore, the initial selection of trustees was strongly contested, although it drew largely on those who had been involved in negotiations with the CSIR, plus South African representatives nominated by Botswana, Namibia and Angola. A trustee explained:

> I got onto the San Council during election at a community meeting, and I got onto the trust as a result of the San Council's and community's decision. This is because we were on the San Council and we were involved in the negotiations and therefore we knew what this was about. We can serve on the trust because we can offer expertise and knowledge gained from the negotiations (San-*Hoodia* Trustee 2, September 2006, personal communication, Upington).

One needs to see this in the context of the previously dispersed power relations among community members (see Table 12.1). The need to select representative decision-makers caused decision-making power to shift from its relatively egalitarian distribution to a Western model, according to which it is held by representatives. Disputes also unfolded at a later stage as leadership positions in particular communities were challenged. In the Khwe community, for example, the position of a trustee elected by the previous Khwe 'regime' was contested by the incumbent leader – to no avail, in the final outcome.

An important lesson to emerge from this process was the need for long-term trustees, and the necessity of divorcing the trust from the ongoing erratic, volatile and often transitory politics in San communities. One of the trustees put it this way:

We decided that we don't want trustees to chop and change as the community politics take place because it could become very vulnerable to a coup taking place among one or other community and then that community saying that well so and so, that trustee must now be deposed and we're putting this one in ... we need to ensure that this trust is kept stable and is not vulnerable to politics and personality disputes and all of those things that happen in the best of circles (Roger Chennells, September 2006, personal communication, Upington).

12.4.3 Receiving and Disbursing Funds

With the trust set up, crucial decisions had to be taken. How were early incoming funds from the benefit-sharing agreement to be divided up – who would be eligible, what policies and criteria would dictate use, and how would the money be distributed? These decisions would set important precedents for how it would be done in the future, once substantial amounts of money were being received. The intention, as articulated by one of the trustees, was that after a few years of implementation the principles would be so clear that the only debate would be 'how'. In contrast to common practice in Western governance models, few of these principles were articulated at the start: the process of developing criteria and principles was viewed as an evolving one.

San communities, it was decided, should work through their official organizational structures, stipulate their needs, and propose a business plan and budget for specific projects. This would then be channelled through the San council to the San-*Hoodia* Trust. As one trustee explained, 'We want the community to tell us what they want to do ... we will measure this against the trust deed which states what one can do with this money. If it doesn't comply with the trust deed then we tell them it can't be done' (San-*Hoodia* Trustee 3, September 2006, personal communication, Upington).

Implementing this proved more taxing. An amount of some 569,000 South African Rand (about US$70,000) was received from the CSIR, based on milestone payments received from Pfizer and Unilever (see Table 6.2 in Chapter 6) and

Table 12.2 Expenditure of Milestone Payments Received from the CSIR by the San-*Hoodia* Trust

Item	Amount (ZAR)
Donation to South African San Council	R200,000
Contribution towards *Hoodia* road show	R10 000
Donation to WIMSA	R60,000
Contribution towards Namibia, South Africa and Botswanan San Councils	R150,000
Bank charges	R5,000
Allowances and meeting expenses	R95,000
Audit fees	R20,000
Total	R540,000[a]

[a] The remaining funds which include interest on the initial R569,000 will likely be used for audit fees, allowances, and meeting expenses

decisions needed to be made urgently with regard to its distribution (see Table 12.2 below for a summary of how this money was allocated). Principles for benefit sharing that would bind the trust had been endorsed by the WIMSA annual general meeting in December 2003 (WIMSA 2004). These included a decision to distribute 75% of all trust income equally to the San councils of Namibia, Botswana and South Africa. The remaining 25% was to be allocated for Trust administration purposes, with preliminary discussions suggesting that 10% could be retained by the trust for internal and administration purposes, 10% allocated to WIMSA as an emergency reserve fund and 5% given to WIMSA to cover the administration of San networks. Aside from this formula, however, WIMSA resolved to reward the South African San Council with a once-off amount of R200,000 (US$25,000) for the work they had done in negotiating the benefit-sharing agreement. A trustee explained it this way:

> The communities decided that R200,000 would go to the South African San Council because they did the negotiations – to make their life easier ... and to do its work. Its activities are the only voice of the San within South Africa so it speaks for all the San here – !Xun, Khwe and ≠Khomani (San-*Hoodia* Trustee 3, September 2006, personal communication, Upington).

This amount was intended not only to strengthen the capacity of the council and support its running expenses, but also to spread knowledge about the benefit-sharing agreement and the opportunities it presented for San development.

The remaining funds were allocated to the three different countries, whose representatives were told that R70,000 (about US$8,750) had been earmarked for release upon receipt of a budget and plan.[3] The existence of a San council was also a prerequisite for the release of funding. This presented a particular challenge. Although there was a South African San Council, this was not the case in Namibia, Botswana and Angola. As one trustee explained:

> It is one of the things we have got stuck on ... what are we going to do to get money to the people, because the money can't stay with the trust forever. Only the South Africans have a San Council. The other San Councils [in Namibia and Botswana] are not yet fully stabilized. They still have disputes among them; they can't get their council members together and can't appoint Councils yet because of these disputes (San-*Hoodia* Trustee 3, September 2006, personal communication, Upington).

The isolation and remoteness of many San communities in these countries further compounded this situation.

A central issue that emerged was the way in which information about the *Hoodia* benefit-sharing agreements would be disseminated to San communities and, linked to this, how trustees would speak for the interests of the communities they represented. Because the trust to a large extent was meant to solicit and respond to proposals to spend income, information sharing was vital. The trust therefore decided unanimously to allocate R10,000 of the funds to support a 'roadshow' in Namibia and Botswana, with the twofold intention of introducing the trust and providing support for the establishment of local councils. In the words of one trustee:

---

[3] In May 2007 a decision was made by the Trust to reduce this amount to R50,000 (US$6,250) per country and to allocate R60,000 (US$7,500) to WIMSA to assist with financial difficulties.

We decided to take co-responsibility ... to support the establishment of councils in Namibia and Botswana. So we decided to undertake a tour to Namibia and Botswana, just to talk, personally visit the communities there, together with the South African San Council ... and to talk about the important process, because we realize now that the advantages must be shared with all, so it was very important for us to make that decision (San-*Hoodia* Trustee 5, September 2006, personal communication, Upington).

Perhaps the trust underestimated the difficulties of implementing this decision. Few of the trustees knew anything about the geography and location of San communities in these countries, and the logistics of ensuring input were sometimes overwhelming. Overall, however, it was considered a useful interim measure while San councils were still establishing themselves.

Information sharing was a lot more straightforward in South Africa because of the existence of the San Council. An identified need for community report-back, for example, led to the allocation of R5,000 (US$625) for each of the three South African San communities, released upon written request from the communal property association. In the !Xun and Khwe areas, communities were also kept informed by the representative trustee reporting back in local languages through the local radio station.

12.5 Challenges and Conclusion

The San-*Hoodia* case is one of the few benefit-sharing cases to have realized income for an indigenous community. The San are a particularly marginalized and impoverished group in southern Africa, yet the predicted amounts they will receive are considerable, potentially amounting to millions of dollars. Add to this that the community is spread across three different countries (South Africa, Namibia and Botswana) with smaller groups elsewhere (e.g. Angola) and it is clear that communal decision-making is a major challenge. The process is still embryonic, but some important lessons for decision-making have already emerged.

12.5.1 Tension of Time Frames

First of all, the rushed nature of negotiations has important implications for other communities engaged in developing benefit-sharing agreements: how can a balance be struck between community processes, which require capacity-building and awareness raising, and thus time, and the economic expectations of companies and shareholders, who require immediacy and certainty? It is here that the discrepancy between the models of democratic decision-making and traditional indigenous decision-making is most pronounced.

In Western countries, communal decision-making is facilitated through long-established groups with legal standing, agreed-upon operating mechanisms and representatives who are educated and conversant with their role. Representatives routinely make decisions without consulting those they represent. By contrast

(Table 12.1), decision-making in traditional indigenous communities often involves all community members, which is time-consuming and thus not feasible in negotiations with industry partners.

In the case of the San, a legally representative group was needed to negotiate with the CSIR, but such groups were (a) in their infancy (having been established partly to administer land claims for the ≠Khomani, !Xu and Khwe) and (b) without the time and resources to enable fast decision-making through majority voting by professional representatives rather than consensus building among unpaid community leaders. Forming such a body was a race against time, and it would seem that the decision-making abilities of the San were compromised by the need for urgent resolution on the part of the CSIR and its commercial partners. Pfizer was clearly anxious about the negotiations and did not want any negative publicity. The San in turn were under pressure to come to an agreement, largely dependent on the resources of the CSIR and were not sure how hard they could push without jeopardizing negotiations.

12.5.2 Lack of Adequate Resources for San Negotiating Team

The tensions about time frames were aggravated by the lack of financial resources to fund meetings, obtain additional advice and hone negotiating skills, all vital elements of effective decision-making in the San's circumstances. But where did responsibility lie for securing these? The CSIR, upon request from the San, invested in facilitating San representation and decision-making capability because it needed an agreement with the San, and getting them to the negotiating table was an essential prerequisite to that. However, one could not have expected a commercially motivated entity to invest freely large amounts of time and money in sustained capacity-building to enable the San to become equal negotiating partners. If the CSIR had done this, the extra time needed might have undermined its chances of a licensing agreement with Pfizer, a very attractive potential licensee. In addition, one might well ask whether capacity-building and education are not a responsibility of national governments. The National Environmental Management: Biodiversity Act (Act 10 of 2004), which was not in place at the time of negotiations, now locates such support firmly with the South African government, to ensure negotiations are on an equal footing when benefit-sharing agreements are negotiated.

12.5.3 Continued Lack of Resources for Trust

The lack of resources to educate the San negotiating team continues to be problematic, even after the conclusion of the negotiations. In fact, one could argue that the mammoth task of distributing funds fairly and in line with the trust deed is a much more difficult one than closing benefit-sharing negotiations. The San-*Hoodia* Trust is made up of strong individuals but it has not been able to develop the expertise required to take on responsibility for a large influx of funds. Capacity development is urgently needed, and it is critical that the criteria and procedures for fund distribution be established before expenditure begins, in order to facilitate effective and fair decision-making. Both are lacking at present: the criteria are very unclear and the absence of stable, well-prepared San institutions in the three countries means that there is currently no coordinated effort to formulate procedures.

For example, one trustee remarked: 'There isn't a who gets and who not ... all the San are entitled to it (San-*Hoodia* Trustee 3, personal communication, September 2006, Upington).' To illustrate this point he asked how the trust would differentiate between a group of ten individuals wanting to open a supermarket in Platfontein, which would be profit-making but also serve the broader community, and a purely developmental initiative, such as a clinic. Another asked how one community requiring schools and clinics would be prioritized over another with similar needs: 'We must allow space for our brothers and sisters in other countries ... because the circumstances of other San people are very much the same' (San-*Hoodia* Trustee 5, September 2006, personal communication, Upington). And what of remote San communities in Namibia and Botswana that have little knowledge of the opportunities *Hoodia* brings? Could a system be designed to ensure their inclusion?

12.5.4 Clarity on Roles and Responsiblities

One factor hampering the formulation of procedures is the fact that the respective roles and responsibilities of the trust and the San Council are not clear. Figure 12.1 illustrates how the trust envisages decision-making and fund distribution, with San councils playing a facilitating and coordinating role, soliciting proposals from community-based organizations, reviewing funding applications to check that they are in accordance with the trust's overall aim and channelling these requests to the trust for final decision. The trust, in turn, evaluates proposals against its objectives and specific criteria, and then either grants the funding, asks for a revised proposal or rejects the application. WIMSA plays a supportive role for both the San Council and the trust ('the father of the trust', in the words of one trustee).

In many ways, however, this represents an ideal scenario. In reality, there is considerable confusion among trustees as to where the final decisions on applications are made. Some understand the San councils to hold this responsibility, while others see it as the role of the trust. Part of this uncertainty might be due to the fact that many leaders in the South African San Council are also trustees, and thus there is some blurring of institutional function – notwithstanding the benefit, of course, of continuity of knowledge between institutions. One trustee noted:

> [T]he San-*Hoodia* Trust supplies money to the various San Councils, and that is why members from communities who want to apply need to do so via the San Council, who will bring their applications to the meeting. The San-*Hoodia* Trust will sift through applications and indicate if these are not correct ... as they have a copy of the trust deed and know what it says and what types of applications are acceptable. Some of the people on the San-*Hoodia* Trust are also members of the San Council, and that is why we do not have problems in receiving the applications ... but the final decision will be by the San-*Hoodia* Trust (San-*Hoodia* Trustee 4, September 2006, personal communication, Upington).

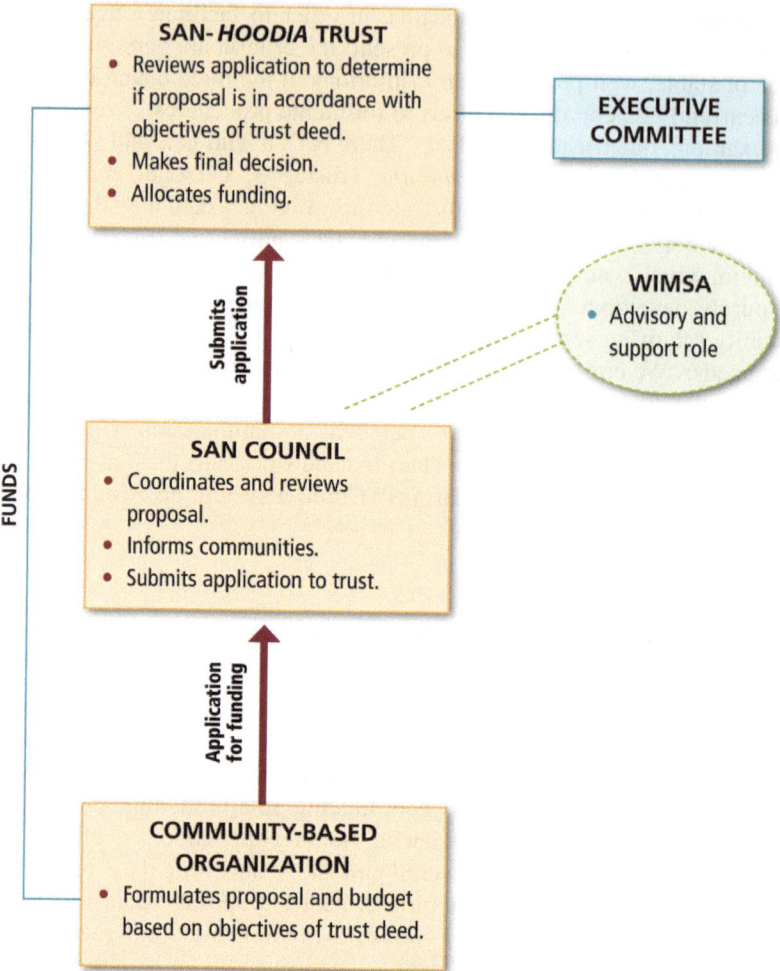

Fig. 12.1 Decision-Making in the Allocation of Funds from the San-*Hoodia* Trust

Giving a San council the role of coordinating proposals, monitoring implementation and ensuring report-back with regard to expenditure could also change the nature of the council fundamentally – from a political structure to something more like a nongovernmental organization (NGO) or service organization, with paid employees. In time, therefore, the trust might well take on more of these responsibilities.

12.5.5 Lack of Success in Local Community Governance

When considering the roles and responsibilities of the trust and the councils and their possible shift to NGO-type governance, it would be helpful if one could learn from earlier successful attempts at governance at community level. However, the instability of local community organizations, such as the communal property associations (CPAs), is cause for concern, with many fraught by suspicion and conflict. One trustee reflected:

> With the ≠Khomani we have had three CPA committees – three. Not one of them performed their duties as they should have, because it was such a hassle. They had to do pension applications, manage the land ... in other words they were like an overloaded *bakkie*[4] riding around in the desert getting stuck in the sand dunes all the time. After ten years the San realized they had used the wrong vehicle (San-*Hoodia* Trustee 2, September 2006, personal communication, Upington).

Another trustee was more positive: 'I'm not saying that there will not be any conflicts, but we will stand strong, those of us working with the money so that we do not spend it recklessly' (San-*Hoodia* Trustee 1, September 2006, personal communication, Upington). But the point remains that the dearth of successful examples of local community governance poses a challenge to the distribution of incoming funds and related activities.

12.5.6 Cross-Country Cooperation

Those representing San in other countries face an especially difficult challenge. While WIMSA offers a useful network for information dissemination and coordination, it is virtually impossible in practice to implement benefit sharing in another country without a receiving organization there that has legal status, management capacity and community support. Even once such an organization is established, major implementation challenges remain. Community organizations and San councils in all three countries are notoriously undercapacitated and require substantial support to develop proposals, manage finances and implement projects.

The significant benefit streams predicted in the future will have a profound impact on the trust, which will have to 'professionalize' itself rapidly into a body equipped to deal with complex reporting, monitoring and financial accountability. Creative solutions may be needed: possibly the employment of staff, or the use of existing NGOs or other service structures identified for this role.

A related challenge is the increasing demand by San representatives to be paid for the time they spend at committee meetings. One trustee explained it like this:

> The people who work on the committee are unemployed and don't have an income. How are they accommodated? People need to feel that they do work, they represent people, and they have worth. You can't always come with empty hands and go away with empty hands because later you may start feeling that you are wasting your time (San-*Hoodia* Trustee 4, September 2006, personal communication, Upington).

Steps to address some of these issues were taken at a 2006 workshop at the Molopo Lodge, near South Africa's border with Botswana, attended by San participants from

---

[4] A South African term for a pick-up truck.

> **Box 12.2 The Molopo (San-!Khoba) Declaration**
>
> Fifty San delegates, from Botswana, Namibia and South Africa, met at Molopo Lodge on the 20 and 21 September 2006, in order to discuss the *Hoodia* case, fair benefit sharing, and San structures.
>
> The following consensus statement was unanimously adopted by the delegates.
>
> - All San structures should include and respect the San traditional values of fair sharing, consensus decision-making, and respect for culture.
> - San structures must actively strive to ensure that a clear majority of funds received should reach and benefit San communities.
> - Administrative costs at all institutions (which include rentals, administrative salaries, communications and related costs) should be kept to a bare minimum.
> - On a projected annual income of R1 million, the approved percentages were determined as 80% for San projects and 20% for administrative expenses.
> - Corruption in any form is totally unacceptable. Professional management of funds, transparency and accountability will be demanded and expected from all San organizations.
> - Priorities are and will be different in Botswana, Namibia and South Africa. San councils are thus required to consult extensively in order to establish the most important priorities in their countries.
> - Projects that are environmentally sustainable, economically viable, and that benefit many San, will be encouraged.

South Africa, Namibia and Botswana. The resulting declaration, reproduced in Box 12.2, describes points of consensus from the meeting and highlights the shared vision that San communities have for ensuring that the benefits from *Hoodia* lead to tangible, effective and equitable community development. This path will undoubtedly be thorny. A member of the trust summarized the prospects eloquently:

> As we all know, there are always problems in the world of money. So we do anticipate obstacles. We know people will fight and there will be people that say, 'we are more, we are a bigger country, we need to get more.' We just expect it, it will happen. But the trust, as I know it, is strong enough to tell people how things work. We cannot allow a [bad] precedent to be set (San-*Hoodia* Trustee 2, September 2006, personal communication, Upington).

The declaration, agreed upon by 50 San leaders, rests upon firm principles that show foresight, integrity and a strong wish to serve their communities. The real test of the principles will only come in the future. As the broader framework of the Convention on Biological Diversity aims to restore justice for traditional knowledge holders (Schroeder, Chapter 2), one hopes that the benefits the San derive from sharing their knowledge will be distributed equitably and to the benefit of all.

12.6 Appendix: Methods and Approach

12.6.1 *Planning and organization*

Semi-structured interviews and focus groups were the main data-gathering tools for the research. They were conducted in parallel with the organization of a workshop in South Africa's Northern Cape province, bringing together San representatives from the Kalahari regions of Namibia, South Africa and Botswana. The apparent simplicity of these arrangements belies the logistical and financial complexity of assembling 50 participants from extremely remote parts of a region poorly connected by roads and public transport. Many participants had never travelled out of their countries and most were not organized through representative structures. Some 5 months were thus spent in a series of planning meetings to set up a process for the selection of participants, to organize their transport and accommodation, to establish relationships with NGOs and others who could support participants, and to resolve the way in which participants could receive financial support for their costs.

12.6.1.1 Selection of Workshop Participants

The process of selecting participants presented particular challenges. Although a San Council existed in South Africa through which nominations could be made, there were no formal San structures in Namibia and Botswana. Moreover, the leadership challenges described in earlier parts of this book (e.g. Chennells et al., Chapter 9) required invitations to go beyond the San councils to include other San opinion makers. Invitations were therefore issued to chairpersons of existing or interim San councils in each country, requesting the participation of ten San representatives, four serving on the country's San council and six others shown to have an interest in San matters or serving in San community organizations. A number of support NGOs were also invited to participate in deliberations.

In South Africa, the South African San Institute (SASI), a San support organization familiar with the areas where San live, was asked to assist with the distribution of invitations to San who could not be reached telephonically or by e-mail. In Namibia and Botswana, WIMSA was asked to perform a similar function.

Despite the lengthy preparations, the process of selection was fraught. Consensus was not easy to reach because of the remote location of many San and the difficulties of consultation. In Namibia and Botswana in particular, the lack of a formal San council significantly hindered the process. In Namibia, an initial list of participants was changed no fewer than five times due to concerns that some groups were over- or under-represented. For both Namibia and Botswana, a final list of participants was submitted just days before the workshop, creating enormous logistical problems.

12.6.1.2 Logistics

The Molopo Lodge in the South African Kalahari was chosen as the workshop venue because of its relatively central location. Transport logistics were coordinated by SASI. Participants travelled by bus or minibus taxi from various locations to the designated meeting places – Windhoek in Namibia and Ghanzi in Botswana – from where they made the journey to the Molopo Lodge. In many cases participants travelled for 2 or 3 days to reach the venue, and the early closing time of the Namibian-South African border post led to the entire Namibian delegation, including children, having to overnight in the bus! South African San were in a better position, generally travelling only 1 day in rented cars to reach the venue.

12.6.1.3 Costs

All travel and subsistence costs were covered by the project, but it was difficult to get cash to participants. Their remote locations precluded the use of bank transfers, and the amounts involved were too large for post office money orders. Eventually, money was transferred to the bank accounts of certain individuals who had to distribute it fairly and accountably. Nonetheless, conflicts arose over the allocation of daily allowances, the balancing of accounts and, in some cases, the lodging of inflated claims. These were resolved after much consultation and deliberation.

Overall, the workshop costs amounted to some R272,000 (US$34,000): R56,000 (US$7,000) for San travel and subsistence, R102,000 (US$12,750) for the accommodation, venue and food, R92,000 (US$11,500) for the project team's international and national travel costs and R22,000 (US$2750) for miscellaneous expenses (e.g. filming the event, see DVD included in this book).

12.6.2 Semi-Structured Interviews with the San-Hoodia Trust

Two days before the workshop, interviews were conducted with members of the San-*Hoodia* Trust, which holds responsibility for distributing benefits among San organizations, and other respondents. The one-on-one interviews were approximately 90 min long and used semi-structured questionnaires as the primary method of data collection. This enabled consistency across interviews, but also allowed space for questions to be posed spontaneously by the interviewer where required.

Interviews were held with all but one of the San-*Hoodia* Trust's nine members and with representatives of the CSIR who had been involved in negotiating the benefit-sharing agreement. All respondents were conversant in English and/or Afrikaans and chose the one they preferred for the interview. All interviews were conducted by the same person (Rachel Wynberg), with the assistance of an Afrikaans translator (Samantha Williams) where necessary. At four interviews one of the other authors of this chapter (Doris Schroeder) was also present. Where permission was given, interviews were audiotaped and later transcribed and translated. Written notes were taken when respondents were not willing to be taped, and were later typed up and transcribed. Respondents were given the assurance that interview material would not be passed on to third parties without their express approval.

Letters had been written to respondents several months earlier, introducing the objectives of the research project and requesting an interview. This followed permission from WIMSA and the trust for the study to proceed (see Preface to the book). Because the trust members are geographically dispersed, in some cases living 1,000 km apart, and often lack access to communication, SASI was asked to help contact trust members and facilitate their travel to a central place (Upington in the Northern Cape) for the interviews. Those interviewed were under no obligation to answer questions and were assured of anonymity.

An emphasis was placed on creating a non-threatening and empathetic interview environment and on encouraging an open and trusting attitude between respondents and interviewers. Respondents were asked introductory questions about the history of their involvement in the case and how they had become involved. This was followed by questions on the negotiation of the benefit-sharing agreement, the establishment of the trust and the way in which trustees made decisions. Concluding questions focused on future developments, in an attempt to elucidate future strategies and key problems. Respondents were also encouraged to put questions to the interviewer.

12.6.3 *Analysis*

Each interview transcript was read through thoroughly, and accompanying notes and diagrams were made to highlight key themes. Once all of this material had been assembled, the information contained in the transcripts was categorized and coded, based on similarity of theme. Information drawn from the different interviews was then clustered into a number of key themes, with phrases and quotations highlighted to illustrate emerging themes. This was expanded and verified by way of secondary data sources and publications such as the minutes of trust meetings, reports and letters.

12.6.4 *Focus Groups*

In addition to the interviews, focus groups were held as part of the Molopo Lodge workshop to assist with the development of ground rules for distributing benefits from the commercial development of *Hoodia* to the San. Three focus groups were created, representing San participants from South Africa, Namibia and Botswana respectively. These groups were asked: 'What would you do if your San council received R1 million?' Scripts were developed to guide focus group discussions and each group was asked to elect a facilitator and a rapporteur. Eight questions were asked:

1. Who will you consult with to spend the money?
2. How will you ensure that all San are represented in your consultation?
3. How will you decide if it is a good project and should be funded?
4. How much money can be spent on administration (rent, travel, meetings, etc.)?
5. How will everyone know who can apply and how and for what?
6. How will you prevent fraud and corruption?
7. List the types of projects that will receive priority.
8. What San-based values should be included in your decision-making?

Groups were asked to report back with their responses to each question, and answers were tabulated for circulation and confirmation.

References

Barnard, A., & Taylor, M. (2002). The complexities of association and assimilation: an ethnographic overview. In S. Kent (Ed.), *Ethnicity, hunter-gatherers, and the 'other': association or assimilation in Africa*. Washington, DC: Smithsonian Institution Press.

Biesele, M. (1978). Sapience and scarce resources: communication systems of the 'Kung and other foragers'. *Social Science Information, 17*, 921–947.

CIPR (2002). *Integrating intellectual property rights and development policy*. Report of the Commission on Intellectual Property Rights, London. www.iprcommission.org/graphic/documents/final_report.htm. Accessed 10 April 2006.

Craig, G. (2004). *The media, politics and public life*. Sydney: Allen & Unwin.

Crucible II Group (2001). *Seeding solutions, volume 2: options for national laws governing control over genetic resources and biological innovations*. International Plant Genetic Resources Institute, Rome; Dag Hammarskjöld Foundation, Uppsala; International Development Research Centre, Ottawa, ON.

Ekpere, J. A. (2001). *The OAU's model law: the protection of the rights of local communities, farmers and breeders, and for the regulation of access to biological resources, an explanatory booklet*. Lagos, Nigeria: Organisation of African Unity, Scientific, Technical and Research Commission.

ETC Group (2001). New enclosures: alternative mechanisms to enhance corporate monopoly and bioserfdom in the 21st century. *Communique*, issue 73. www.etcgroup.org/upload/publication/230/01/newenclosuresfinal.pdf. Accessed 11 October 2008.

GRAIN (2001). No patents on rice! No patents on life! Statement from peoples' movements and NGOs across Asia, revised August. www.grain.org/briefings/?id=172. Accessed 11 October 2008.

Guenther, M. (2002). Independence, resistance, accommodation, persistence: hunter-gatherers and agropastoralists in the Ghanzi veld, early 1800s to mid-1900s. In S. Kent (Ed.), *Ethnicity, hunter-gatherers, and the 'other': association or assimilation in Africa*. Washington, DC: Smithsonian Institution Press.

Hayden, C. (2003). *When nature goes public: the making and unmaking of bioprospecting in Mexico*. Princeton, NJ: Princeton University Press.

Huntington, S. P. (1996). *The clash of civilizations and the remaking of world order*. New York: Simon & Schuster.

Lee, R. (2003). *The Dobe Ju/'hoansi*. Toronto, ON (first published in 1983): Thomson Learning.

Locke, J. (1960). *Two treatises of government*. Cambridge: Cambridge University Press.

Mill, J. S. (1910). *Utilitarianism, liberty and representative government*. London: J. M. Dent & Sons.

Mitchell, J. (1987). Women and equality. In A. Phillips (Ed.), *Feminism and equality*. Oxford: Basil Blackwell.

Plato. (1935). *The Republic, vol II*. London: William Heinemann.

Robins, S. (2002). NGOs, "bushmen", and double vision: the ≠Khomani San land claim and the cultural politics of "community" and "development" in the Kalahari. In T. A. Benjaminsen, B. Cousins, & L. Thompson (Eds.), *Contested resources: challenges to the governance of natural resources in South Africa*. Programme for Land and Agrarian Studies, School of Government, University of the Western Cape, Cape Town.

Rousseau, J.-J. (1973). *The social contract and discourses*. London: J. M. Dent & Sons.

Sen, A. (1999). Democracy as a universal value. *Journal of Democracy, 10*(3), 3–17.

Silberbauer, G. (1982). Political process in G/wi bands. In E. Leacock & R. Lee (Eds.), *Politics and history in band societies*. Cambridge: Cambridge University Press.

Then, C. (2004). The true cost of gene patents: the economic and social consequences of patenting genes and living organisms. Greenpeace, Hamburg. http://weblog.greenpeace.org/ge/archives/1Study_True_Costs_Gene_Patents.pdf. Accessed 11 October 2008.

Ury, W. (1990). Dispute resolution notes from the Kalahari. *Negotiation Journal, 6*(3), 229–238.

Ury, W. (1995). Conflict resolution among the Bushmen: lessons in dispute system design. *Negotiation Journal, 11*(4), 379–389.

Vierich, H. (1982). Adaptive flexibility in a multi-ethnic setting: the Basarwa of the southern Kalahari. In E. Leacock & R. Lee (Eds.), *Politics and history in band societies*. Cambridge: Cambridge University Press.

WIMSA. (2004). *WIMSA Annual Report, April 2003 to March 2004*. Windhoek, Namibia: Working Group for Indigenous Minorities in Southern Africa.

WIPO. (2001). *Intellectual Property Needs and Expectations of Traditional Knowledge Holders: WIPO Report on Fact-Finding Missions on Intellectual Property and Traditional Knowledge (1998–1999)*. Geneva: World Intellectual Property Organisation.

Part III
Reflections

Chapter 13
The Role of Scientists and the State in Benefit Sharing: Comparing Institutional Support for the San and Kani

Sachin Chaturvedi

Abstract This chapter compares access and benefit-sharing (ABS) arrangements among San and Kani communities and in the process identifies prerequisites for effective ABS.

The experience of the Kani suggests that a strong institutional framework with committed staff is essential for the success of any ABS arrangement. It is also important that those managing the system be well aware of the predicaments and apprehensions of indigenous communities vis-à-vis formal research and development processes and administrative systems. The cultural ethos and values of indigenous communities may be very different from those in a more formal regime.

Both San and Kani cases have their own strengths and weaknesses. The Kani case is strong in terms of its institutional support, with the Tropical Botanic Garden and Research Institute (TBGRI) providing the skills required for the cultivation and harvesting of plants, the adoption of good manufacturing practices by the manufacturing firm, and the protection of intellectual property through a patent (Chaturvedi 2007). The TBGRI also employed a lawyer for the community, who drafted the trust deed and facilitated the opening of an account.

In this regard, the San had no institutional support from the South African Council for Scientific and Industrial Research, against which they actually had to assert their rights, nor did they receive any support from related government departments. The San were supported by a non-governmental organization, the South African San Institute, which had assisted the regional San organization WIMSA (Working Group of Indigenous Minorities in Southern Africa) since 1996, and which provided a lawyer and other forms of development assistance. This could ensure better returns for the San.

The two cases bring out different scenarios in which institutional structures or the lack thereof may emerge as barriers to the transfer of indigenous knowledge.

S. Chaturvedi
Research and Information System for Developing Countries (RIS), Core IV-B, Fourth Floor, India Habitat Centre, Lodhi Road, New Delhi-110 003, India
e-mail: sachin@ris.org.in

Keywords access and benefit sharing • India • Jeevani • Kani indigenous community • research • scientific institutions • traditional knowledge

13.1 Introduction

In modern economics, knowledge is considered a panacea for all the ills of industrialization and economic growth. A strategy increasingly favoured by economic development practitioners is building programmes to strengthen knowledge-generating capacities. However, it seems the linkage between indigenous knowledge systems (IKS) and the market is not as straightforward as is the case with knowledge emanating from formal research and development institutions. There are several scenarios in which the structures of institutions (or lack of them) may emerge as barriers to the transfer of indigenous knowledge. Additionally, if indigenous knowledge is used productively, it may not lead to the most equitable distribution of benefits.

This chapter compares the access and benefit-sharing (ABS) experiences of the San with those in the Kani case. There are several similarities in the development experiences of San communities in South Africa and the Kani from Kerala in India, with some major differences which eventually influenced the growth trajectory of ABS arrangements in those communities. These differences also seem to determine the future growth prospects.

The main aim of this chapter is to draw attention to the need for stable and simple institutions which inspire confidence in indigenous knowledge holders, so as to facilitate equitable conduct in the economy and establish trust. If agents in formal institutions conduct themselves in a selfless manner, they may help to establish an atmosphere which is not self-seeking and corrupt. The formal arrangements of modern societies, which are highly complex and rely on legal support, are completely alien to indigenous knowledge holders.

In such a situation, formal institutions and people, particularly scientists linked to these institutions, have an important role to play. They may help create better opportunities for indigenous knowledge holders by making systems more predictable and appropriate to the needs of indigenous communities. Industrial structures may also be organized in such a way that they help move indigenous knowledge holders up the value chain. Such structures may eventually unleash the enterprise hidden in a community, enabling its members to concentrate on their own aspirations and capabilities without having to wait for development assistance from government or other such agencies. For example, San organizations have worked on differentiating their products and identifying specific niche markets, while the Kani have remained confined to one product without envisaging other avenues for product development.

A brief overview of the Kani case is followed by a detailed comparison of the San and Kani cases in Section 13.3. The last section draws conclusions.

13.2 Brief Overview of Kani Case

The Kani ABS experience began in April 1987, when a scientist from the All India Coordinated Research Project on Ethnobiology (AICRPE) (see Section 13.3.1) arrived in the forests of the Agasthyar hills in southern India to seek permission from the Mottu Kani (head of the Kani tribe) to launch an expedition into the forests. Within the first few days, the scientists realized that the Kanis who were accompanying the team as guides did not feel as tired and fatigued as the scientists. The scientists found that the fruit the tribal group members were chewing had imparted this vitality and rejuvenation. Members of the Kani tribe considered this sacred knowledge and did not want to disclose it to others. However, after much persuasion the members of the Kani tribe agreed to share details about the plant with the scientists.

The plant was identified as *Trichopus zeylanicus subsp. travancoricus*, which the Kanis describe as Arogyappacha (meaning 'source of evergreen health'). In November 1996 the formulation and technology for production of the drug were transferred by the Tropical Botanic Garden and Research Institute (TBGRI) to the Coimbatore-based Arya Vaidya Pharmacy (AVP), one of the largest Ayurvedic manufacturing companies in India, against a licence fee of Rs.1 million (approximately US$25,000) and royalties of 2% at ex-factory sale. The TBGRI proposed to share the licence fee and royalty with the Kanis on a 1:1 ratio. The Kerala Kani Samudaya Kshema Trust[1] was registered in November 1997, to regulate and direct the inflow of money (Equator Initiative, 2002). The product developed by TBGRI was called Jeevani.

The product is a polyherbal drug[2] in a granular form. The members of the Kani tribe were actually chewing the fruit of the Arogyappacha plant, but since the fruit is available in limited quantities, the TBGRI team scientifically validated all parts of the plant for possible leads, including the roots and leaves. Eventually they found leaves with the necessary chemical and pharmacological properties. The final product includes three other medicinal plants apart from Arogyappacha (*Trichopus zeylanicus subsp. travancoricus*) or Jeevani, namely *Withania somnifera* (Ashwagandha), *Piper longum* and *Evolvulus alsinoides*.

The ABS arrangement, established through a trust fully owned by the community, entered a second phase in 2006 when the TBGRI invited the Kani tribes on board and constituted a Business Management Committee (BMC). The BMC decided to set minimum conditions for the ABS arrangement. It suggested the licence fee be doubled to Rs.2.1 million (US$52,000) and that the royalty payment also be doubled to 4% (*The Hindu*, 2006).

---

[1] *Samudaya* means 'community' and *kshema* means 'welfare'.
[2] Polyherbal drugs are plant-based drugs with constituents from more than one plant.

13.3 Key Issues for Comparison

As discussed in Chapter 6 by Wynberg and Chennells, the San were historically a nomadic group of hunters and gatherers, but currently reside in conditions of abject poverty in rural areas. This precisely mirrors the Kani experience. The Kani tribe is a small, previously nomadic, but now settled community of almost 25,000 members, based in the Agasthyar hills in southern India. Under the modern administrative system in India, this tribal group is spread over six gram panchayats.[3] The requirements imposed on the tribal communities by the Forest Department have increased over the years, and this has adversely affected their ability to make decisions. For instance, the individual areas which the Kanis now occupy are on long-term lease from the Forest Department. Their choices for cultivation thus depend on a list of non-timber forest products issued and amended from time to time by the Forest Department (Gupta 2004).

Various institutional structures have influenced the different outcomes in the San and Kani cases. This is of particular significance as in both communities there is a decline in traditional societal arrangements. Among the San, their former egalitarian and consensus-based hunter-gatherer lifestyles have had to adapt to rapid sedentarization. Similarly among the Kanis, over the years legal requirements for tribal communities have adversely affected their ability to make decisions and live as they were accustomed to living. The consequence of this has been the decline and eventual collapse of institutions which used to deal with community matters. For instance, in the Kani tribe the customary rights to transfer and use certain traditional medicinal knowledge were held by tribal healers called Plathis. This practice has almost disappeared, as most of the tribe are now settled and are no longer supported by the earlier social security system provided by the community, in which Plathis played a key role.

13.3.1 Sensitivity of Scientists

In benefit-sharing cases the individual commitment of scientists and their sensitivity towards indigenous knowledge holders seem far more important than any existing rule, regulation or guideline. The fact that the South African Council for Scientific and Industrial Research (CSIR) worked on the molecular structure of *Hoodia* from 1986 to 1995, a period during which there was much debate about indigenous rights at the international level, eventually culminating in the Convention on Biological Diversity (CBD) in 1992, is a major factor distinguishing the San case from the Kani one. The South African CSIR did not consider the option of prior informed

---

[3] The gram panchayat is the smallest administrative unit at village level in India. The six panchayats belonging to the Kani are Amboori, Kuttichal, Vithura, Peringamala, Kulathupuzha and Aryankavu.

consent arrangements or engage the community in the eventual outcome of the product. As a result, the San were excluded from all stages of product licensing, development, and final production and marketing. It required substantive intervention from outsiders to persuade the South African CSIR to consider the 'existence' and knowledge of the San and enter into negotiations with them.

By contrast, in the Kani case the scientists from the TBGRI were already aware of the historical and commercial importance of IKS. They were part of national efforts to conserve medicinal plants and related indigenous knowledge that were being undertaken as part of the AICRPE. The necessary prior informed consent was obtained by the AICRPE team when they first arrived in the forests of the Agasthyar hills in southern India to seek permission from the Mottu Kani to launch an expedition into the forests.[4] There is a practice among the Kerala tribes that any outsider is first supposed to meet the tribal chief before entering their settlements. Accordingly, Adichan Kani, the head, deputed a team of three Kanis to accompany the expedition as guides. The full team, led by the chief coordinator of AICRPE, Dr. P. Pushpangadan, arrived in the forests in December 1987.

It was also significant that Dr. Pushpangadan, as head of the AICRPE, was closely involved in national and global discussions on the protection of IKS at the time. The mandate of the AICRPE and the background of the Regional Research Laboratory (RRL) in Jammu[5] had framed his approach in such a way that he was actively involved at international fora for the cause. He provided inputs for the Declaration of Belém of 1988, which eventually led to the establishment of the International Society of Ethnobiology, which later provided key inputs for the contents of the CBD text. This declaration[6] explicitly recognized that indigenous peoples have been stewards of much of the world's genetic resources and that biological diversity would decrease significantly if knowledge underlying the resource management practices of the world's indigenous peoples depreciated due to the forces of rapid social change in the societies in which this knowledge was reposited.

As a result of these special circumstances, the scientists in the Indian team forfeited their own share and gave it to the indigenous knowledge holders. Once the transfer of technology and production was finalized, other issues related to the modalities for transfer emerged. The director of the TBGRI proposed that the proceeds be shared with the tribal community, which, to the executive committee of the TBGRI at the time, was an alien concept and without precedent. The executive

---

[4] The project coordinator, Dr. S. Rajasekharan, met the Mottu Kani from the Chonampara tribal settlement in Kootur, in the Thiruvananthapuram district, in April 1987 (personal communication with Dr, S. Rajasekharan).

[5] A constituent national biological research institute under the umbrella of India's CSIR, the RRL was initially called a drug research laboratory. It was established in November 1941 by Colonel Sir Ram Nath Chopra to gainfully exploit the biodiversity of the Himalayas. The Indian CSIR took it over in 1957 and made it part of the network of regional laboratories, and it became RRL. In December 2006, the Indian CSIR renamed RRL Jammu the Indian Institute of Integrative Medicine, or IIIM.

[6] See http://ise.arts.ubc.ca/global_coalition/declaration.php (accessed 15 April 2008).

committee decided to follow the Indian CSIR's model of benefit sharing, according to which 60% would go to the scientists responsible for product development and 40% to the institutions. At the meeting of the executive committee of the TBGRI in September 1995, it was resolved that the proceeds would be shared on a fifty–fifty basis. The scientists chose to forego their share in favour of the tribal community. The fact that, by then, India had signed the CBD and that articles 8(j) and 15.7 (UNEP 1992) were directly applicable helped the director pursue this case with the executive committee. As a result, the arrangement was worked out at 1:1 – that is, 50% to the tribal community and 50% to the institute. After the TBGRI decision was made, the institute approached the community and discussed the plan. Apart from the three guides who had initially accompanied the scientists, ten more members of the community were invited to be present.

13.3.2 Role of the State

In each of the two cases, the state has played (or failed to play) its role in very different ways and at different times. In the case of the South African CSIR, the South African government and the Department of Science and Technology did not act until Biowatch South Africa and Action Aid had intensively intervened, attracting huge international media attention. It was only after these developments that the South African CSIR and the San entered into negotiations. For the San, matters were complicated as the community also had members residing in the neighbouring countries of Namibia and Botswana, with whom the South African CSIR had no mandate to negotiate. As a result, as Chennells et al. describe in Chapter 9, the Working Group of Indigenous Minorities in Southern Africa (WIMSA) mandated the South African San Council to work on its behalf.

In the Kani case the Kerala state government was involved right from the beginning. The case was the centre of attention of the political system, particularly because India had just introduced a process of liberalization and opened markets to foreign investment, so some domestic entities felt threatened. These apprehensions, raising the wider spectre of biopiracy, meant a major political alarm was raised by the opposition parties.

The negotiation process at the TBGRI has been intense and very interesting. The Kanis initially participated in an informal manner, more as bystanders, but entered the process in a formal way in the second phase in 2004. However, the negotiations within the TBGRI reveal more about how the actual ABS regime emerged. When the product was first developed, the TBGRI invited companies to bid for the product's commercial production. AVP was shortlisted for production of the drug after it agreed to establish a good manufacturing practice (GMP) facility according to World Health Organization standards. It was decided to sign the agreement in the presence of the chairman of the governing body of the TBGRI (who is *ex officio* the Chief Minister of Kerala State) on 22 July 1995. However, the Chief Minister did not witness the signature on that day, as a letter written to him by the then leader of the opposition, Mr. V.S. Achuthanandan, (the current Chief Minister) argued that the lump sum amount offered by the private company was inadequate and that public limited companies owned by the government should be given priority over private companies (*The Hindu*, 1995).

The TBGRI appointed a committee of scientists to look into both points. The committee found that there was no GMP-standard production and marketing capacity in either Kerala State Drugs and Pharmaceuticals Ltd., based in Alappuzha, or the Pharmaceutical Corporation (I.M.) Kerala Ltd. (Oushadhi), located in Trichur, the two major public sector organizations. Neither agency was producing or marketing herbal drugs. With regard to the lump sum licence fee of Rs.1 million (approximately US$25,000), it was noted that no institute had earned any better amount on any herbal product. It was pointed out that the Central Drug Research Institute at Lucknow, an institute under the Indian CSIR, had earned only Rs.0.5 million (approximately US$12,500) through its memory-enhancing drug *Bacopa moneri* (brahmi).

At a subsequent meeting on 20 October 1995, the agreement with the AVP was cleared by the governing body (*The Hindu*, 1995), empowering the TBGRI, on behalf of the Kerala government, to proceed with negotiations with the AVP. The agreement proposed that after 7 years the AVP would have no right over the drug and that the TBGRI would be free to negotiate with any other company.

As mentioned earlier, in this phase the Kanis had no formal presence in the process, though informally the two guides who had accompanied the original expedition remained involved.

13.3.3 Institutional Framework

There are striking similarities and differences between the San and Kani experiences when it comes to the institutional arrangements. In both cases the key stakeholders decided to put in place very similar institutional structures. The San-*Hoodia* Benefit-Sharing Trust was established in February 2005 and the Kerala Kani Samudaya Kshema Trust was registered in November 1997. The difference is that the South African CSIR is represented on the board of the San-*Hoodia* Trust, while the TBGRI has remained separate from the activities of the Kani Trust.

The objective of the San-*Hoodia* Trust is to raise the standards of living and well-being of the San peoples of southern Africa. At a meeting in Upington in 2003, the San-*Hoodia* Trust decided to distribute 75% of all trust income between the San-*Hoodia* councils of Namibia, Botswana and South Africa. The trust would retain 10% for internal administration expenses and legal support mechanisms, while 10% would go into an emergency fund and 5% to WIMSA-related expenses. The trust decided not to pay money to individuals. This is very different from the Kani Trust, which is authorized to spend money for schemes not only for group benefit, but also for individual benefit – for instance, supporting women in distress. The provisions of the trust also facilitate the conservation of biodiversity and the adoption of best practices for plant collection. Both trusts have provisions in this regard.

Kani trust members have the freedom to identify their priority areas. This ability of trust members to distribute funds according to their own priorities, limited only by the trust's legal mandate, is an important part of the case. Some decisions have been questionable, from an outsider's perspective. For instance, presenting the three Kani guides with a total of Rs.50,000 (approximately US$1250) was believed to be out of proportion to their relatively low levels of effort. The knowledge the three tribe members shared belonged to the whole community. Sharing traditional knowledge, particularly related to plants, is considered a 'sin' in the community. Therefore the honouring of the three members who divulged knowledge with considerable monetary benefit was highly questionable. However, it seems that the community is gradually becoming more sensitive to its priorities and able to articulate its expenditure accordingly.

In the Kani case the emergence of institutional arrangements was complicated, but viable mechanisms were established. The executive committee of the TBGRI initially suggested transferring the money to the Scheduled Caste Development Department and the Scheduled Tribe Development Department of the Kerala government. The Kanis vehemently opposed the idea. However, the TBGRI was reluctant to transfer the money directly to the Kanis, because of serious levels of alcohol abuse in the community. The director of the TBGRI then contacted the leading experts in traditional knowledge. Some of them, including Professor Anil Gupta, proposed the idea of a trust for the tribal community (Gupta 2004). The TBGRI also used the expertise of other individuals and social workers to educate the Kani people in organizing themselves to form a society or trust.[7] Eventually, the TBGRI engaged the services of Advocate Kariyam B. Vijayakumar, who developed the trust deed. According to the Indian Societies Registration Act of 1860,[8] a trust can have either six or nine members. In this case the advocate suggested nine members. Accordingly, an amount of US$11,000 (50% license fee and 50% royalties) was transferred to the registered trust (No 109/97) on 22 February 1999. Provision was made for elections every third year.

In February 1999 a bank account was opened for the trust at the Union Bank of India in Kuttichal, the nearest town to the tribal settlement, exclusively under the control of the office bearers of the trust. The first funds were scheduled to arrive within a year. After transfer of the technology for manufacturing Jeevani to the AVP, in 1996, the TBGRI earned US $50,000. Half the license fee and half the royalties from sales went to the Kani tribes. It is interesting to note that following the transfer of money to the trust, the first meeting of the trust was not held until 19 March 1999. This, one could argue, shows that the trust was not well prepared for the arrival of the funds. At the meeting, it was decided that the three Kanis who had passed on the information to the scientists would be rewarded with cash prizes (Khwaja 2001).

One particularly interesting feature of the Kani case is that in the second phase of the ABS agreement, the TBGRI rendered the process more democratic and transparent. This phase also formalized the presence of the trust representatives in the

---

[7] Mr. P.R.J. Pardeep was one of the leading figures in this group.
[8] www.orissagov.nic.in/p&c/ngo/societies%20registration%20act.pdf.

Table 13.1 Comparison of San and Kani ABS Agreements Between Stakeholders

	San agreement, 2003	Kani first agreement, 1996	Kani second agreement, 2006
Parties	The South African CSIR and the South African San Council	The TBGRI and the AVP	Kanis, the TBGRI and the AVP
Entry into force	Entered into force in March 2003	Entered into force on 10 November 1996	Yet to be implemented
Validity	Valid for the royalty period of patents, being 20 years from 1996, or as long as South African CSIR receives financial benefits from patents	Valid for a period of 7 years	Valid for a period of 7 years
Licence fee or milestone payment	Milestone payment of R560,000 ($95,000)	Licence fee of Rs.1,000,000 ($25,000)	License fee Rs.2,000,000 ($50,000)

new negotiation process. In 2004, the new director at the TBGRI constituted a BMC, as noted above, with a membership of seven: two members of its faculty, three outside experts and two representatives of the Kani Trust. The role of the BMC was to negotiate fresh bids with companies interested in the commercial production of the drug. The BMC placed advertisements in leading newspapers, on the basis of which they received a number of proposals. As the table shows, the BMC decided to set minimum conditions for the ABS arrangement. It suggested the licence fee be doubled to US$52,000 and that the royalty payment also be doubled to 4% (*The Hindu*, 2006) (Table 13.1).

13.4 Concluding Remarks

It is clear from this discussion that the successful implementation of ABS regimes requires certain preconditions. These include robust and strong institutional frameworks with sensitive and committed staff. It is essential that the people managing these institutions be well aware of the predicaments and apprehensions of indigenous communities vis-à-vis formal research and development processes and administrative systems.

The San and Kani cases clearly demonstrate that the absence of these prerequisites may adversely affect the outcome of the ABS regime. Legal dispensations in several countries have exhibited only token consideration for indigenous communities and apportioned only limited forest and land rights to tribes and other traditional communities that have been living in forests and other areas for generations. Such limitations have adversely affected the economic well-being of these communities. It is therefore not very surprising that both San and Kani people are living in abject poverty, although they are long-term recipients of development assistance.

Both cases have their strengths and weaknesses. The Kani case is strong in terms of its institutional support, and the TBGRI has always provided the skills required for the cultivation and harvesting of plants, the adoption of GMP by the manufacturing firm and the protection of intellectual property through a patent. The TBGRI also employed a lawyer for the community, who drafted the trust deed and facilitated the opening of an account. In this regard, the San had no institutional support from the South African CSIR, against whom they actually had to assert their rights, nor did they receive any support at all from related government departments. The San were supported by an NGO, the South African San Institute, which had assisted the regional San organization WIMSA since 1996, and which provided a lawyer and other forms of development assistance.

Because the San case became an internationally known case, and because the San negotiations were based upon a patent with high commercial value, the strong externally assisted intervention could ensure far better returns on indigenous knowledge to the San community than were enjoyed by the Kani people. As is clear from the earlier discussion, the Kani case was settled between two mutually agreeing parties, so whatever returns were offered by one were readily accepted by the other. The San agreement was the result of negotiations in which the San had external assistance. Since there were no outsiders involved in the Kani negotiation, the valuation of Jeevani overlooked the potential global market for the product. Had that market been taken into account, the price settled on for the Kani community's product might well have been just as good as that negotiated for the San.

References

Chaturvedi, S. (2007). Kani case: a report for GenBenefit. RIS, New Delhi. www.ris.org.in/Kani_Case.pdf. Accessed 15 April 2008.

Equator Initiative (2002). The Innovative Partnership Awards for sustainable development in tropical ecosystems. Equator Prize Announcement, UNDP, New York.

Gupta, A. K. (2004). *WIPO-UNEP Study on the role of intellectual property rights in the sharing of benefits arising from the use of biological resources and associated traditional knowledge.* Study No. 4, WIPO and UNEP, Geneva.

Khwaja, R. H. (2001). Report to CoP-MoP 4 from National Focal Point. Ministry of Environment and Forests, Government of India, New Delhi.

The Hindu (1995). TBGRI Pact with Private Firms Put Off, 25 July.

The Hindu (2006). Tribals to Benefit from Renewal of Licence from Herbal Drug, 28 March.

UNEP (1992). Convention on Biological Diversity. http://www.cbd.int/doc/legal/cbd-un-en.pdf. Accessed 15 April 2008.

Chapter 14
The Law is not Enough: Protecting Indigenous Peoples' Rights Against Mining Interests in the Philippines

Rosa Cordillera A. Castillo and Fatima Alvarez-Castillo

Abstract Increased mining activities in indigenous peoples' lands in the Philippines have brought to the fore questionable free and prior informed consent processes despite the legal protection of indigenous peoples' rights to autonomous decision-making regarding the use of their lands and resources. Inadequacies in the implementation of the law, as well as the complicity of state agencies in circumventing its requirements, are among the major causes of this problem. This reflects the distribution of resources and power in a country where indigenous peoples are among the most marginalized in terms of influencing policy decisions and implementation. Given this situation, Philippine indigenous peoples and advocates have resorted to direct political action to assert their right to autonomous decision-making over their lands.

The ways in which indigenous peoples wage their struggle for respect of their rights are often influenced by contextual specificities. This explains the differences in the methods of political action between the San's assertion of their right to a fair distribution of benefits from the use of their plant genetic resources on the one hand, and the Philippine indigenous peoples' struggle with the mining industry on the other. Nonetheless, similar insights can be drawn from both cases, such as the importance of collective, participatory action and the delicate roles that civil society advocates can play. In the ultimate analysis, advocates for indigenous peoples' rights should learn to take a supportive role enabling indigenous peoples to speak with their own voice and actualize their autonomy.

Keywords Decision-making • free and prior informed consent • indigenous peoples • mining • participatory action • Philippines

---

R.C.A. Castillo (✉)
Department of Behavioural Sciences, University of the Philippines-Manila, Manila, Philippines
e-mail: rosacordillera@yahoo.com

F. Alvarez-Castillo
Department of Social Sciences, University of the Philippines-Manila, Manila, Philippines
e-mail: fatima.castillo@up.edu.ph

Violations of the rights of indigenous peoples continue in the age of the modern state. In the Philippines, these violations are often linked to large-scale development projects like dams, logging and mining in indigenous peoples' lands. Such development projects are accompanied by numerous cases of fraudulently obtained 'free and prior informed consent' (FPIC), militarization, displacement and killing of indigenous peoples and their advocates (Tebtebba et al. 2006). These occur despite the recognition of the rights of indigenous peoples in both the Constitution and national laws, including the right to give or withhold FPIC. The law's implementation leaves much to be desired.

Like the San, Philippine indigenous peoples have existed for a long time outside the consciousness of society at large and have been virtually forgotten by the state. Very little has been done to improve their lives, resulting in high rates of poverty (by cash economy standards), illiteracy and poor health. The largest concentrations of indigenous peoples are in the most economically poorest regions of the country (Asian Development Bank 2002).

Over time indigenous peoples in the Philippines have been pushed by land-grabbing and development projects to the geographic margins of the country, which happen to be rich in mineral deposits. The irony today is that they are becoming increasingly visible as the subjects of news reports in the popular media precisely because the remote physical locations they inhabit (which signify their social and political marginalization) are attractive to the mining industry.

Increased interest in indigenous peoples' lands and the resources these contain (e.g. mineral resources and biodiversity) is being driven by a high demand internationally for mineral and biodiversity products – a demand fueled in part by the expanding markets for medicines, food and cosmetic products derived from plants that are in many instances found only in indigenous peoples' lands. This is clearly seen in the case of the San in southern Africa, whose knowledge of the appetite-suppressing properties of the *Hoodia* plant led to the plant's exploitation and appropriation and eventually to the patenting of these properties (initially without the consent of the San) (Wynberg and Chennells, Chapter 6).

The Philippine government's vigorous implementation of its export-oriented economic growth strategy for national development, primarily through the opening of the country's mineral resources to foreign investors, is a key factor in the increased incursion into indigenous peoples' lands.[1] This strategy is in line with the free trade policy of the World Trade Organization (WTO) and the World Bank via the liberalized global trade regime (LRC-KSK/FoE Philippines 2001). The Mining Act of 1995 is an implementation of the strategy. It allows transnational companies to own 100% of the mines, to move communities away from mine areas and to obtain complete water and timber rights over mineral-rich lands (Tartlet 2001).

This paper focuses on the problems experienced by indigenous peoples in the Philippines related to FPIC issues in mining projects. It delineates the socio-economic and political contexts and dynamics of these problems. It argues that in a situation

---

[1] NGOs and indigenous peoples' organizations call these kinds of development projects 'development aggression'.

of extreme power imbalances among the parties, where the state acts as an agent of corporate interests, the legal requirement for FPIC is hugely inadequate to protect the rights of indigenous peoples, safeguard their interests and minimize harm. FPIC thus is a political problem that reflects larger, structural power distributions, not only nationally but also globally. One way of addressing these problems is by political action.

A comparative analysis with the San-*Hoodia* case is made to draw attention to the ways in which context influences indigenous peoples' campaigning in defence of their rights.

14.1 Violation of Indigenous Peoples' Human Rights

Philippine society today is rich in examples of how human rights violations follow society's fault lines of inequity[2] in economic and political power. Those who bear the burden of such violations are impoverished, marginalized and powerless. Those who perpetuate such violations are rich and powerful. This is clearly seen in the experience of the Alangan community with mining, as described in Box 14.1.

Box 14.1 The Alangan People (Helle [2007])

The Alangan Mangyan is a tribal community[3] living in a forest on the island of Mindoro. Typical of societies that depend on the forest for their needs, the Alangan have evolved a way of life and belief system in which nature is respected and cared for. They believe that spirits dwell in nature, that the god Alulaba protects the rivers and the god Kapwambulod watches over the forest and its resources.

In the latter part of the 1990s, the mining company Mindex set up its operations in the forest. This alarmed the Alangan people. A tribal leader, Ramil Baldo, voiced his concern:

> When I saw that the mining company had started
> working in the forest, I became frightened. I felt the
> forest's own fear. This is not what our ancestors
> wanted for the forest ...The mining operations will just
> make rich people richer.

---

[2] The phrase 'social faultlines' referring to social inequities was used by Paul Farmer in his discussion of the distribution of the burden of infectious diseases in society and inequities in income and power (Farmer 1999).

[3] There are no tribes in the Philippines in the technical anthropological sense. However, most indigenous peoples have appropriated the term to refer to themselves.

> **Box 14.1 (continued)**
>
> When drilling was also done on the burial site of their ancestors, considered sacred ground by the people, they decided to oppose the project.
>
> On their informed consent, Ramil Baldo said: 'We did not know that it involved an agreement about mining operations. That we only found out afterwards. We object to such trickery and misleading'

14.1.1 Background to the Case

In March 1997, the regional office of the Department of Environment and Natural Resources – Mines and Geological Bureau gave Mindex Resources Development Inc. a permit to explore a 9,720 ha concession in Sablayan, Occidental Mindoro, for nickel and cobalt. Mindex is a public Norwegian company engaged in the exploration and extraction of mineral resources in several countries. In 1998, its subsidiary Aglubang Mining Corporation applied to the Philippine government for a mineral production sharing agreement[4] for the same area being explored by Mindex.

The mining area is located within the ancestral domain claim of the Alangan and Tadyawan indigenous Mangyan communities. It is also a watershed area. Granting a mining permit to Mindex/Aglubang implied displacing the Mangyan communities and destroying the environment through strip mining (ALAMIN n.d.).[5]

The Mangyan organizations Samahan ng Nagkakaisang Mangyan Alangan Inc. (Organization of United Mangyan Alangan Inc., or SANAMA) and Kaisahan Mangyan Tadyawan Inc (Organization of United Mangyan Tadyawan Inc., or KAMTI) opposed the project. At the same time, another Mangyan organization was formed, Lupaing Ninuno Kabilogan Mangyan (Ancestral Land of All Mangyan, or Kabilogan). Some Alangan leaders suspected that the emergence of this new Mangyan organization, which was sympathetic to the mining project, was the result of the machination of Mindex/Aglubang with help from the National Commission of Indigenous Peoples (NCIP) (personal interview with Alangan leaders, 26 May 2008).

In a meeting called by the regional office of the NCIP, without the presence of SANAMA and KAMTI, the members of Kabilogan were asked to raise their hands if they were in favour of the mining project, which they did. However, the officers of Kabilogan had been elected inside the Mindex compound, and only 14 individuals signed the memorandum of agreement. They were given watches and a monthly

---

[4] In the agreement, the Philippine government grants the contractor the right to mineral resources in the area, while the contractor provides the capital, infrastructure and personnel.

[5] Strip mining is very similar to open pit mining. It involves the removal of the topsoil to get to the subsoil resources. In Mindoro, Mindex/Aglubang will remove about 10 m of the topsoil, then transport the nickel-containing soil to a facility which will separate out the nickel.

allowance of 1,000 pesos (about US$24) by the company. The NCIP then issued a certificate to the company to operate (Nettleton et al. 2004). This was despite the following facts:

- The Mangyan do not traditionally vote by raising of hands.
- A majority of the people were opposed to the project.
- The Mining Act prohibits granting a permit to operate when there is a pending ancestral domain claim by the people.

When a national newspaper, the *Manila Bulletin*, reported that the group had given its support for the mine, the leader of Kabilogan clarified that their support of the mine was on condition that their ancestral domain rights be recognized first (Eraker 1999).

In February 2000, Crew Development of Canada bought up 97% of the shares of Mindex, which then changed its name to Crew Minerals Philippines Inc. In December 2000, the Department of Environment and Natural Resources granted the newly named company rights to the mineral resources in the area.

Alyansa Laban sa Mina (Alliance Against Mining, or ALAMIN), a broad coalition of the people of Mindoro who were opposed to the mine, was formed in May 1999. ALAMIN initiated huge public rallies against the mining project, filed formal protests at different levels of authorities and collected 25,000 signatures against Mindex mines. In July 2001, the Department of Environment and Natural Resources cancelled the permit issued to Crew Minerals. However, the company filed a petition for reversal of the cancellation. In March 2004, the government reversed its decision and allowed Crew to resume operations (Crew Gold Corporation 2004).

The Alangan village is now a heavily militarized area. (See also below for military operations in indigenous peoples' areas.)[6]

14.2 Open Season for Large-Scale Mining

The Philippines is among the world's biggest producers of copper, nickel, chrome, zinc, gold, and silver. The government estimates the country's metallic mineral reserves at about 7 billion metric tons, valued at between US$840 billion and US$1 trillion (Tujan and Guzman 2002).

The constitutionality of the Mining Act has been raised in the Philippine Supreme Court by indigenous peoples and their advocates. While the Constitution expressly prohibits wholly foreign-owned companies from controlling, managing or engaging in the exploitation of the country's natural resources, the Mining Act allows foreign mining companies to have full equity and control of mining projects in the country. The petitioners also raised concerns over the potential environmental

---

[6] Arne Isberg, Crew country manager, claimed that ALAMIN was connected to the New People's Army (a rebel group that has been waging a Maoist insurgency in the country since 1969). He requested the Armed Forces of the Philippines to provide security after a series of attacks on Crew's facilities by armed groups (Gariguez et al. 2005).

effects of large-scale mining activities (which the law favours over small-scale mining), based on environmental disasters in the past (Cruz 1999). In 2005, the Supreme Court ruled that the Mining Act was consistent with the Constitution.

A single company can be awarded thousands of hectares for exploration and/or exploitation. As a result, mining projects have encroached into 17 important biodiversity areas, 35 national conservation priority areas and 32 national integrated protected areas. If all applications were approved, 41% of the country's total land area would be covered by mining claims (Cruz 1999).

It is in this context that FPIC to projects is applied for in mining areas. Of the 119 certificates of compliance with the FPIC process issued by the NCIP between January 2004 and February 2008, 70 are for mining-related projects such as exploration and surveying (NCIP 2008). This shows the extent of mining applications in indigenous peoples' areas.

The mining industry has become notorious for environmental disasters and health hazards (Tujan and Guzman 2002). Dominated by foreign capital, the industry is mostly in the extractive stage (Tujan and Guzman 2002). Many companies do not put expensive health and safety standards in place despite legal requirements for safe procedures and infrastructure.

14.3 The Legal Framework of FPIC

Two pieces of domestic legislation are directly relevant to obtaining FPIC from indigenous peoples. They are the Mining Act of 1995 and the Indigenous Peoples Rights Act (IPRA).

Enacted on 29 October 1997,[7] the IPRA implements the constitutional provisions regarding the rights of indigenous cultural communities. The Constitution provides that the state 'recognizes and promotes the rights of indigenous cultural communities within the framework of national unity and development'. Furthermore, the state, 'subject to the provisions of this Constitution and national development policies and programs, shall protect the rights of indigenous cultural communities to their ancestral lands to ensure their economic, social and cultural well-being'.

Following the enactment of IPRA, the NCIP was formed, to be headed by seven commissioners from major groupings of indigenous peoples. The NCIP, with its regional offices, is the main implementing agency for the law.

---

[7] Mining companies vehemently opposed IPRA. They argued that it violated the Regalian Doctrine. They also claimed that foreign mining investors could be discouraged from investing despite the liberalized mining law (Cruz 1999). The Regalian Doctrine, also known as *'jura regalia'*, is a fiction of Spanish colonial law that has been said to apply to all Spanish colonial holdings. It refers to the feudal principle that private title to land must emanate, directly or indirectly, from the Spanish crown with the latter retaining the underlying title. Lands and resources not granted by the Crown remain part of the public domain over which none but the sovereign holds rights (MacKay 2004).

The IPRA has clear and specific provisions on the requirements of FPIC, defining it as

> the consensus of all the members of the indigenous peoples to be determined in accordance with their respective customary laws and practices, free from any external manipulation, interference, and coercion, and obtained after fully disclosing the intent and scope of the activity, in a language and process understandable to the community.

FPIC is required for the following: the exploration, development and use of natural resources; research and bioprospecting; displacement and relocation; archaeological explorations; policies affecting indigenous peoples; and the entry of military personnel.

Furthermore, it provides that concessions, licences and leases cannot be issued, renewed or granted by any government agency without the NCIP's certification that the areas involved do not include any ancestral domain.[8] It also gives indigenous peoples the right to stop or suspend projects that do not satisfy the consultation process required by the law.[9]

Despite the legal requirements, violations of the law on FPIC are rampant. A study of such violations (Tebtebba et al. 2006)[10] found the following in regard to disclosed information:

- Information is not provided in a language that community members can understand.
- Information provided relates only to the potential benefits, leaving out potential adverse impacts.

There are problems with the process as well. These include:

- Requirements for the consultation process and short time frames are not in accordance with customary indigenous peoples' practices.
- Customary practices on consensus building are not followed when people are made to vote.
- When the community tends to disapprove the project, fake tribal councils are set up with fake tribal leaders.
- The process often excludes women and youth.

Furthermore, there are no mechanisms to ensure compliance and accountability. For instance, there are no mechanisms for the revocation of FPIC and the project

---

[8] IPRA defines ancestral lands or domains to include concepts of territories that cover not only the physical environment but the total environment, including spiritual and cultural bonds to areas which the indigenous peoples possess, occupy and use and to which they have claims of ownership.

[9] The full text of IPRA is at www.ncip.gov.ph/downloads/philippines-ipra-1999-en.pdf.

[10] The Tebtebba Foundation, an NGO having special consultative status with the Economic and Social Council of the United Nations, with the Asian Indigenous Women's Network and the Cordillera People's Alliance, submitted a report based on their study on the experience of indigenous peoples with the FPIC provision of IPRA to the UN Permanent Forum on Indigenous Issues in 2006. See also www.tebtebba.org/tebtebba_files/ipr/ipr.html for more on Tebtebba's case studies on FPIC issues.

contract, even where it has been shown that consent was fraudulently obtained and violations of the contract have been committed (Tebtebba et al. 2006).

14.4 Resistance

A national movement of environmentalists, indigenous peoples' organizations and local church people has emerged to oppose the vigorous, full-scale drive of the national government to entice foreign investment in mining. As a result, as shown in the Mindex case, the government occasionally has had to revoke a licence, although only temporarily.

Many indigenous communities have broken their silence and are beginning to speak for themselves – and to the powerful through various forms of direct political action. For example, they have formed their own organizations and are gaining the capacity to document violations, to engage with the media, external supporters and politicians, to organize public demonstrations and, when necessary, to form and sustain physical blockades to prevent mining equipment from entering their territory. Often this entails using their bodies to block the entry of bulldozers and other mining equipment. For example, in Kasibu, Nueva Vizcaya, independently and without any initial assistance from non-governmental organizations (NGOs), indigenous people formed human barricades and organized an around-the-clock watch to prevent the entry of mining equipment into their community (Galvez 2007).

Resistance can be extremely risky, however. In communities where strong opposition prevails there is a heavy military presence characterized by military operations, the setting up of military detachments and the recruitment of paramilitary forces. The militarization of indigenous peoples' community areas has resulted in various human rights violations such as the harassment and intimidation of indigenous peoples' leaders, illegal arrests and detention, threats and extrajudicial killings.[11] The military often conduct 'clearing' operations in communities to ensure the operation of the mines. Of the more than 800 victims of extrajudicial killings in the country since 2001, 18 were environmental advocates (Kalikasan-PNE 2006). Eleven of these victims (indigenous and non-indigenous persons) had actively protested against mining in their areas (Kalikasan-PNE 2006). Their cases remain unsolved (Bulatlat 2007).

---

[11] The UN Human Rights Council sent special rapporteur Philip Alston to the country in 2007 to investigate hundreds of reported extrajudicial killings and enforced disappearances. In his report, he concluded that the state military forces were involved but the government was in a state of almost total denial (Alston 2007). Also in 2002, Rodolfo Stavenhagen, UN special rapporteur on the situation of human rights and fundamental freedoms of indigenous people, arrived in the country to investigate the plight of indigenous peoples. He found out that many indigenous peoples, especially those who resisted projects deemed development aggression, were victims of state harassment (Bulatlat 2007; Tebtebba 2002).

14.5 Comparative Insights with the San-*Hoodia* Case

Essentially, both the San-*Hoodia* case and the numerous cases of violations of FPIC procedures in Philippine indigenous communities are issues of self-determination – that is, the right of indigenous peoples to control the direction and pace of development, and to practise their cultural, economic and political rights over their territories.

There are, however, important differences between their situations. In a valid FPIC agreement, the granting or denial of consent presupposes a discussion of potential benefits as well as risks or adverse impacts among the parties involved. It is in regard to risks or adverse impacts that FPIC issues in mining are fundamentally different from those in biopiracy. While both can be seen as forms of mining (i.e. the extraction of resources, physical or ideational) the scale of extraction and the damage to environment and cultural life are larger in mining for subsoil resources than in bioprospecting for plants. While the San case is an issue of the fair sharing of benefits from the exploitation of the *Hoodia* plant, the case of indigenous peoples in mining areas in the Philippines is an issue of cultural and physical survival. This is because mining not only involves a specific resource such as the *Hoodia* and the knowledge around its use, but implicates larger issues of land rights. Rights to land are directly related to the indigenous peoples' cultural, economic and political rights (see Vermeylen, Chapter 8).

Given the unique relationship that indigenous peoples have to their lands, the issue of fair benefit sharing and compensation in mining takes on dimensions other than those found in the San-*Hoodia* case. How could, for example, the loss of sacred ground to mining ever be compensated for? How could the sense of uprootedness and loss of identity be realistically compensated for? The loss of traditional relationships to land and of the traditional utilization of resources, with the eventual loss of indigenous knowledge and practices rooted in those lands, is not something an exchange value can easily be placed upon. The extent of such losses has been termed **ethnocide**, the destruction of a group's culture (IPHR Watch et al. 2006). Since the indigenous peoples' culture is rooted in their lands, how could fair benefit sharing ever be possible in cases involving potential ethnocide?

Given that the continued survival of entire indigenous communities is at stake in mining, the presumption of freedom to decide (to consent or to say no) is an urgent matter. While the San have experienced a more brutal history than the Philippine indigenous peoples, in that there is so far no evidence of large-scale genocidal huntings and killings of Philippine indigenous peoples in the colonial past, the political context in South Africa today is more encouraging of free and open negotiation among parties to a conflict than is the case in the Philippines. For example, while the South African Council for Scientific and Industrial Research (CSIR) erred originally in not obtaining the prior informed consent of the San, when challenged it eventually behaved with more respect for their rights. In the Philippines, besides instances of government agencies wilfully manipulating informed consent, the use of armed force and violence by the state does not provide the atmosphere necessary for fair negotiations. Militarization, on top of manipulation and deceit, prevents indigenous communities from exercising their right to free decision-making on matters important to their survival as peoples.

14.6 Contextualizing the FPIC Political Problematique

A primary function of the modern state is the distribution of public resources. There is an assumption in liberal political theory that the modern state, born out of a popular mandate, is rational and fair; that because it is detached from narrow vested interests, it adjudicates wisely on contending interests in society to promote the common good (Axford et al. 1997).[12]

In reality, government is often not detached from narrow, vested interests. Its policies are influenced by interest and pressure groups (Laski 1967). In the Philippines, the government's distribution of public goods takes place within a context of sharp power imbalances among groups that try to influence policies from both within and outside the country. The government itself is dominated by families from the local economic elite, many of whom have joint venture arrangements with foreign companies (Simbulan 2005).

FPIC can be viewed within the framework of distributing public resources. It is a tool, as defined in the IPRA, that is designed to protect indigenous peoples' rights to their ancestral lands and the resources these contain.

In the Philippines, the distribution of public resources follows the grid lines of power distribution and the fault lines of inequity, where the most powerful get the most and the powerless get the least.

What does FPIC within the framework of the distribution of public resources in a context such as the Philippines reveal? Theoretically (by law) it is a tool that indigenous peoples can use to protect their rights to their ancestral domains. In reality, however, indigenous peoples are powerless vis-à-vis very powerful forces allied to the state that are able to manipulate FPIC provisions to further their economic interests, as exemplified in the case of the Alangan Mangyan. It is in this context that the legal requirement of FPIC has become useless, simply because, in many instances, despite that requirement, the FPIC of indigenous peoples to mining activities is not being sought in accordance with the law. This implies that the law is being ignored by those who are supposed to abide by it. Metaphorically, therefore, the law on FPIC is a scarecrow that the birds of prey (i.e. the violators) ignore. This happens mainly because the state, which is supposed to execute the law, is an active party to its violation.

In this situation, where the indigenous peoples cannot rely on the state and the law to protect their rights to their ancestral domains against big mining interests, they have no option but to resort to direct political action, exemplified in public rallies to demonstrate their opposition and petition signing to put pressure on government officials to act more responsibly.

---

[12] For a critical and historical analysis of the state, see Hall (1994).

14.7 Common Lessons

14.7.1 Collective Action

Both the San-*Hoodia* and the Philippine mining cases illustrate the fundamental role of collective action. Because the San, including those in Botswana and Namibia, are able to organize themselves and make decisions through participatory mechanisms, the potential of dissension during negotiations, which could be exploited by unscrupulous parties, has been largely avoided. The strategy of divide and rule, which is successfully employed by mining companies and government agents in the Philippines when they pit the families of indigenous peoples against one another, indicates that some Philippine indigenous peoples' communities need to improve their efforts at collective action. The fact that there are more than a 100 indigenous peoples' ethnolinguistic groups in the country further complicates this matter.

Nonetheless, communities that have a history of organized resistance to development aggression are better able to prevent manipulation by mining companies. For example, the peoples of Cordillera were able to stop the World Bank-funded Chico Dam project in the early 1980s, during the Marcos dictatorship (CPA n.d.). These communities have continued to oppose mining projects. They are among the most organized indigenous communities in the country.

Both the San and Philippine cases also underscore the importance of advocates and of building advocacy networks. This is especially so because indigenous peoples are virtually powerless, without experience in dealing with dominant groups, formal or legal institutions and the mass media. As shown earlier, in the Philippines the indigenous peoples and their advocates often risk their lives when they oppose mining projects. Because of this, many community members eventually give their consent to such projects.

However, there are also many examples of resilient resistance by indigenous communities and their support groups. Unity and willpower within the community, solidarity and logistical support from NGOs and church groups, particularly in providing paralegal training and food – because lobbying, organizing forums, picketing and barricading take people away from their livelihood – sustained media coverage and international lobbying usually make the difference between successful and failed resistance. And such resistance might have to be kept up for years.

The story of the San is rich with insights on how relationships between indigenous peoples and their advocates evolve during struggle, and how roles and boundaries develop, hopefully towards the indigenous peoples taking on greater leadership roles while advocates assume supportive and facilitative roles. The evolving relationship in the San case has not been conflict-free (nor should one expect it to be), but these lessons will help Philippine indigenous peoples and advocates define how they should relate to one another as partners in their common struggle for the protection of indigenous peoples' rights.

14.7.2 Innovation in Democratic Work for Indigenous Peoples' Rights

Both the San and Philippine indigenous peoples have developed skills in building national and international networks of solidarity to bring their issues to a wider audience. In the Philippines, because the state is perceived as a party to the violation of FPIC, indigenous peoples and their advocates have sought help from other quarters. Activists write to the governments of the countries where foreign mining companies are based about violations committed by those companies. They also ask citizens of those countries to help give the issue a higher public profile and to put pressure on their governments to make the companies withdraw.

Some of these efforts to draw international attention have borne fruit. For example, persistent lobbying from indigenous peoples' groups and NGOs brought the United Nations special rapporteur on the situation of human rights and fundamental freedoms of indigenous people, Rodolfo Stavenhagen, to the Philippines in 2002 to investigate the plight of indigenous peoples (Tebtebba 2002). Public campaigns are becoming more comprehensive as NGOs protest not only against mining companies and the violation of FPIC provisions, but also against the neoliberal trade policies that fuel such violations. The discourse is thus global in scope.

14.8 Conclusion

We have shown how differences in the context and experiences of the San and the Philippine indigenous peoples differentiate their actions and options in asserting their right to give or withhold FPIC. In the Philippines, where the state fails to implement the law, conspires with big corporations to circumvent it or attacks those who try to use the law for their protection, more than the law is required to uphold this right. When the victims of fraudulent FPIC are impoverished and powerless, what is needed is direct political action such as organizational and advocacy actions.

In most cases, this is met with armed force by the government or by continued mining activities. That is not surprising. The fact that indigenous peoples continue to organize and use legal protest is significant because it could have more strategic impact in building a sense of agency and empowerment among indigenous peoples.

The San experience is instructive. FPIC can easily be bypassed. It may have been serendipitous that the failure of the CSIR to obtain the San's consent reached public attention. But of more significance is what the San did after learning that external groups would profit from their knowledge without a fair sharing of benefits with them.

There was a fortuitous coming together of circumstances: the San in South Africa had an existing lobbying organization working for their land rights which became the organizing platform in their struggle for fair benefit sharing from *Hoodia*; a legal advocate with whom they had won their land rights case was available and willing to assist them; San leaders were ready to take active leadership roles; and, eventually, civil society and the mass media came together to support them.

Central to both the San and Philippine experiences is the evolving relationship between indigenous peoples and their advocates. This relationship is formed 'on the job', without predefined structures, form, shape or content. Differentials in power and skills as well as differences in culture and background make this a complex terrain to navigate. Commitment to the principles of justice alone is not enough. Sensitivity to cultural, linguistic, gender and behavioural nuances and a readiness to listen and learn, especially among the advocates, might eventually prove the most important factors sustaining solidarity and challenging injustice.

References

ALAMIN (n.d.). *Case brief: the Mindoro nickel project*. Alyansa Laban sa Mina/Alliance Against Mining. www.mangyan.org/current/environment/casebrief-nickelproj.html. Accessed 1 February 2008.

Alston, P. Press Statement. Professor Phillip Alston, Special Rapporteur of the United Nations Human Rights Council on extrajudicial, summary or arbitrary executions, Manila, 21 February. http://www.inquirer.net/verbatim/philip-alston-statement02222007.pdf. Accessed 12 August 2008.

Asian Development Bank. (2002). *Indigenous peoples/ethnic minorities and poverty reduction – Philippines: Chapters 6 and 7*. http://www.adb.org/Documents/Reports/Indigenous_Peoples/PHI/chapter_6.pdf. Accessed 10 May 2008.

Axford, B., Browning, G. K., Huggins, R., Rosasmond, B., & Grant, A. (1997). *Politics: an introduction*. London: Routledge.

Bulatlat (2007). UN Rep says killings hurting Arroyo's credibility abroad, 11–17 February. www.bulatlat.com/news/7-2/7-2-un.htm. Accessed 1 March 2008.

CPA (n.d.). *A history of resistance: the Cordillera mass movement against the Chico dam and Cellophil Resources Corporation*. Cordillera Peoples Alliance. www.cpaphils.org/campaigns/A%20History%20of%20Resistance.rtf. Accessed 21 June 2008.

Crew Gold Corporation (2004). Mindoro Nickel Project MPSA reinstated, 24 March. www.minesandcommunities.org/Action/press300.htm. Accessed 1 March 2008.

Cruz, I. (1999). Setting the scene: an overview of the mining industry and regulatory climate in the Philippines. In *Minding mining! Lessons from the Philippines*, The Philippine International Forum. Philippine: Philippine-European Solidarity Centre (PESC-KSP). www.philsol.nl/pir/v2/RegClimate-99b.htm. Accessed 1 February 2008.

Eraker, H. (1999). Go home to Norway, Mindex! *Philippine International Review*, 2(1). www.philsol.nl/pir/v2/Mindex-99b.htm. Accessed 1 March 2008.

Farmer, P. (1999). *Infections and inequalities: the modern plagues*. Berkley, CA: University of California Press.

Galvez, J. K. (2007). Bloody mining conflict looms – NGO. *The Manila Times*, 17 August. www.manilatimes.net/national/2007/aug/17/yehey/prov/20070817pro6.html. Accessed 28 April 2008.

Gariguez, E., Sarmineto, J., & Aguilar, A. L. (2005). *Rapid field appraisal: Mindoro nickel project crew minerals (Philippines)–crew gold A-S Aglubang Mining Corporation, Victoria, Oriental Mindoro*. Calapan City, Mindoro: Peasant and Advocacy Network (PEASANT-NET).

Hall, A. J. (1994). *The state: critical concepts* (Vols. 1–3). London: Routledge.

Helle, K.-E. (2007). *The Alangans: forest tribe on the verge of mining*. www.justmake.no/kunder/norwatch/index.php?artikkelid=1649&back=1. Accessed 14 October 2007.

IPHR Watch, KAMP and NCCP (2006). *A primer on indigenous peoples and human rights*. Quezon City: Author.

Kalikasan-PNE (2006). *Environmental advocates from Kalikasan pay tribute to 18 slain colleagues*, Kalikasan-People's Network for the Environment, 31 October. www.arkibongbayan.org/2006-10Oct31-kalikasanBantayog/kalikasanvskillings.htm. Accessed 1 February 2008.

Laski, H. J. (1967). *A grammar of politics*. London: Allen & Unwin.
LRC-KSK/FoE Philippines (2001). Philippine case study. In *Globalization, trends, and the peoples of Southeast Asia and the Pacific: case studies from Australia, Bangladesh, Indonesia, Japan, and the Philippines*. Manila: Friends of the Earth International and Legal Rights & Natural Resources Center–Kasama sa Kalikasan.
MacKay, F. (2004). *Indigenous peoples' rights to lands, territories and resources: selected international and domestic legal considerations*. FAO Corporate Document Repository. www.fao.org/docrep/007/y5407t/y5407t0g.htm. Accessed 15 June 2008.
NCIP (2008). *List of issued compliance certificates as of February 2008 (Certificate of compliance to FPIC process and certification that the community has given its consent)*. Quezon City: Ancestral Domains Office, National Commission on Indigenous Peoples.
Nettleton, G., Whitmore, A., & Glennie, J. (2004). *Breaking promises, making profits: mining in the Philippines*. Manila: Christian Aid and PIPLinks.
Simbulan, Dante (2005). *The modern principalia: the historical evolution of the Philippine ruling oligarchy*. Quezon City: University of the Philippines Press.
Tartlet, R. K. (2001). The Cordillera people's alliance: mining and indigenous rights in the Luzon highlands. *Cultural Survival Quarterly, 25*(1). www.culturalsurvival.org/publications/csq/csq-article.cfm?id=645. Accessed 15 August 2006.
Tebtebba (2002). *Visit to the Philippines of Professor Rodolfo Stavenhagen, UN special rapporteur for the human rights of indigenous peoples*. Press release, Tebtebba Foundation, 11 December. www.tebtebba.org/tebtebba_files/ipr/stavenhagenpress.html. Accessed 10 March 2008.
Tebtebba, CPA and AIWN (2006). Recent experiences and recommendations on the concept and implementation of the principle of free, prior and informed consent. Tebtebba Foundation, Cordillera Peoples Alliance and Asian Indigenous Women's Network. http://www.sarpn.org.za/documents/d0002040/index.php. Accessed 12 August 2008.
Tujan, A., & Guzman, R. B. (2002). *Globalizing Philippine mining (2nd ed)*. Manila: Ibon Foundation Inc. Databank and Research.

Chapter 15
Benefit Sharing is No Solution to Development: Experiences from Mining on Aboriginal Land in Australia

Jon Altman

Abstract This chapter looks at what happens in Australia when indigenous people who are landowners need to negotiate with multinational corporations engaged in mineral exploration and production on their lands. Focusing on a number of significant benefit-sharing agreements, the chapter explores some of the broad fundamental tensions that arise when the interests of indigenous minorities in commercially valuable resources are belatedly recognized in post-colonial circumstances.

The chapter begins with a brief background of the situation of indigenous people in Australia. A synoptic historical and statutory overview of the relationship between miners and indigenous people, as mediated by the state, follows. Next is an analysis of five important issues that have arisen in Australia: How do relatively powerless and marginalized groups gain leverage for commercial negotiations? On what basis are benefit-sharing agreements made? To whom should payments made under benefit-sharing agreements be distributed? How should payments made under benefit-sharing agreements be utilized? Who should be responsible for decision-making?

The chapter identifies a range of emerging issues of equity and effectiveness that have created problems in the Australian situation. Elements of these problems resonate with the circumstances of the San and their negotiations over the utilization of *Hoodia*, and these are discussed.

Ultimately, there are no easy solutions to the development problems faced by indigenous peoples. While scarce capital generated by benefit-sharing agreements should help to ameliorate these problems, it is important to acknowledge that any one agreement will only provide a partial solution. Managing expectations while sustainably implementing agreements is clearly an emerging challenge. This is especially the case over the life cycle of a long-term agreement. Recognizing the inevitable challenges posed in agreement implementation should, at the very least,

J. Altman
Centre for Aboriginal Economic Policy Research (CAEPR), Hanna Neumann Building #21, The Australian National University, Canberra ACT 0200, Australia
e-mail: jon.altman@anu.edu.au

ensure early investment in capacity-building that might allow adaptive and informed management of agreement implementation.

Keywords Ancestral land rights • benefit sharing • Australian Aboriginals • managing expectations • mining

15.1 Introduction

In this chapter, I briefly explore the relationship between indigenous people, mining corporations and the state in liberal democratic Australia. In the last 30 years, with land rights and native title laws, indigenous people have significantly increased their land holdings to now cover over 20% of the continent, mostly in the remotest parts (Altman et al. 2007). Today, indigenous people number just over 500,000 or 2% of Australia's population.[1]

Australia is a rich First World nation, and mineral production and export, much from Aboriginal-owned land, are fundamental to its wealth. Increasingly, benefit-sharing agreements are completed between resource developers and indigenous parties – the Minerals Council of Australia (MCA 2006) recently estimated that there were over 300 agreements between mining companies and indigenous communities throughout Australia.[2] And yet, according to all available social indicators, indigenous people remain economically marginalized and socially disadvantaged (Altman et al. 2005). Clearly they do not share equitably in the mineral wealth that is extracted from their lands.

This chapter does not explore the political economy of such inequity in any great detail. Rather it focuses on some of the broad fundamental tensions that arise when the interests of indigenous minorities in commercially valuable resources are belatedly recognized in post-colonial circumstances. In the Australian case, the focus is on land rights and native title, and on negotiation rights in respect of minerals. This focus reflects the particularities of the Australian situation, in which almost all significant benefit-sharing agreements have occurred at the interface between miners and indigenous landowners. Paradoxically, this is not because indigenous landowners enjoy property rights in minerals (which they invariably do not), but because they do have variable rights to influence land access, and this provides the lever in commercial negotiations.

Interestingly, despite the scale of the indigenous estate, covering over 1.5 million square kilometres, and ongoing bioprospecting, there has been no significant discovery

---

[1] In Australia, there are two indigenous minorities, Aboriginal people and Torres Strait Islanders. It is current convention to refer to them as either 'Aboriginal people' and 'Torres Strait Islander Australians' or 'Indigenous Australians' with initial capital letters.

[2] In the past these agreements were referred to as 'mining agreements', but increasingly they are being referred to as 'benefit-sharing agreements' in accordance with global practice.

with commercial application to date and no major benefit-sharing agreements with respect to a plant species. The resources exploited in the Australian Aboriginal and San cases are thus fundamentally very different. In Australia, minerals are a non-renewable resource whose extraction invariably results in environmental damage to a landscape that, from an indigenous perspective, is imbued with religious significance, sentience and ancestral spirits. In the San case, *Hoodia* is a wild-harvested plant that can also be cultivated to ensure sustainability.

This chapter is structured as follows. I begin with a very brief background of the situation of indigenous people in Australia to provide some comparative context. Then I give a synoptic historical and statutory overview of the relationship between miners and indigenous people as mediated by the state. Next I address the following five important issues that have arisen in Australia:

1. How do relatively powerless and marginalized groups gain leverage for commercial negotiations?
2. On what basis are benefit-sharing agreements made?
3. To whom should payments made under benefit-sharing agreements be distributed?
4. How should payments made under benefit-sharing agreements be utilized?
5. Who should be responsible for decision-making?

Finally, in the last part of the chapter, I outline some issues that arise from the Australian situation that might be instructive for the San case addressed in this book.

It should be noted that my focus is on remote Australia and most specifically on the Northern Territory. This is primarily because most benefit-sharing agreements have been signed in relation to remote Australia. The Northern Territory is especially instructive for a variety of reasons, but mainly because it was here in the 1950s that the notion of benefit sharing between indigenous Australians and miners was first conceived and legally enshrined.

15.2 Brief Background on Indigenous People in Australia

In 1788, when the British colonists arrived at Sydney Cove, it is estimated that there were about 500,000 indigenous people in Australia speaking 200 languages and 600 dialects. Like the San they lived as hunter-gatherers, and the European Enlightenment took a dim view of this as an inferior, Hobbesian mode of living. Such views conveniently justified the wholesale and illegal alienation of land and resources.

Keen (2004) has carefully examined the early colonial and ethnographic records to highlight some commonalities across the continent in three broad areas – ecology, institutions and economy. He concludes (2004) that the fundamentals of material culture and technology were similar; kin relations extended to the whole social universe and structured social roles; cosmologies demonstrated similar relations between ancestors, the living and the dead; and everywhere the landscape was imbued with totemic significance. The all-encompassing Dreaming or Dreamtime, referring to stories and myths that have a primordial character, is conceived as the period in which the original totemic ancestors shaped the material landscape and left traces of themselves in it. The law produced by these ancestors incorporates the explicit rules and regulations that governed rights over land, water and resources. Anthropologist Stanner (1965) provided a sense of Aboriginal cosmology conceiving the land as animate or 'sentient'. People lived in a reciprocal relationship with the land: they nurtured the land through proper observance of ritual relations at sacred places, and the land in turn nurtured them through healthy reproduction of natural species essential for survival.

The colonization of Australia from 1788 saw indigenous hunter-gatherer societies displaced from their customary lands – at worst destroyed, at best radically transformed. The gradual European occupation of the Australian continent from 1788 saw the colonial state and settler capitalism ignore indigenous interests in land for nearly 200 years. In the process the advancing colonial frontier destroyed indigenous economies and societies and alienated land and resources, but in a highly variable manner contingent on the time when the colonial frontier arrived.

Since first contact the state has loomed large in the lives of indigenous Australians. The history of state policy is greatly complicated by the emergence in the nineteenth century of an Australia comprising six colonies, each of which developed its own policies for dealing with its indigenous inhabitants. Generally, special laws set indigenous Australians apart from other colonial citizens for their 'protection and preservation'. The purpose of this approach can be interpreted positively as a means to prepare Aborigines for future full citizenship or negatively as a way to 'smooth the pillow for a dying race' (Altman and Sanders 1991). Subsequently, when it became clear that the Aboriginal population was not disappearing, the policy changed to assimilation, the dominant paradigm from 1952 to 1972.

It is only since 1972 that the state's approaches to Australia's indigenous minority have undergone radical change, with self-determination becoming the central term of indigenous affairs policy and the failed official policy of forced assimilation being abandoned. Suddenly there was a rapid escalation in federal government involvement in indigenous affairs, including a dedicated government department, an elected indigenous representative organization, indigenous-specific programmes, the establishment of thousands of community-based organizations to administer programs locally and a bold initiative in the creation of laws to enshrine land rights for indigenous peoples. But the historical legacy has made it difficult to reverse the entrenched indigenous marginality that remains today.

In the 2006 census, just over 500,000 indigenous Australians were enumerated, with fewer than 30% resident in remote Australia. Both census-based and special survey social indicators indicate that according to most criteria, indigenous people are very badly off compared to other Australians. This situation has been evident in social indicators measuring relative health, education, housing and employment in every five-yearly census since 1971 (Altman et al. 2005), when indigenous people were first comprehensively included in this statistical instrument. Hunter (2006) has documented that more than 40% of indigenous people live in poverty irrespective of geographic location.

Yet the use of such statistics overlooks the enormous diversity in the contemporary situation of indigenous people, besides being Eurocentric and reflecting the values of the dominant society. In recent times, indigenous identity has re-emerged as strongly linked to land, language and distinct customary practice, but in only some situations have people been in a position to reclaim their ancestral lands, a crucial anchor for cultural revival. Likewise, only in a minority of situations do Aboriginal languages remain in use. The same is the case with continuities in customary practices such as religion and ritual and hunting and gathering for a livelihood.

15.3 Historical Genesis: Miners, Indigenous People and Benefit Sharing

The focus in this chapter is on benefit-sharing agreements. These are most evident between miners and indigenous people, mainly in remote regions where an estimated 120,000 indigenous people continue to live in approximately 1,200 small, discrete communities mainly on what is now Aboriginal-owned land.

Relations in Australia between miners and indigenous people have a relatively long and complex history. In effect, colonial law ignored indigenous interests in land, although there was some rare recognition prior to the era of land rights and native title from the 1970s. In the first half of the twentieth century, under the broad policy umbrella of protection and preservation, some reserves were gazetted for Aboriginal people, most extensively in the Northern Territory, which the Commonwealth of Australia[3] controlled from 1911. Miners were specifically excluded from reserves and people required permits to enter.

Even such exclusion was overlooked when commercial imperatives loomed, as in the 1930s at Warramanga in central Australia, where reserve boundaries and people were shifted when gold was discovered. Similar amendments to reservation conditions occurred in Cape York in Queensland in the 1950s when mining company Comalco discovered bauxite and, most famously, at Gove when another mining company, Nabalco, gained a licence to mine bauxite within the Arnhem Land reserve in 1968 (Altman 1983).[4]

It is frequently overlooked today that in Australia the notion of benefit sharing was first introduced by the conservative Minister for Territories, Paul Hasluck, in 1952, when post-war Australia was considering mining bauxite for 'strategic national purposes' (to produce aluminium to build warplanes) in north-east Arnhem

---

[3] In this chapter, 'Commonwealth' refers to the Commonwealth of Australia, not the Commonwealth of Nations.
[4] Much of this history is covered in Jon Altman, *Aborigines and mining royalties in the Northern Territory*, Australian Institute of Aboriginal Studies, Canberra, 1983.

Land, an area reserved for Aborigines. By now, after the native welfare conferences of 1951 and 1952, Commonwealth policy had shifted from protection and preservation to assimilation, and Hasluck wanted to see reserves used for Aboriginal economic benefit.

In an unprecedented move that still has implications today, Hasluck enshrined three broad measures in the law by amending the national Northern Territory (Administration) Act and the Northern Territory's Mining and Aboriginals Ordinances. First, while reserves were to be opened on a controlled basis for mining, any royalties paid by mining companies were to be earmarked for Aboriginal people. Second, a trust fund was to be created, named the Aboriginal Benefits Trust Fund, into which all royalties would be paid. Third, and perhaps most innovatively, in 1953 Hasluck determined that the statutory royalty paid on Aboriginal reserves would be double the normal rate stipulated in the Mining Ordinance. This was mainly intended as a means of discouraging marginal operations, as the 'national interest' had been the original reason for legal changes.

The Hasluck reforms had one major oversight: there was no requirement to share the benefits from mining with Aboriginal communities or groups adjacent to mines and there was certainly no reference to Aboriginal traditional owners of mine sites, as these people were not legally recognized. This partly reflected the fact that much of this statutory change was driven by a proposal to mine the Wessel Islands, a part of the Arnhem Land reserve that was deemed uninhabited, and it was not until a decade later, when mining at Gove began, that the interests of a directly socially impacted community, in this case the mission at Yirrkala, were considered.[5]

In 1968, the Nabalco mine was approved under a special ordinance that ignored all Hasluck's earlier progressive requirements about benefit sharing. This precipitated the famous legal case *Milirrpum and others v Nabalco and the Commonwealth*, the Gove land rights case that challenged the right of the Commonwealth to issue mining leases irrespective of whether traditional owners of the land consented and without the payment of any compensation. Another aspect of Hasluck's law was blatantly overlooked but not pursued in the Gove case: the payment of a double royalty was never considered in the sweetheart deal between the Commonwealth and Nabalco, although it had been paid at Groote Eylandt in the agreement struck in 1965. The Commonwealth was keen to see remote Arnhem Land commercially developed and believed that such development would be in the interests of Aboriginal people, irrespective of their perspectives and aspirations. The Aboriginal challenge in the Northern Territory Supreme Court was lost and mining at Gove proceeded.

This somewhat detailed historical introduction is provided for a number of reasons, and not just to demonstrate that history shows we do not learn from history. First, the issue of benefit sharing with Aboriginal interests has a far longer history than

---

[5] In 1965 an agreement was completed for manganese mining at Groote Eylandt, also within the Arnhem Land Aboriginal reserve. In this case a special royalty was paid by the miner, GEMCO, a subsidiary of BHP, to the Church Missionary Society, which held these monies in trust for Aboriginal people residing on Groote Eylandt (Altman 1983).

generally recognized in contemporary Australian policy discourse and discussions relating to the Convention on Biological Diversity (CBD). Second, the Gove case shows quite clearly that the state and mining interests colluded to override Aboriginal opposition. Here was a case of power asymmetry accentuated beyond any notion of ethical fairness, and ultimately the judiciary supported the state and miners. The negotiation playing field was non-existent in 1968: Aborigines were not even represented at the table. And third, despite good initial policy intent by Hasluck in the 1950s and apparently progressive laws, changes in policy thinking, commercial pressure and corporate advocacy demonstrated that statute law could always be overridden.

15.4 Indigenous Leverage in Mining Agreements

The first issue that I address is how relatively powerless and marginalized groups gain leverage for commercial negotiations. The answer in Australia is provided by the workings of Commonwealth and state land rights and native title laws, of which there have been many since the late 1960s. Rather than examine all of them, I focus on the two iconic events that stand out as institutions to reverse the processes that resulted in indigenous land alienation: a political commitment to land rights in 1972 for social justice and economic reasons that resulted in the Aboriginal Land Rights (Northern Territory) Act of 1976 (referred to in this chapter as the Land Rights Act), and the Mabo High Court legal judgment in 1992 that overturned the legal fiction of *terra nullius* and resulted in the passage of the Native Title Act of 1993.

In 1973 the government of Prime Minister Gough Whitlam instructed Mr Justice Edward Woodward to examine ways to provide both land and mineral rights to Aboriginal people in the Northern Territory. However, his commission of inquiry decided that Aboriginal people should not be vested with mineral rights, as this was deemed too radical a measure and was strongly opposed by the Australian Mining Industry Council. Instead, Woodward (1974) recommended that Aboriginal people be granted a right to veto mineral exploration and production on their lands: this is now termed 'free and prior informed consent'. Such a right-of-consent provision constituted a de facto property right in minerals because it could be traded away. Woodward was well aware of the Hasluck legacy and so recommended the retention of the double royalty requirement and of a trust fund arrangement for Northern Territory Aboriginal people, both of which were embodied in the creation of the statutory Aboriginals Benefit Trust Account (ABTA).

Woodward modified the Hasluck scheme in two important ways. First, the statutory royalty was regarded as a minimum that would be paid to the ABTA, but Aboriginal landowners would be at liberty to negotiate additional benefits above this minimum. Second, and more significantly, Woodward recommended that the statutory royalties be divided according to a formula: 30% would be paid as compensation to people affected by a mine; 40% would be paid to Aboriginal land councils, statutory authorities created to represent landowner interests, as an independent source of funding; and the remaining 30% would be retained by the ABTA and distributed more widely to, or for the benefit of, Aboriginal people throughout the Northern Territory. This scheme was by and large adopted in the Land Rights Act, which remained relatively unchanged until 2006, when the conservative government of John Howard made statutory amendments that reduced the independence of the Aboriginal land councils and enhanced the powers of the Minister for Indigenous Affairs. Nevertheless, under this system some significant mining agreements have been completed, with several 100 million dollars being paid into the ABTA to date.

The Native Title Act of 1993, which applies Australia-wide, not just to the Northern Territory, provides somewhat different negotiation leverage. Under this law, Aboriginal people determined to hold native title rights in land do not enjoy a right of veto, but a right to negotiate, which is without question a weaker form of property. Furthermore, since the Act was amended in 1998, this 'right to negotiate' has been limited to situations where people have 'exclusive possession' determinations and to a period of six months.[6] Importantly, though, the right to negotiate is also extended to native title claimants whose claim has passed a registration test administered by the National Native Title Tribunal and whose case for determination has not yet been heard by the Federal Court. Following amendments to the Native Title Act in 1998, the situations where the right to negotiate could be exercised were greatly reduced. The extraordinary procedural complexities embedded in the native title legal framework have resulted in growing recourse to negotiated and expedited agreements and arbitration. Ritter (2002) and Corbett and O'Faircheallaigh (2006) argue that new institutional arrangements are biased against indigenous interests and that forced arbitration in particular has favoured miners.

The native title framework is heavily skewed in favour of resource developers and lacks some of the important and progressive institutional arrangements introduced in the 1970s in earlier land rights laws. For example, there is no trust account arrangement and there is no earmarking of statutory royalties (or their equivalents) raised by the state for the benefit of Aboriginal people. Agreements that are concluded are strictly between mining companies and native title parties, although in fact the state has played a role in overseeing the making of the agreement. If agreements cannot be reached within the stipulated period of 6 months then mining can proceed and an arbitral process is instigated to determine compensation. This arbitral process cannot consider the value of minerals in its determination of compensation, a mechanism established to hasten agreement-making, again in favour of miners.

---

[6] Exclusive possession is the strongest form of native title possession and has similarities to inalienable freehold title. However, it is a misnomer as it does not exclude mining exploration and production. Lesser forms of native title that are only partial might see native title rights co-exist with the rights of other interests such as those of pastoralists. In such situations native title rights might be limited to a right to forage and perform ceremonies, and there may only be a requirement for miners to consult or notify native title groups.

Despite these hurdles, a large number of agreements have been completed in the native title era over the past 15 years, with many described in the comprehensive Agreement Treaties and Negotiated Settlements database.[7] Some major agreements have been signed, most notably at the Century zinc mine and the Comalco bauxite mine in Queensland, in the Pilbara 'iron province' and at the Argyle diamond mine. These agreements were comprehensive and included benefits such as financial payments, generally in the region of several million Australian dollars per annum, and also employment concessions, training and other non-monetary benefits. The willingness of multinational corporations to make such agreements mainly reflects the mining boom of the last decade or so and the high profitability of mining in Australia by global standards.[8] Whether such agreement-making has been equitable or made a real difference to Aboriginal socio-economic status is an issue to which I turn now.

15.5 Emerging Issues of Equity and Effectiveness

The development outcomes from engagement between multinational corporations and indigenous peoples in Australia have been highly variable, being dependent on many factors including regional histories of colonization, the nature of mines, the value of negotiated benefits packages and the forms of indigenous engagement with the mine economy. Most importantly, perhaps, notions of development are culturally constructed and are never easy to measure objectively. In the absence of comprehensive frameworks to monitor development outcomes independently, one is generally limited to either official statistics that have not been purpose-designed for measuring socioeconomic impacts or to case studies.

Statistical social indicators that formally measure health, education, income, employment and housing status have been applied to assess the impact of mining in eight remote regions with major mines identified and clustered together (Altman 2006). The analysis compares the socio-economic status of indigenous people in the mining aggregations with that of people in a number of other aggregations. A major qualification on this analysis is that it is crude, as in some cases other opportunities besides mining might be available. Nonetheless, mining does appear to make some difference, at least according to mainstream social indicators, although quite clearly the economic status of indigenous people in mining areas does not approach that of non-indigenous Australians at the national level. The information also suggests that there is enormous variation among mining regions.

---

[7] www.atns.net.au/ accessed 28 February 2008.

[8] For example, in terms of the major agreement for mining at Century, Aboriginal people receive payments of $A3 million per annum for 20 years, but in 1 year alone (2006–2007) the mining company Zinifex reported profits of over $A1 billion.

Arguably, it is impossible to adequately summarize the benefits and costs of a large number of diverse benefit-sharing agreements. This is partly because very few comprehensive assessments are transparently available. One notable exception is the Kakadu Region Social Impact Study (1997a, 1997b) completed over a decade ago.[9]

Even this comprehensive year-long study was ambiguous in its findings, showing that there had been positive effects from the comprehensive Ranger uranium mine agreement, as well as many negative effects from the arrival of a mining township and the swamping of the local Aboriginal population by miners (and by tourists to Kakadu National Park). The most negative impact was easier access to alcohol, which, combined with the additional income from mining, led to increased alcohol consumption, with a detrimental impact on health and an increase in mortality.

In Australia, benefit-sharing agreements have proven to be poor instruments for the delivery of development, no matter who defines development – the state, mining companies or intended Aboriginal beneficiaries. This is primarily due to the absence of a cogent policy framework to ensure that beneficial agreements generate positive outcomes. The following four broad issues that remain fundamentally unresolved exacerbate this situation.

15.5.1 On What Basis Are Benefit-Sharing Agreements Made?

The 50 year history of benefit sharing outlined above has seen a shift in purpose from Hasluck's vision that payments should be a financial pool for economic development to Woodward's model encompassing both compensatory payments and sharing mineral rent (profits). The latter view is the one that dominates today, but it is inherently problematic. On the one hand, if benefits are compensatory, then arguably they should be clearly earmarked to offset the negative impacts of mining, not to ensure development. On the other hand, if these payments are a form of rent-sharing, then it is unclear why indigenous people have not been vested with mineral rights or why there is no external accountability for how such payments are utilized.

These ambiguities generate two further issues. The first is the diverse forms that agreement payments take, ranging from an *ad valorem* royalty (linked to value of production) to a straight quantum-based royalty (on tonnes of ore extracted) to a share of profits to commercially negotiated lump-sum payments. Clearly each form of payment has a different logic. The second is the crucially important issue of state responsibility. There is evidence in Australia that where indigenous communities benefit from mining agreements, there is a tendency for the state to reduce its public sector funding commitments to indigenous individuals as needy Australian citizens. Such cost-shifting either to agreement beneficiaries or to mining companies can see reduced public investment offset increased benefit payments, as occurred in the Kakadu region in the 1990s (Kakadu Region Social Impact Study 1997a, 1997b).

---

[9] In the interests of transparency it should be noted that I was appointed the 'independent expert' to the study advisory group and subsequently to the UNESCO Kakadu Mission in 1998.

If payments are intended as compensation, then such state abdication of responsibility is unacceptable in a rich, liberal, multicultural state. If payments are a form of profit-sharing, such action makes more sense, but needs to be based on evidence that such profits are being equitably distributed and are sufficiently large to justify reduced state expenditures.

15.5.2 To Whom Should Payments Under Benefit-Sharing Agreements Be Distributed?

In Australia, gaining legal rights to land involves an institutional codification of 'traditions and customs' for making claims over unalienated Crown land. For example, Section 3 of the Land Rights Act requires Aborigines to demonstrate that they are a local descent group with primary spiritual responsibility for sacred sites and for land and are entitled 'as a right to forage over the land claimed'. And Section 223 of the Native Title Act requires claimants to demonstrate continuity of rights and interests under traditional laws acknowledged and traditional customs observed, and the maintenance of connection with lands and waters since colonization. Through these requirements indigenous Australians have become trapped in a Western legal definition of authenticity, and the onus of proof is on them to prove entitlement to their ancestral lands. This process, aptly captured by Wolfe's (1999) notion of 'repressive authenticity',[10] has important implications in determining who benefits from agreements, especially as it is landownership that provides the leverage for negotiations.

Because mining is geographically bounded and has environmental impacts that also encompass cultural impacts for people who retain their traditions and customs, benefits should arguably be primarily earmarked for traditional owners. Such allocation of benefits to a strictly defined and contained group would accord with a mineral rent-sharing principle that the owners of the land should be those compensated. However, the Australian state is not comfortable with confining payments to traditional owners, and there is constant pressure to distribute benefits more widely in accordance with a compensatory principle that any people socially impacted should receive ameliorating compensation.

This creates two major problems. First, there is a logical tension between those who are asked to either consent to (under land rights law) or negotiate over (under native title law) mine development and those who will benefit. Second, this generates inevitable conflict within the indigenous domain, between traditional owners whose land is desecrated by mining and other Aboriginal people who might be socially impacted in a less immediate way. In reality there is a hierarchy of rights clearly recognized under customary law, and the inconsistency between Western and customary law can create conflict. Paradoxically, there is an inverse relationship

---

[10] 'Repressive authenticity' is the term Patrick Wolfe (1999) uses to describe the history of anthropology's codification of particular forms of Aboriginality.

between the number of potential agreement beneficiaries and the likelihood of positive development outcomes. If benefits are spread too thinly, as occurred in the case of the Naberlek mine, they cannot have a positive impact, and this in turn can encourage irresponsible expenditure (Altman and Smith 1994).

15.5.3 How Should Payments Made Under Benefit-Sharing Agreements Be Utilized?

The issue of how payments should be utilized is clearly complex and highly contentious. As a general rule, benefit-sharing agreements seek to curtail the autonomy of Aboriginal beneficiaries in making decisions about the use of agreement monies in accordance with local aspirations. Increasingly, agreements impose controls over expenditure, with most agreements limiting access to cash payments for individuals. These controls have been over influenced by negative incidents in the early days of the land rights era (Altman and Smith 1994) and clearly are inconsistent with notions of self-determination, local empowerment and profit-sharing. It appears that the neoliberal obsession with the individual and individual income maximization currently favoured by the state does not extend to indigenous individuals.

Clearly a balance is needed between payments to individual landowners and pooled benefits for sustainable community benefit. There have certainly been examples in Australia of indigenous groups adopting policies that have seen benefits spent in line with local aspirations, and very beneficially. For example, the Gagudju Association invested its agreement payments in successful commercial ventures and in the provision of services, while also making some individual payments to its members (Kakadu Region Social Impact Study 1997a). The Ngurratjuta Aboriginal Corporation adopted similar policies, but would not allow cash payments to individuals. Instead it allowed purchases to be made via purchase orders in accordance with set guidelines (Altman and Smith 1999).

What is clear from the Australian situation, and should be reiterated, is that at times agreement benefits have been used to purchase goods and services that should, under normal circumstances, be provided from the public purse. At times this has been done as a conscious strategy because the needs-based queue is too long. The danger, though, is that if the need always exceeds the resources available, benefit payments may always be used in this way.

15.5.4 Who Should Be Responsible for Decision-Making?

Ultimately agreement payments should be used beneficially, but who defines the word 'beneficially'? In Australia, there has been a growing tendency for the state, multinational corporations and powerful indigenous spokespeople to decide what is 'beneficial', often to the exclusion of local groups of those impacted by mining, who should make such decisions. Again this logic assumes that payments are a sharing of mineral rent rather than compensation for negative impacts.

Governance of the implementation of agreements has emerged over time as a critically important but somewhat neglected issue. Typically, when a major agreement is completed, an Aboriginal organization is established as a legal entity, but with limited expertise and little experience in managing large sums of money. There is clearly a need to ensure that such incorporated bodies are properly structured to suit local circumstances (in terms of membership and decision-making powers) and properly resourced. In some cases, all-Aboriginal boards have been appointed with very limited powers. A key example here is the board of the Aboriginals Benefit Account (as the ABTA is now called), which oversees a trust fund with reserves of nearly $A150 million. Unfortunately this board is only advisory, and it is the Minister for Indigenous Affairs (or her/his delegate) that has decision-making authority. In the past decade, the minister has dominated decision-making more and more, with many decisions based on government policy positions rather than Aboriginal priorities. At other times, mixed boards have been appointed, but these have been dominated by non-Aboriginal members who have a better understanding of Western corporate law. Ultimately Aboriginal people should be empowered to make decisions, but that requires appropriate structures as well as investment in local capacity-building for effective decision-making that is principally accountable to local, rather than external, parties.

In Table 15.1, potential hurdles that need to be crossed for leveraged benefits to make a difference are summarized schematically across a spectrum from the optimal to the problematic. It is far more likely that benefits will be delivered where positive features are evident than when situations are problematic. A crucial factor is whether property rights are well defined, so that benefits can be delivered to clearly

Table 15.1 Schematic Representation of Development Challenges for Indigenous Beneficiaries of Agreements (adapted from Altman [2001])

Issue	Positive extreme	Problematic extreme
Property rights	Well-defined and strong	Poorly defined and weak
Beneficiaries	Clearly defined and geographically bounded, based on traditional ownership	Poorly defined and spatially scattered, based on historical association
Development strategy	Shared uncontested vision, based on community cohesion	Diverse and contested, based on individual action
Financial policy	High emphasis on accumulation and investment	High emphasis on distribution and expenditure
Investment policy	Asset forming, strategic formation of a sustainable corpus	High risk, speculative
Expenditure policy	Focused on group or community; social benefit	Focused excessively on a few individuals or families; private benefit
Time frames	Long term, strategic	Short term, immediate
Accountability	Transparent and rigorous	Opaque and conflict-ridden

defined beneficiaries. Another important issue is whether beneficiaries share a development vision, although it important to emphasize that this vision does not need to be homogeneous: it can allow for diverse aspirations and outcomes. A vital element of sustainable outcomes is that agreement beneficiaries should be able to productively accumulate and financially invest benefits.

Clearly the terms 'positive' and 'problematic' used in this schema contain value judgements; the criteria used in the table employ standard 'good' governance principles over which there is room for debate. The greatest debate, as already noted, arises in financial and expenditure policies, where there is a tension between making payments for group benefit and distributing payments to individuals and families to supplement their income. Unfortunately, if leveraged benefits are small (usually due to a poor property basis for negotiation) then there is little incentive to save and invest.

15.6 Comparative Implications for the San *Hoodia* Case

The benefit-sharing agreement between the San and the CSIR is fundamentally different from the agreements signed between Aboriginal people and multinational corporations in Australia. As noted earlier, the San *Hoodia* agreement is an innovative contract with respect to intellectual property in a plant species. There have been no similar bioprospecting agreements in Australia. Secondly, the San agreement utilized international leverage provided by the CBD, whereas agreements between Aboriginal and mining interests rely on Australian law. Finally, it is clear that the leverage for the San agreement is based on an acknowledgment that San intellectual property (traditional knowledge) in *Hoodia* requires remuneration. In Australia there are strong negative environmental and cultural impacts associated with mining that require compensation by law, even if only for surface disturbance on Aboriginal-owned land, that is commensurate with the surface disturbance compensation that other Australians can expect. Arguably in the San case there is less tangible negative impact from the commercial use of *Hoodia*,[11] so the benefit-sharing agreement that has been struck is an important legal precedent for future bioprospecting agreements and a crucially important victory for the San.

Agreement-making in Australia over the past three decades clearly indicates that measures are needed to address the asymmetric power relations between indigenous interests and multinational corporations. To a limited extent this issue has been addressed by state funding of indigenous land councils and native title representative bodies, new institutions that have played a major role in representing

---

[11] It is recognized, however, that there certainly are costs, if less tangible ones, associated with the cultural and heritage losses that accompany the commodification of *Hoodia*.

indigenous interests in negotiations. Without a level negotiations playing field, multinational corporations will always be exposed to charges of exploitation. The Minerals Council of Australia recognizes this and recently lobbied the federal government to change course on amendments to land rights and native title laws that would reduce the independence and capacity of indigenous representative bodies. This reflects the industry's sensible realization that more robust indigenous organizations result in more effective and efficient agreement-making (MCA 2006). To date its calls have not been heeded. In the San case, consideration needs to be given to whether better benefits might have been negotiated if San representation had been better resourced, thus reducing power imbalances between the San and the CSIR in negotiations. This needs to be kept in mind for future negotiations.

In Australia there is a lack of clarity about the purposes for which agreement benefits are paid. The options include: compensation for loss of land and its desecration; profit-sharing with landowners; an arbitrary source of development finance; and a mix of all three. Similarly in the San case, it is far from certain whether benefit sharing is just profit-sharing with traditional knowledge holders, compensation for the loss of intellectual property rights over *Hoodia*, an arbitrary source of reparation for colonialism and apartheid or an arbitrary source of development aid. This lack of clarity could have implications for the clear specification of appropriate outcomes from the utilization of agreement payments.

In Australia the intended beneficiaries of benefit-sharing agreements are rarely precisely defined, and this creates tensions and contestation in the indigenous domain. In particular it is often unclear if it is landowners (who do not own minerals) or those who are socially impacted that should share in benefits. If it is both, then how are benefits to be equitably shared? All too often, under the guise of self-determination, payments have been made to Aboriginal regions affected by mining, and then other indigenous interest groups have been left to compete, usually quite unproductively, if not destructively, over the division of the spoils. In the San case, the question is whether the geography of the plant's distribution, the knowledge of its medicinal properties or the original habitation pattern identifies beneficiaries. In the former two cases, the Nama peoples would also qualify for benefits, which they have not received to date.

Both in Australia and in the San case, the identification of beneficiaries is one of the ongoing challenges in the governance and implementation of benefit-sharing agreements. To some extent in the San case this issue has been addressed with a high level of inclusiveness (if not in respect of the Nama peoples), but as a result considerable administrative effort and resources will be required to distribute limited benefits among a very large number of potential beneficiaries (100,000 plus) in several countries.

The Australian experience indicates that the state looms very large in setting the legal framework for benefit sharing and defining the terms on which indigenous people can take part in negotiations. There is undoubtedly a marked difference in Australia between an earlier progressive statutory framework that provided free prior informed consent provisions under land rights law in the 1970s and more recent and less benign native title law allowing commercial exploration and mining to take precedence over native title rights. Surprisingly, the much weaker negotiating power of native title parties has not resulted in less lucrative agreements, but that is primarily because of the superprofitability of mining in the past 15 years. However, the assumption of the state appears to be that indigenous people will utilize native title leverage as a pathway to mainstream development, whereas indigenous beneficiaries frequently seek to use agreement benefits to meet their own very distinct aspirations – often to return to look after their 'country' that they perceive as threatened by mining. Similarly with the San, various nation states appear unwilling to support their interests in traditional knowledge through the passage of robust domestic biodiversity legislation and enforcement mechanisms, preferring instead to privilege national and international commercial interests over the customary interests of indigenous people. This antipathy to the San has been clearly demonstrated in Botswana by the draconian and illegal efforts of the Botswana government to bring the San into the mainstream, irrespective of their aspirations and inherent rights to pursue alternate livelihood pathways (Hitchcock and Babchuk 2007).

The benefit-sharing agreement between the San and the CSIR is in its early days, and it remains to be seen if the division of royalties will provide a sufficient pool of development capital to make a difference to the San. Clearly the availability of discretionary finance to be placed in the San *Hoodia* Benefit-Sharing Trust under the effective control of the San is important. This also creates a series of challenges for the trustees, who will need to ensure that these scarce financial resources are utilized beneficially. Already important decisions have been made that will preclude payments to individuals and require potential beneficiaries to provide sound business plans and to demonstrate financial capacity to manage and account for any grants from the trust.

Other issues will now arise. For example, will the percentages committed for internal and administrative purposes (10%) and for the administration of San networks (5%) be adequate? And will the 10% allocated to the emergency reserve fund allow for the growth of a sufficient corpus? The range of issues raised in Table 15.1 above might prove a useful checklist for the trustees to consider regularly as they strategically assess the trust's performance against the aspirations of San beneficiaries.

There is no doubt that this will be a challenging task for two main reasons. First, the distribution of the San population and its overall size will make it extremely difficult for the trustees to establish cost-effective member feedback mechanisms. Second, a development breakthrough like the agreement between the San and CSIR always raises high, possibly unrealizable expectations. The experience in Australia indicates that managing expectations can be very difficult, especially where a range of stakeholders – the state, multinational corporations and indigenous peoples themselves – have diverse, and at times divergent, development expectations and aspirations. This in turn brings political pressures about whether to favour internal over external accountability: in other words, should trustees be principally accountable to beneficiaries or should they also take into account the expectations of external stakeholders, especially the state? And in seeking internal accountability should the views of some beneficiaries take precedence over others?

15.7 Conclusion

The development challenges facing indigenous Australians, especially those living on the remote indigenous estate, are enormous. Mining is an increasingly dominant form of highly profitable commercial activity in remote regions. And land rights and native title laws give indigenous people negotiation rights that generate significant benefit-sharing agreements. Nevertheless, even where significant multimillion multiyear agreements have been negotiated, development outcomes have frequently been disappointing for all parties. This partly reflects the extent of the historic legacy of marginalization facing indigenous people and partly indicates more recent state neglect. But it also reflects an extreme divergence between, on the one hand, the shared project of the neoliberal state and mining corporations to assimilate indigenous people into the mainstream and, on the other, the goals of indigenous peoples to have their different world views and diverse aspirations recognized.

In Australia, a rich First World country, the state looms large in the lives of indigenous people, and a high level of state dependence is a big part of the development problem. Under such circumstances access to private sector finance from benefit-sharing agreements should provide a mechanism for beneficial development. Unfortunately this is rarely the case, for many complex reasons outlined in this chapter. Sometimes the negative impacts of development outweigh compensatory benefits; sometimes the state reneges on its responsibilities; and sometimes new institutions established to manage significant financial flows are inadequately structured and resourced. Where there has been success, it has occurred because Aboriginal groups have been empowered to take control and to tailor agreement implementation to address their particularities and aspirations. This has often required innovative approaches that recognize fundamental differences between mainstream Australian and remote indigenous livelihood options and preferences, and diversity within regional indigenous domains.

There are doubtless elements of the Australian indigenous situation that resonate with the circumstances of the San. Ultimately, there are no easy solutions to the development problems faced by the world's indigenous peoples. While scarce capital generated by benefit-sharing agreements should help to ameliorate these problems, it is important to acknowledge that any one agreement will only provide a partial solution. Managing expectations while sustainably implementing agreements is clearly an emerging challenge. This is especially the case over the life cycle of a long-term agreement. Recognizing the inevitable challenges posed in agreement implementation should, at the very least, ensure early investment in capacity-building that might allow adaptive and informed management of agreement implementation.

References

Altman, J. (1983). *Aborigines and mining royalties in the Northern Territory*. Canberra: Australian Institute of Aboriginal Studies.

Altman, J. (2001). Economic development of the indigenous economy and the potential leverage of native title. In B. Keon-Cohen (Ed.), *Native title in the new millennium*. Canberra: Aboriginal Studies Press.

Altman, J. (2006). Indigenous peoples, the state and multinational corporations: Australian perspectives on contested notions of property, power and sustainable development. Unpublished paper presented at the UNRISD workshop: Identity, Power and Rights: The State, International Institutions and Indigenous Peoples, Geneva, 26–27 July.

Altman, J., Biddle, N., & Hunter, B. (2005). A historical perspective on indigenous socio-economic outcomes in Australia, 1971–2001. *Australian Economic History Review*, *45*(3), 273–295.

Altman, J., Buchanan, G., & Larsen, L. (2007). The environmental significance of the indigenous estate: natural resource management as economic development in remote Australia. *CAEPR Discussion Paper No. 286/2007*. www.anu.edu.au/caepr/Publications/DP/2007_DP286.pdf. Accessed 29 February 2008.

Altman, J., & Sanders, W. (1991). From exclusion to dependence: aborigines and the welfare state in Australia. *CAEPR Discussion Paper No. 1/1991*. www.anu.edu.au/caepr/Publications/DP/1991_DP01.pdf. Accessed 29 February 2008.

Altman, J., & Smith, D. (1994). The economic impact of mining moneys: the Nabarlek case, Western Arnhem Land. *CAEPR Discussion Paper No. 63/1994*. www.anu.edu.au/caepr/Publications/DP/1994_DP63.pdf. Accessed 29 February 2008.

Altman, J., & Smith, D. (1999). The Ngurratjuta aboriginal corporation: a model for understanding Northern Territory royalty associations. *CAEPR Discussion Paper No. 185/1999*. www.anu.edu.au/caepr/Publications/DP/1999_DP185.pdf. Accessed 29 February 2008.

Corbett, T., & O'Faircheallaigh, C. (2006). Unmasking the politics of native title: the National Native Title Tribunal's application of the NTA's arbitration provisions. *UWA Law Review*, *33*(1), 153–177.

Hitchcock, R. K., & Babchuk, W. (2007). Kalahari San foraging, land use and territoriality: implications for the future. *Before Farming*, March, article 3.

Hunter, B. (2006). Revisiting the poverty wars: income status and financial stress among indigenous Australians. In B. H. Hunter (Ed.), *Assessing the evidence on indigenous socioeconomic outcomes: a focus on the 2002 NATSISS*. Canberra: ANU E Press.

Kakadu region social impact study (1997a). Report of the Aboriginal Project Committee, Supervising Scientist, Canberra, June.

Kakadu region social impact study (1997b). Report of the Project Advisory Committee, Supervising Scientist, Canberra, June.

Keen, I. (2004). *Aboriginal economy and society: Australia at the threshold of colonisation*. Melbourne: Oxford University Press.

MCA (2006). *Minerals Council of Australia 2007–08 Pre-budget submission*. Minerals Council of Australia, December. www.minerals.org.au/__data/assets/pdf_file/0006/18816/FINAL_Pre-Budget_sub.pdf. Accessed 13 April 2008.

Ritter, D. (2002). A sick institution? Diagnosing the Future Act Unit of the National Native Title Tribunal. *Australian Indigenous Law Reporter*, *7*(2), 1–11.

Stanner, W. (1965). Religion, totemism and symbolism. In R. Berndt & C. Berndt (Eds.), *Aboriginal man in Australia: essays in honour of Emeritus Professor A.P. Elkin*. Sydney: Angus and Robertson, 207–237.

Wolfe, P. (1999). *Settler colonialism and the transformation of anthropology: the politics and poetics of an ethnographic event*. London and New York: Cassell.

Woodward, E. (1974). *Aboriginal land rights commission: second report, April 1974*. Canberra: Australian Government Publishing Service.

Chapter 16
Human Research Ethics Guidelines as a Basis for Consent and Benefit Sharing: A Canadian Perspective

Kelly Bannister

Abstract While no access and benefit-sharing policy is yet in place in Canada, consent, benefit sharing and other issues relevant to bioprospecting and biodiversity research are important points of discussion at national as well as institutional and community levels. Although some plant species in Canada that are heavily exploited for commercial purposes have suffered serious decline, which has affected cultural uses by Aboriginal peoples, a primary difference from the San-*Hoodia* case is that much of the Canadian biodiversity debate has taken place in the abstract, based largely on rights and responsibilities emerging from potential scenarios rather than real instances of successful commercial products derived from traditional knowledge. New national human research ethics guidelines for research involving Aboriginal peoples are likely to provide a key reference point for addressing issues raised in the biodiversity context when the appropriation of traditional knowledge is involved.

Canadian biodiversity policy will likely include good ethical practice standards and promote collaboration and adherence to community-level protocols for current and future biodiversity research and development. However, traditional knowledge appropriation from already published literature is unlikely to be addressed, even though such knowledge may not have made its way into the published record by ethical means. This significant policy gap should be informed by the San-*Hoodia* case, which sets an important precedent by enabling Aboriginal communities and biological populations to be supported through benefits and capacity-building, even when bioprospecting is based on the published literature.

Keywords Benefit sharing • biodiversity policy • Canadian indigenous peoples • research ethics • traditional knowledge

K. Bannister
POLIS Project on Ecological Governance, University of Victoria, University House 4, P.O. Box 3060, Victoria, British Columbia, Canada, V8W 3R4
e-mail: kel@uvic.ca

16.1 Introduction

Domestic interest in genetic resources and the traditional knowledge of Canada's Aboriginal peoples[1] has risen significantly over the past couple of decades. The increased scientific and corporate interest has likely been influenced in part by the evolving regulations and restrictions on the movement of genetic resources across national borders (an impact of the 1992 Convention on Biological Diversity), making it seem less burdensome in terms of time and cost to study and access biodiversity and associated traditional knowledge at home rather than abroad. Moreover, a fairly extensive published literature has accumulated on traditional plant uses by Aboriginal peoples in Canada and associated biological properties (e.g. Turner et al. 1990; Kuhnlein and Turner 1991; McCutcheon et al. 1992; McCutcheon et al. 1994; Moerman 1998; Marles et al. 2000; Turner 2004), making it easier for academic and corporate parties to access traditional plant knowledge second-hand, without the need to invest time and incur expenses for fieldwork or building relationships with source communities.

Biodiversity studies undertaken today, however, are subject to higher levels of scrutiny than in the past, and are governed by different ethical expectations, particularly relating to Aboriginal rights and interests. Not surprisingly, consent, benefit sharing and other ethical and legal issues relevant to bioprospecting and biodiversity research have become important points of discussion in developing national law and policy, as well as institutional and community-level policies and procedures in Canada. This paper compares the San-*Hoodia* case with the Canadian context and discusses, in particular, the relevance of human research ethics guidelines to emerging Canadian biodiversity policy where the appropriation of traditional knowledge is involved.

16.2 Similarities and Differences

The San-*Hoodia* case offers an important concrete example of the complex issues at stake and furthers our understanding of both generalizable and non-generalizable aspects. While there are numerous similarities between the issues raised by the San-*Hoodia* case and those in the Canadian context, there are also important differences. For example, like the San-*Hoodia* case, some traditional plant uses in Canada may be attributed exclusively to a given Aboriginal community or cultural group,

---

[1] 'Aboriginal peoples' refers collectively to the descendants of the original inhabitants of Canada. The Constitution of Canada recognizes three separate peoples with unique heritages, languages, cultural practices, and spiritual beliefs. The three Aboriginal peoples of Canada are: Inuit, Métis and Indian (noting that the term 'Indian' has largely been replaced with 'First Nation' in common usage). Inuit are northern Aboriginal peoples. Métis are of mixed European and First Nations ancestry. First Nation people comprise the largest and most diverse group of Aboriginal peoples in Canada with over 600 First Nations bands and over 50 languages.

but many plant species are widely distributed or traded and used in similar ways by different communities and cultural groups, making it difficult to determine who is the original source of knowledge and who has the rights to benefits derived from its commercial development.[2] A primary difference is that, to date, much of the Canadian debate has taken place in the abstract, based largely on rights and responsibilities emerging from 'what if' scenarios rather than any real instances of successful commercial products derived from traditional knowledge.

While there are no Canadian examples that parallel the San-*Hoodia* case, an example of some relevance is that of the blockbuster anti-cancer agent paclitaxel (Taxol®), used in the treatment of ovarian, breast, lung and other cancers. Taxol was originally isolated from the bark of the Pacific yew tree (*Taxus brevifolia* Nutt.) in the 1960s as part of a random screening programme for anti-cancer activity headed by the US National Cancer Institute (Cragg et al. 1994; Goodman and Walsh 2001). Years later, it was commercially developed by Bristol-Myers Squibb. Originally Pacific yew was collected in the state of Washington, USA, but subsequently massive collections totaling thousands of kilograms of bark were made throughout the tree's range in the Pacific Northwest, including regions of both Canada and the USA. The Pacific yew is no longer used as the source of Taxol due to concern about the environmental impact on the slow-growing tree and the discovery of alternative sources of Taxol and related anti-cancer compounds.

There are numerous documented medicinal uses of Pacific yew by Aboriginal peoples in Canada and the USA (e.g. Moerman 1998), including the treatment of cancer by Tsimshian people (Compton 1993) and internal ailments by Salishan people (Turner and Hebda 1990) in British Columbia, Canada. However, it is impossible to know for certain whether or not the traditional knowledge of Pacific yew influenced the original collection – and, if so, whose knowledge. The media continue to cite Taxol as an example of bioprospecting and infer that Aboriginal intellectual property rights were ignored (e.g. Ramsay 2005) while the National Cancer Institute has always maintained that its screening programme through which Taxol was isolated was random (e.g. Cragg et al. 1994).

What is clear is that the environmental impacts of research and development involving Pacific yew have had accompanying cultural impacts, as noted by renowned Canadian ethnobotanist Nancy Turner (2001):

> Within a short time, yew trees all along the Pacific Coast were being cut down for their high value bark – in some cases, trees were poached from private lands and parks – with little consideration for the other values of the yew tree (Hartzell 1991; Foster 1995). In particular, little recognition was given to the high cultural values that Pacific yew has for First Peoples, both for its medicinal use (see Turner and Hebda 1990) and for its tough, resilient wood.

The Pacific yew is one of several North American species with commercial value that raise biocultural issues – that is, where overharvesting for commercial markets has limited the supply and decreased the accessibility of a species for cultural uses. Other well-known examples are goldenseal, echinacea and American ginseng.

---

[2] With over 600 recognized First Nations bands in Canada, multiple claims to knowledge of a particular plant would be likely and claims could be extremely complex.

Canada is a signatory to the Convention on Biological Diversity and committed to the '2010 target' of significantly reducing the rate of biodiversity loss through domestic policy and/or legislation. Unlike South Africa, however, to date there is no specific or overarching Canadian policy or legislation on biodiversity that addresses traditional knowledge issues. Efforts are under way to develop a national access and benefit-sharing policy as a formal means to grant access to biological resources and associated traditional knowledge and maximize benefits derived from their use for Canadians. The overarching policy goals include:

- Facilitating the conservation and sustainable use of Canada's biodiversity, seen as 'the raw material of the bio-based economy'
- Enhancing economic productivity through research and innovation
- Contributing to the health and well-being of rural and Aboriginal communities (Environment Canada 2005)

The protection of traditional knowledge is seen as a component of the latter goal.

While a number of awareness-building activities have taken place[3] progress on policy development has been slow. In 2008, a Federal/Provincial/Territorial Task Group was established to examine ABS policy in Canada. The Task Group identified a series of options on an ABS policy framework for Canada and released a discussion paper for public feedback in Spring 2009 (Environment Canada 2009). According to the Canadian ABS Portal the results are proposed to be discussed at a possible meeting of Federal/Provincial/Territorial Ministers in the fall of 2009.[4] Access and benefit sharing is a complex, cross-cutting topic in Canada that includes multiple jurisdictions involved in the management of natural resources (federal, provincial/territorial and Aboriginal), as well as a range of stakeholders from different sectors. The approach promoted for policy development is to balance the interests of users, providers, stakeholders and the Canadian public. As noted in a recent scoping paper: 'Finding the right balance between the respect for cultural practices and spiritual beliefs inherent in traditional knowledge and knowledge sharing is crucial for advancing scientific research that can support the health and well-being of Canadians' (Environment Canada 2005). The adequate protection of

---

[3] Awareness-raising activities have included the following workshops and meetings:

- Access and Benefit Sharing of Genetic Resources: The Science and Technology Agenda Experts' Workshop (December 2004)
- Northern Workshop on Access to Genetic Resources and Associated Traditional Knowledge and Benefit Sharing (March 2005)
- Genetic Resources in Agriculture: Their role, Their Governance – Implications for Access & Benefit Sharing (November 2005)
- Access to Forest Genetic Resources and Benefit Sharing: Potential Opportunities and Challenges for Governments and Forest Stakeholders (February 2006)
- National Meeting on ABS and Certificates (November 2006)

[4] See http://www.ec.gc.ca/apa-abs/index.cfm?lang=eng. Accessed 1 August 2009

traditional knowledge, at least from the perspective of Aboriginal peoples, may be a key challenge in this context.

16.3 Researchers as Agents of Appropriation

Canada's approach to ABS policy development assumes academic researchers are the primary accessors and users of biodiversity resources and associated traditional knowledge, often serving directly (e.g. through university-corporate partnerships) or indirectly (through publishing) as intermediary providers of these to third parties who are interested in commercial development. The idea of researchers as potential agents of cultural appropriation is not new to biodiversity research. It was articulated by ethnopharmacologist Elaine Elisabetsky (1991) in the early years of the controversy generated by bioprospecting ventures aimed at discovering new pharmacueticals:

> Usually, indigenous knowledge was crucial to the development of such products; nevertheless, indigenous groups tend not to benefit from the achievements of research. ... As a result, such research efforts are perceived as scientific imperialism: scientists are accused of stealing plant materials and appropriating traditional plant knowledge for financial profit and/or professional advancement.

Elisabetsky speaks to the significant role researchers can play – and have played for decades – in facilitating the appropriation and commodification of traditional knowledge and genetic resources, even if that is not their intention. The San-*Hoodia* example is a case in point: the anthropological record of *Hoodia* use by San peoples and their descendents served as a key contribution to the discovery of appetite-suppressing properties with commercial potential (see Wynberg and Chennells, Chapter 6). It is impossible to assess retrospectively what ethical standard was upheld by researchers who created the early published literature on San use of *Hoodia*. For example, was San knowledge recorded and published with San people's consent or even awareness? Even if consent was given, could those who originally shared (or those who documented) the knowledge have had any inkling that it would become so valuable in later years?

Certainly it is not reasonable to hold researchers of the past to today's ethical standards. Yet there is clearly a pressing need for those currently involved in accessing and using biodiversity and associated traditional knowledge (whether for bioprospecting, basic biodiversity research, or other purposes) to be acutely aware of the complex and interrelated ethical, legal and political issues, not the least of which are the implications of using traditional knowledge from the published literature, and the potential consequences of placing traditional knowledge in the public domain in the first place (e.g. in academic literature and open-access databases). Use of traditional knowledge from the academic literature may raise difficult questions to be sorted out after the fact, as the San-*Hoodia* case has shown, about who ought to benefit – and how – from the use of traditional knowledge already considered part of the public domain. Making traditional knowledge public may raise concerns about diminishing the context in which the knowledge evolved (and therefore the sense of responsibility and accountability to source communities according to their world views) and making the knowledge accessible for 'free and unfettered use' by third parties.

While the San-*Hoodia* case seems to have become an exception, the past two decades of intensive literature-based biodiversity prospecting have indicated that all too often third parties from the commercial sector (e.g. biotechnology, pharma, herbal, floral) lack sufficient awareness or incentive to address the inequities and potential harms, direct and indirect, that their enterprises might bring for source communities and ecosystems. Lawyer Brendan Tobin (2004) notes:

> While the international debate has tended to focus primarily on the question of biopiracy, there are many more immediate threats to traditional knowledge which require attention if it is to be conserved and strengthened. These include loss of land and language, insensitive educational and health policies, agriculture and fisheries extension programs, and the impact of organized religion, amongst others. Development of any effective global program for protection of [traditional knowledge] should, therefore, include not only a means for the recognition of ownership rights but also a system for strengthening the continued use and development of [traditional knowledge] as part of the global body of science, and a mainstay of the populations in developing countries, where local sustainability and development opportunities are closely linked to the integrity of [traditional knowledge] systems.

Ultimately, a balance must be found between the need to document traditional knowledge and make it more widely available and the need to ensure protections against the unfair or harmful exploitation of that knowledge and interrelated biological and cultural resources. One could argue that the dilemma is partly a consequence of the academic enterprise itself – a system based predominantly on a linear, extractive model of knowledge acquisition that, in biodiversity-related fields, has typically sought to take useful information and biological resources from source communities and channel these into the academic chain of knowledge production, leading to new knowledge, publications, patents or commercial products.

A key question, then, is: how can traditional knowledge be gathered in respectful and culturally appropriate ways that benefit (rather than harm) source communities and support (rather than sever) interrelationships with biodiversity and cultural heritage? The San-*Hoodia* case offers a unique example of some degree of community benefit being achieved retrospectively. In the Canadian context, proactive solutions are being encouraged through the identification of new ethical frameworks and methodologies that are based on equity, partnership and power-sharing – that is, the evolution of models of research and knowledge production that are more circular than linear and more self-sustaining than extractive.

One mechanism that has been proposed is the development of a national code of conduct to offer voluntary guidance to researchers on access to traditional knowledge associated with genetic resources and to establish best-practice standards (Bannister and Haddad 2006). The code of conduct would build upon existing research policies and institutional structures, specifically those involving human research ethics, as described below.

16.3.1 Human Research Ethics as a Basis for Biodiversity Policy

Canada has in place a national ethics policy for all academic research involving humans (CIHR, NSERC and SSHRC 1998, currently under revision),[5] as well as specific guidelines for health research involving Aboriginal people (CIHR 2007). Although not specific to biodiversity research or bioprospecting and more limited in scope, the guidance provided by these research ethics policies offers the most relevant framework at this point to approach the topic of Aboriginal peoples, consent and benefit sharing in Canada, as discussed below.

The overarching national research ethics policy in Canada is articulated in the *Tri-Council Policy Statement: Ethical Conduct for Research Involving Humans*. It requires compliance by individuals and institutions receiving funds from any of Canada's three federal research granting bodies: the Canadian Institutes of Health Research (CIHR), the Natural Sciences and Engineering Research Council of Canada (NSERC) and the Social Sciences and Humanities Research Council of Canada (SSHRC). Section 6 of the statement is specifically about research involving Aboriginal peoples. It acknowledges that research with Aboriginal communities involves extra complexities and outlines additional requirements to ensure that the rights and interests of the community as a whole are respected. These requirements include:

- Consideration of past harms to individuals and communities incurred through the expropriation of cultural properties
- Respect for the culture, traditions and knowledge of the Aboriginal group
- Consideration of the interests of the Aboriginal group when property or private information belonging to the group is studied or used
- The conceptualization and conduct of the research as a partnership with the Aboriginal group
- Adjustment of the research to address the needs and concerns of the Aboriginal peoples involved
- Willingness to deposit data and other research outcomes in an agreed-upon repository
- An opportunity for the community to react and respond to research findings and publications (as summarized in Bannister, 2009)
- Substantial changes to strengthen and clarify Section 6 are anticipated in the revised second edition of the Tri-Council Policy Statement.

---

[5] Canada's national ethics policy governing academic research involving humans, called the *Tri-Council Policy Statement: Ethical Conduct for Research Involving Humans*, was adopted in 1998. In December 2008, a substantially revised draft second edition was released for public consultation. The release of a revised version of the draft to the public for further comment is anticipated in Fall 2009 and, following a second consultation period, a final draft is expressed in February 2010 (see http://pre.ethics.gc.ca/eng/index/). Accessed 1 August 2009).

While specifically focused on health (broadly defined), the *CIHR Guidelines for Health Research Involving Aboriginal Peoples* (CIHR 2007) are the most progressive and comprehensive yet of the national ethics policies in regard to Aboriginal peoples, consent and benefit sharing in Canada. The guidelines address a wide spectrum of difficult issues from both philosophical and practical perspectives, including:

- Aboriginal jurisdiction over the conduct of health research within Aboriginal communities
- Requirements for community project approval
- The promotion of research partnerships and participatory research methodologies
- Requirements for both collective and individual consent, as well as collective and individual confidentiality and privacy
- Respect for individual autonomy and responsibility
- The importance of, and responsibilities involved in, using indigenous knowledge in research
- The protection of cultural knowledge as a shared responsibility of communities and researchers
- Expectations for benefit sharing, community empowerment and capacity development
- Rights to control the collection, use, storage and potential use of data (where the use of data and biological samples by researchers is based on 'loaning' and 'researcher as steward' concepts that vest ownership in Aboriginal individuals and communities)
- Community involvement in the interpretation and dissemination of results
- Explicit support for cultural protocols and Aboriginal communities' own research ethics guidelines and processes where they exist, including local Aboriginal ethics review boards

The last issue mentioned above represents an important new direction in Canadian research ethics policy approaches, recognizing that overarching national guidelines must intercalate in meaningful and effective ways with existing and yet-to-be-developed community-level guidelines and research protocols articulated by Aboriginal communities. The full implications of attempting to integrate these parallel processes and share decision-making power in research have yet to be understood, but the principle is a significant step towards decolonizing research and an important learning opportunity for all involved.

Another innovative feature of the CIHR guidelines is the explicit attempt to address a key problem of both health and biodiversity research: promoting an understanding of what 'traditional knowledge' is from an Aboriginal perspective and of what it means to access and use traditional knowledge in research. Article 1 of the guidelines states:

> A researcher should understand and respect Aboriginal world views, including responsibilities to the people and culture that flow from being granted access to traditional or sacred knowledge. These should be incorporated into research agreements, to the extent possible (CIHR 2007).

This article underscores a reciprocity in being granted access to traditional knowledge, and a broader sense of accountability and responsibility that emerges from entering into a research relationship with Aboriginal people. As the guidelines go on to explain:

> Within Aboriginal cultures, the notion of accountability may imply responsibility across a temporal dimension that is foreign to western notions of accountability (for example, accountability to past and future generations may take primacy over accountability to community authorities for certain types of knowledge). Accountability may also involve a sacred dimension such as a sense of relational accountability to a recognized spiritual entity or to the Land. Researchers should understand these broader practices of accountability in order to understand the responsibility that they have once they enter into the research relationship (CIHR 2007).

Amid the increasing interest by scientific and corporate communities alike in incorporating traditional knowledge into research (whether related to plant medicines, conservation, climate change or other areas), it is often not clear what level of understanding Western-trained scientists have about the nature of traditional knowledge. Clearly there is no universally agreed definition, and traditional knowledge is not a concept uniformly held by all indigenous peoples, so it is problematic to generalize, especially using Eurocentric frameworks. Understandings necessarily depend on specific contexts and applications, but it is clear that instrumental and reified notions are common in academia, industry and government alike.

As indicated above, Canadian ethics policy does not simply see traditional knowledge as the identification of a plant species or a plant use for medicine, but recognizes it as a complex *system* of knowledge and relationships, arising from Aboriginal cosmologies and based on Aboriginal epistemologies. It is not surprising, then, that misunderstandings and clashes of world view can arise in the discussion of traditional knowledge and how it ought to be treated by those outside a given cultural group, especially when knowledge appropriation and commercialization are (or are perceived to be) involved.

16.4 Conclusion

Canada's national research ethics policies involving Aboriginal people, as described above, will likely influence the future development of domestic policy and/or legislation that explicitly addresses Aboriginal peoples, consent and benefit sharing in a biodiversity context. At this stage, it is only possible to make an educated guess at what Canadian access and benefit-sharing policy will look like in the future. The rationale for change, however, is as much practical as moral: eventually the supply of starting materials (e.g. traditional knowledge and biological resources) that feed the knowledge production process will be depleted if the outcomes are not fed back in useful ways to support and sustain these living systems. Some conceptual shifts needed include:

- Moving beyond the sense of individualism that has long been ingrained in human research ethics models towards recognizing the collective decision-making, legal rights and cultural responsibilities of Aboriginal communities

- Recognizing biocultural interdependence by extending ethical considerations beyond humans to include the surrounding environment upon which human well-being depends
- Acknowledging the importance of broader time frames needed for research projects, as well as considering the implications of projects beyond a single generation

These elements are premised on collaborative research and the co-creation of knowledge. If adopted, they are likely to result in new requirements to proactively establish mutually agreed terms for research and development, including obtaining appropriate consent or permissions to access knowledge and/or biological resources; informing and involving local communities and knowledge-holders directly in the research process as collaborators; sharing research outcomes and benefits in meaningful and useful forms; and ensuring that adequate protections (legal or administrative) are in place to prevent the misrepresentation, misappropriation or unwanted commodification of traditional knowledge and cultural property.

Protection mechanisms have already begun to emerge from the community level. For example, a number of Aboriginal communities require initial community review of research projects (e.g. through community research or ethics committees)[6]; compliance with community protocols and guidelines that address consent, benefits and intellectual property rights[7]; and community review of findings prior to publication. Template protocols and toolkits have also been developed by Aboriginal organizations to assist communities in developing their own capacity in these areas.[8]

In the process of assuming more decision-making and control at the community level, a major challenge in store for many Aboriginal communities is likely to be capacity – for example, developing administrative infrastructure and expertise within the community, or even finding sufficient individuals to be involved in implementing community-based mechanisms to protect traditional knowledge (e.g. community review committees, monitoring approved projects). In many cases, other priorities such as health, education, housing, and employment may compete for limited community resources, personnel and capacity. Investment in community

---

[6] Examples are the Nuu-chah-nulth Tribal Council Research Ethics Committee in British Columbia and the Mi'kmaq Ethics Watch in Nova Scotia.

[7] Here are some examples of Aboriginal community research protocols and guidelines in Canada:

- Mi'kmaq Research Principles and Protocols (Mi'kmaw Ethics Watch 2000)
- *'Namgis First Nation Guidelines for Visiting Researchers/Access to Information* ('Namgis First Nation n.d.)
- *Code of Ethics for Researchers Conducting Research Concerning the Ktunaxa Nation* (Ktunaxa Nation 1998)
- *Tl'azt'en Nation Guidelines for Research in Tl'azt'en Territory* (Tl'azt'en Nation 1998)
- *Protocols and Principles for Conducting Research in a Nuu-chah-nulth Context* (Nuu-chah-nulth Tribal Council Research Ethics Committee 2004)

[8] Examples are the *Template Traditional Knowledge Protocol* by the First Nations Technology Council (2005) and *Negotiating Research Relationships: A Guide for Communities* by the Inuit Tapiriit Kanatami and the Nunavut Research Institute (1998).

capacity-building by the Canadian government is clearly necessary for the success of any new access and benefit-sharing policy.

Evolving Canadian policy is aimed at setting good ethical practice standards for current and future biodiversity research and development. It is unlikely to address issues arising from the appropriation of knowledge found in works already published, even though such knowledge may not have got there by ethical means. Thus bioprospecting based on literature that predates any new Canadian policy may represent a significant omission. It is in this context that the San-*Hoodia* case sets an important precedent that the rest of the world, including Canada, will have trouble ignoring, in that it enables Aboriginal communities and biological populations to be supported through benefits and capacity-building, even where bioprospecting is based on pre-existing literature.

References

Bannister, K. (2009). Non-legal instruments for protection of intangible cultural heritage: key roles for ethical codes and community protocols. In C. Bell & R. K. Paterson (Eds.), *Protection of first nations cultural heritage: laws, policy, and reform*. Toronto, ON: UBC Press.

Bannister, K., & Haddad, P. (2006). *Roadmap to developing a code of conduct for researchers accessing genetic resources and traditional knowledge*. Report prepared under contract for Environment Canada, Ottawa, ON.

CIHR (2007). *CIHR guidelines for health research involving aboriginal people*. Ottawa, ON: Canadian Institutes of Health Research. www.cihr-irsc.gc.ca/e/documents/ethics_aboriginal_guidelines_e.pdf. Accessed 23 October 2008.

CIHR, NSERC and SSHRC (1998, with 2000, 2002 and 2005 amendments). *Tri-council policy statement: ethical conduct for research involving humans*. Ottawa, ON: Interagency Secretariat on Research Ethics (on behalf of the Canadian Institutes of Health Research, Natural Sciences and Engineering Research Council of Canada and Social Sciences and Humanities Research Council of Canada). www.pre.ethics.gc.ca/english/policystatement/policystatement.cfm. Accessed 23 October 2008.

Compton, B. D. (1993). Upper North Wakashan and Southern Tsimshian ethnobotany: the knowledge and usage of plants and fungi among the Oweekeno, Hanaksiala (Kitlope and Kemano), Haisla (Kitamaat) and Kitasoo peoples of the central and north coasts of British Columbia. PhD dissertation. Vancouver, BC: Department of Botany, University of British Columbia.

Cragg, G. M., Boyd, M. R., Cardellina II, J. H., Newman, D. J., Snader, K. M., & McCloud, T. G. (1994). Ethnobotany and drug discovery: the experience of the US National Cancer Institute. In D. J. Chadwick & J. Marsh (Eds.), *Ethnobotany and the search for new drugs*. Ciba Foundation Symposium 185. Chichester: Wiley.

Elisabetsky, E. (1991). Sociopolitical, ecological and ethical issues in medicinal plant research. *Journal of Ethnopharmacology, 32*(1–3), 235–239.

Environment Canada (2005). *ABS policies in Canada: scoping the questions and issues*. Prepared by the Federal/Provincial/Territorial Working Group on Access and Benefit Sharing of Genetic Resources. www.ec.gc.ca/apa-abs/documents/ABS_policies_e.pdf. Accessed 22 October 2008.

Environment Canada (2009). *Access to Genetic Resources and Sharing the Benefits of Their Use in Canada: Opportunities for a New Policy Direction*. http://www.ec.gc.ca/apa-abs/utilisant-using2/default.cfm?lang=eng. Accessed 1 August 2009.

FNTC (2005). *Template traditional knowledge protocol, First Nations Technology Council*. www.fntc.info/files/documents/Traditional%20Knowledge%20Protocol%20Template%20Feb%202005.doc. Accessed 23 October 2008.

Foster, S. (1995). *Forest pharmacy: medicinal plants in American forests*. Durham, NC: Forest History Society.

Goodman, J., & Walsh, V. (2001). *The story of taxol: nature and politics in the pursuit of an anticancer drug*. New York: Cambridge University Press.

Hartzell, H., Jr. (1991). *The yew tree: a thousand whispers, biography of a species*. Eugene, OR: Hulogosi.

Inuit Tapiriit Kanatami and Nunavut Research Institute (1998). Negotiating research relationships: a guide for communities. Reproduced in Pimatisiwin, *A Journal of Aboriginal and Indigenous Community Health (2003), 1*(1), 19–27. http://www.itk.ca/system/files/Negotitiating-Research-Relationships-Community-Guide.pdf. www.itk.ca/system/filesNegotitiating-Research-Relationships-Community-Guide.pdf or http://www.pimatisiwin.com/online/?page_id=101. Accessed 1 August 2009.

Ktunaxa Nation (1998). *Code of ethics for researchers conducting research concerning the Ktunaxa nation*. www.law.ualberta.ca/research/aboriginalculturalheritage/casestudies/Ktunaxa%20Code%20of%20Ethics.pdf. Accessed 23 October 2008.

Kuhnlein, H. V., & Turner, N. J. (1991). *Traditional plant foods of Canadian indigenous peoples: nutrition, botany and use, food and nutrition in history and anthropology (Vol. 8)*. Philadelphia, PA: Gordon and Breach Science.

Marles, R. J., Clavelle, C., Monteleone, L., Tays, N., & Burns, D. (2000). *Aboriginal plant use in Canada's northwest boreal forest*. Vancouver, BC: UBC Press.

McCutcheon, A. R., Ellis, S. M., Hancock, R. E. W., & Towers, G. H. N. (1992). Antibiotic screening of medicinal plants of British Columbian native peoples. *Journal of Ethnopharmacology, 37*, 213–223.

McCutcheon, A. R., Ellis, S. M., Hancock, R. E. W., & Towers, G. H. N. (1994). Antifungal screening of medicinal plants of British Columbian native peoples. *Journal of Ethnopharmacology, 44*, 157–169.

Mi'kmaw Ethics Watch (2000). *Research principles and protocols*. http://mrc.uccb.ns.ca/prinpro.html. Accessed 23 October 2008.

Moerman, D. E. (1998). *Native American ethnobotany*. Portland, OR: Timber Press.

Namgis First Nation (n.d.). *Guidelines for visiting researchers/access to information*. www.law.ualberta.ca/research/aboriginalculturalheritage/casestudies/namgisform.pdf. Accessed 23 October 2008.

Nuu-chah-nulth Tribal Council Research Ethics Committee (2004). *Protocols and principles for conducting research in a Nuu-chah-nulth context*. www.clayoquotbiosphere.org/science/ntcethics.xls. Accessed 23 October 2008.

Ramsay, H. (2005). Who owns the healing secrets of plants? *The Tyee*, 20 January. http://thetyee.ca/News/2005/01/20/WhoOwnsHealingSecretsofPlants/. Accessed 23 October 2008.

Tl'azt'en Nation (1998). *Tl'azt'en nation guidelines for research in Tl'azt'en territory*. http://cura.unbc.ca/governance/CEM-Tlazten%20Guidelines.pdf. Accessed 23 October 2008.

Tobin, B. (2004). Towards an international regime for protection of traditional knowledge: reflections on the role of intellectual property rights. Paper presented at Bioethical Issues of Intellectual Property in Biotechnology, Tokyo, Japan.

Turner, N. J. (2001). "Doing it right": issues and practices of sustainable harvesting of non-timber forest products relating to first peoples in British Columbia. *BC Journal of Ecosystems and Management, 1*(1), 1–11. www.forrex.org/JEM/ISS1/vol1_no1_art6.pdf. Accessed 23 October 2008.

Turner, N. J. (2004). *Plants of Haida Gwaii*. Winlaw, BC: Sononis Press.

Turner, N. J., & Hebda, R. J. (1990). Contemporary use of bark for medicine by two Salishan native elders of southeast Vancouver island. *Journal of Ethnopharmacology, 229*, 59–72.

Turner, N. J., Thompson, L. C., Thompson, M. T., & York, A. Z. (1990). *Thompson ethnobotany: knowledge and usage of plants by the Thompson Indians of British Columbia, Memoir No. 3*. Victoria, BC: Royal British Columbia Museum.

Chapter 17
The Limitations of Good Intent: Problems of Representation and Informed Consent in the Maya ICBG Project in Chiapas, Mexico

Dafna Feinholz-Klip, Luis García Barrios, and Julie Cook Lucas

Abstract The Maya International Cooperative Biodiversity Group (Maya ICBG) research project began in 1998 in the central highlands of Chiapas, Mexico, in a difficult and contentious legal, social and political climate. The researchers' good intentions were that the indigenous Maya people would both contribute to the project and benefit from it. However, gaps in the way local communities were included became a focus for international resistance to the project, which was abandoned in 2001.

No single actor should bear the total responsibility for what happened to the Maya ICBG, but none is devoid of it. Through a comparison with the San-*Hoodia* case we discuss how parties on all sides implicitly understood 'collaboration' and 'benefit sharing', which can easily become controversial due to conflicting assumptions about how and to what extent different groups of people should benefit from the potential royalties, and who should make these decisions.

Like the San peoples, the Maya stood to receive a very small proportion of any profit that might come from the development of commercial products. These benefits, whether realized or not, are never ethically neutral, so the transparent, full and free prior informed consent of communities to accept the risk of going along this path is absolutely essential. Both cases played out in a domestic legal and policy vacuum. Questions about the legitimacy of processes and decisions emerge as fundamental.

The failure of the Maya ICBG was due largely to the lack of an appropriate prior informed consent process built on trust and adequate representation. The question of Maya identity and self-representation through forms that are 'credible' to outside

D. Feinholz-Klip (✉)
National Commission of Bioethics Mexico, Bosque del Castillo, 36, Col. La Herradura, Huixquilucan, Edo. Mex, CP 53920, Mexico
e-mail: dafna.feinholz@gmail.com

L.G. Barrios
Departamento de Agroecología, División de Sistemas de Producción Alternativos, El Colegio de la Frontera Sur (ECOSUR), Apartado Postal, 63, San Cristóbal de las Casas, Chiapas, 29200, Mexico
e-mail: lgarcia@ecosur.mx

J.C. Lucas
UCLAN, Centre for Professional Ethics, Brook, 317, Preston, PR1 2HE, United Kingdom
e-mail: jmlucas@uclan.ac.uk

bioprospectors is an ongoing issue. The pan-Mayan identity currently under construction in Chiapas faces similar challenges to those of the San people.

Keywords Benefit sharing • biopiracy • indigenous Maya people • Mexico • policy • representation • International Cooperative Biodiversity Group

17.1 Introduction

The saga of the Maya International Cooperative Biodiversity Group (Maya ICBG) is the story of a research project that began in late 1998 in the central highlands of Chiapas, Mexico, one of the richest and most endangered biodiversity regions on earth. It had the bold purpose of excelling as a model of transparent, legal and ethical plant bioprospecting in an indigenous territory in a very difficult and contentious legal, social and political climate. The researchers intended the indigenous Maya people both to contribute to the project and to significantly benefit from it. However, many difficulties were encountered, and the project was abandoned in 2001.

Many familiar reasons can be identified for the failure of the Maya ICBG: an inadequate regulatory framework, the general political situation, and divisions among the indigenous people and international concerns about biopiracy and commercial exploitation. However, an interesting perspective on the rise of the conflict relates to how parties on all sides implicitly understood 'collaboration' and 'benefit sharing'.

17.2 The MAYA ICBG

'Drug Discovery and Biodiversity among the Maya of Mexico' was a 5 year research project awarded a grant of US$2.5 million from the International Cooperative Biodiversity Groups (ICBG) programme in 1998. The ICBG is a consortium of United States federal agencies including the National Institutes of Health (NIH) and the US Department of Agriculture.

The four major goals of the Maya ICBG were

1. To discover, isolate and preclinically evaluate bioactive agents from pharmacologically important species of vascular plants found in the state of Chiapas;
2. To discover, isolate and evaluate bioactive species of immediate health significance and potential economic value to the local Maya populations – which included targeting species for increased local use and species with commercial production potential – and to develop local capacity for the sustainable management and production of medicinal plants, both for local use and for national and international markets (e.g. public local medicinal gardens, phytomedicines, crop protection and ornamentals);
3. To initiate ecologically sophisticated biodiversity surveys aimed at comprehensive coverage of the vascular flora of the highlands of Chiapas, significantly enriching

the holdings of local herbaria and producing an innovative *Ethnoflora of the Highlands of Chiapas* to be published in the major Maya languages; and
4. To enhance infrastructure through technology transfer and support research training relevant to the goals of the host-country sponsoring institution, by developing a modern natural products laboratory and strengthening existing academic exchanges – in parallel with which academic preparation the Maya ICBG would engage in extensive training and capacity-building for Maya collaborators as well as community development (Berlin et al. 1999).

The project was led by Brent Berlin, professor of anthropology at the University of Georgia, USA (UGA), who had been conducting research among the Maya for 40 years. Partners in the project were a Mexican multidisciplinary research and graduate teaching centre, El Colegio de la Frontera Sur (ECOSUR), and Molecular Nature Limited (MNL), a small for-profit natural products discovery company based in Wales, UK (UGA 1999). The project was subdivided into three associate programmes, one of which, 'Conservation, Sustained Harvest and Economic Growth', was led by one of the authors of this chapter, Luis Garcia Barrios.

The Maya ICBG project viewed the indigenous highlanders as major stakeholders since plant collection would occur in their lands. One of the project's motivations was that it would be more reliable and generate more incentives, both economic and for biodiversity conservation, if it was guided by the previous knowledge of the Maya inhabitants (Berlin et al. 1999). They collectively possess a complex ethnopharmacopoeia comprising hundreds of species of plants in an ancient system of traditional medicine which they are rapidly losing as a consequence of current social and economic processes and policies.

However, despite the crucial role of the indigenous people, the initial consortium did not include any Maya representation. For the Maya ICBG, bringing the Maya people to the table as 'a full partner in our activities' (Berlin et al. 1999) was not a precondition but a *goal* of the project. The Maya ICBG intended highland people to have an equal share of the benefits, and the project included a mechanism to secure this, in the form of PROMAYA (Promotion of Intellectual Property Rights of the Highland Maya of Chiapas, Mexico), 'an innovative non-profit organization that will hold in trust and administer the indigenous community's portion of any financial returns resulting from the activities of the Maya ICBG'(Berlin et al. 1999). However, as we outline below, this became a core focus for resistance to the project.

17.3 Geographic, Cultural and Sociodemographic Political Context

The Maya civilization was at its peak in southern Mexico and Central America around the year 1000 before it suffered Spanish colonization in the 1500s. Through capitalist development of the land, indigenous people were stripped of their territories and chased into marginal areas. Large ventures such as haciendas were created on previously indigenous territories and the indigenous people used as servants. This mistreatment of the indigenous people led to caste wars which culminated in the revolution of 1910. The new constitution of 1917 (article 27) recognized the land rights of the original occupants of these territories.

Highland Chiapas today is a region comprising 8,000 or more indigenous Maya villages or *parajes* representing 900,000 people who speak one of the four Maya languages (Rosenthal 2006). Most of these rural people live in conditions of extreme poverty and experienced or participated in the 1994 Zapatista uprising with its unresolved demands for indigenous rights, relative autonomy and better living conditions. In the aftermath of the uprising, there is a governance crisis in the region, where a significant number of municipalities have been declared autonomous by the Zapatista and pro-Zapatista movements. The latter govern such territories de facto, sometimes in conflict with constitutional authorities.[1] Chiapas is highly militarized, with many displaced indigenous people.

'The concept of a "community" in Chiapas today is the subject of intense debate', states Rosenthal (2006), and most commentators on this case agree. For some, concepts of community are fictions imposed from outside by anthropologists (Nigh 2002), colonists or governments (Ayora Diaz 2002),[2] mere administrative conveniences that may or may not bear any relationship to local residential patterns or social organization. Berlin et al. (1999) argue that in Chiapas ethnic identity is first and foremost defined by the person's membership of a *municipality*, followed closely by identification with a particular *community*. While each of these communities is located within the boundaries of a municipality (*municipio*) politically recognized by Mexican law, Maya traditions of village autonomy, combined with the lack of municipal structures, mean that municipal government authorities generally have no power over decisions on local governance or about individual communities' use of their natural resources.

The Maya population is not formally represented in the region by a single sociopolitical body. Rather, they have a number of very dynamic and sometimes conflicting forms of territorial, economic, political and religious organization. Many of these are increasingly connected with national and international urban society organizations. The establishment of PROMAYA by the research project was partly 'a viable solution' (Berlin and Berlin 2006) to this perceived problem of representation. However, within this volatile political context, the Maya ICBG was to encounter a range of opposition.

---

[1] For a fascinating discussion on obtaining research permits from the leaders of autonomous municipalities in Chiapas, see Simonelli and Earle 2003 ('Yes, they're illegal. But they're legitimate'). Some commentators see the armed conflict as 'the most important limiting factor' for the Mayan people (Field 2006).

[2] Consuelo Sánchez argues that the Mexican government attempted to destroy ethnic identity by 'deliberately [fomenting]... the atomization of Indian peoples into fragmented communities' (Sánchez 1999).

17.4 Indigenous Partnership and Prior Informed Consent: Legitimacy vs Legality

17.4.1 International Regulation: Convention on Biological Diversity

The 1992 Convention on Biological Diversity (CBD) gives states sovereign rights over their biological resources and clearly states: 'Access to genetic resources shall be subject to prior informed consent of the Contracting Party providing access to such resources' (CBD 1992, article 15). It also requires that the sustainable use and wider application of biological diversity should 'respect, preserve and maintain knowledge, innovations and practices of indigenous and local communities ... with the approval and involvement of the holders of such knowledge, innovations and practices and encourage the equitable sharing of the benefits arising from the utilization of such knowledge, innovations and practices' (CBD 1992, article 8j; see also, Wynberg and Laird, Chapter 5)

There is clear international consensus, then, that in this context, consent for the use of genetic resources must be obtained from *the local community*. Working within the requirements of the CBD, the Maya ICBG proceeded on the basis that '[i]n the context of gaining permission and prior informed consent for research, "community" clearly refers to a specific, geographically bounded socio-political unit that forms a recognized subdivision of the municipality' (Berlin and Berlin 2002).[3] However, this strategy caused difficulties due to the contingent and shifting nature of 'community' in Chiapas.

17.4.2 Mexican National Regulations

Mexican governmental institutions are considered by many to be socially illegitimate, particularly in Chiapas. However, Mexican national regulations stipulated that plant collection for scientific purposes could only be permitted when 'supported by the previous, expressed and informed consent of the legal owner(s) of the land(s) on which the biological resource is found' (SEMARNAP 1997, as cited in Nigh 2002). It was this understanding of community consent that was operationalized by the Maya ICBG (Rosenthal 2006). However, this law proved inadequate for regulating *biotechnological* research, in which, as Nigh points out, samples are collected not simply for academic purposes, but for potential development into commercial products that embody indigenous knowledge (Nigh 2002). The Maya ICBG made the first

---

[3] Steffan Igor Ayora Diaz (2002) questions this definition of 'community', to which Nigh (2002) replies.

permit application for such 'biotechnology collections' in Mexico (Rosenthal 2006), but this was never granted, and so the proposed bioprospecting activities of the project never began.

17.5 Negotiating Prior Informed Consent with Communities

The Maya ICBG study area comprised 28 constitutional municipalities and as many as 8,000 hamlets. In line with the CBD, the researchers wanted the information provided to the Maya people to be as comprehensible as possible (Berlin and Berlin 2004), and there was a clear recognition of the need for iterative consent as an ongoing process. The team made a considered decision to go back to communities later to discuss consent for issues such as patenting if and when these arose, rather than trying to address such complexities from the beginning (Berlin and Berlin 2003). This strategy was later subjected to sustained criticism.

Initially the project focused on communities where the researchers had existing contacts. Groups of constitutional authorities at the sub-municipality level (hamlets of 50 to 100 houses) and community members were invited to ECOSUR. During an 8 hour visit, a play in native languages was peormed which dramatized the aims and objectives of the project and its potential benefits. The 200 visitors were taken on a tour of the herbarium, library and laboratory facilities and informed about all aspects of the project. Project brochures and documents in a variety of languages were also provided.

The researchers arranged to visit those villages that expressed an interest, where the play was presented again to the hamlet assembly. Communities that agreed to participate were asked to indicate their approval by signing a model agreement consenting to the project collecting plants and fungi.[4] The project followed individual community norms and practices. In most instances, elected community representatives signed. In some communities, individuals who were not elected leaders, but who wished their names to be recorded, also signed. In one community, the heads of households of each family signed. Other communities developed their own statements of agreement. During a three-month period, 46 out of 47 hamlets decided to sign up. The communities would later discuss a written proposal and accept or reject it. This process formed the basis of ECOSUR's April 2000 application to the government to issue permits to start plant collection for bioprospecting purposes (Berlin and Berlin 2003, 2004; Rosenthal 2006).

---

[4] Maya ICBG developed an agreement form modelled on guidelines from the US Department of Health and Human Services, the National Institutes of Health and the University of Georgia Institutional Review Board.

17.6 Intellectual Property Rights and Benefit Sharing

The theatrical presentation included details of 'a fair and just benefit-sharing plan', while written materials were provided to potential participants about the project to 'indicate how monetary benefits would accrue to communities as a result of any commercial applications' (Berlin and Berlin 2003). This demonstrates the attitude of the Maya ICBG towards benefit sharing – that while a share of the benefits would flow back to the Maya people, the *process* by which this would take place would be top-down – that is, conceived and constructed by the research community. 'Near-, medium-, and long-term benefits for Mayan communities were outlined explicitly' (Rosenthal 2006) in the *Benefit-Sharing and Protection of Intellectual Property Agreement*, and the details of PROMAYA had already been decided before the Maya communities were approached about the project. Neither was ever taken to them for consultation or negotiation, and this later became a huge problem.

The initial appointments of external experts to the PROMAYA board of directors had already been made. Yes, there was a stated intention to give the communities a voice in PROMAYA: 'When fully functioning the general membership of PROMAYA will consist primarily of members of Highland Maya communities represented in the study area, all of whom will have voting and governance rights' (Berlin et al. 1999). But the major decisions had already been taken: 'The members of the Maya ICBG [that is, the research institutions] have reached agreement governing terms of benefits sharing and protection of intellectual property by the participating parties of the Maya ICBG vis a vis the collaborating Highland Maya communities of Chiapas' (Berlin et al. 1999).

In line with the International Society of Ethnobiology (ISE) Code of Ethics (ISE 2006), the intended intellectual property rights arrangements were that Maya medicinal knowledge would be carefully documented to classify it, among other things, as 'prior art' in order to prevent attempts to patent it. Patents over discoveries, inventions, novel molecular structures and bioactive compounds would be held in property by all partners, with PROMAYA having a veto over their use. In no case were plants to be patented or research products used for the development of transgenic organisms. Most economic benefits were expected to be locally generated and not dependent on bioprospecting results or eventual agreements with pharmaceutical companies. In the event of any bioprospecting results, the three parties agreed to share monetary benefits in four equal parts, including PROMAYA, with the latter having power of approval over the decision (Berlin et al. 1999; Berlin and Berlin 2003).

In the course of the project, ECOSUR decided to transfer any eventual monetary benefits to PROMAYA. UGA researchers also decided to recover as much as possible from the UGA research fund for this purpose. For the reasons outlined below, potential PROMAYA members were never actually invited to meet, and the NGO never got off the ground during the life of the project.

No one doubts the good intentions of the project (Nigh 2002),[5] which proposed to administer the fund widely and equitably to share benefits across the entire region, where all communities, whether they had cooperated or not, would be able to apply for financial support for local health, production and conservation projects proposed by the communities themselves. However, there were glaring gaps in the way in which the local communities were included in the development of the benefit-sharing agreement, and these contributed to the subsequent difficulties.

17.7 Opposition to the Project

Concerns were first voiced by the local healers and midwives organization, the Organization of Indigenous Physicians of the State of Chiapas (*Organización de Médicos Indígenas del Estado de Chiapas* – OMIECH), which had an existing conflict with Dr Berlin over a previous project.[6,7] OMIECH had originally been invited to participate in PROMAYA in 1998. They declined, and were given assurances that the project would not proceed until legal requirements were fulfilled (Nigh 2002). They therefore objected when, in June 1999, they were shown a contract already signed by the three existing research partners (RAFI 1999a). OMIECH mandated the Chiapas Council of Traditional Indigenous Doctors and Midwives (*Consejo Estatel de Organizaciones de Médicos y Parteras Indígenas Tradicionales de Chiapas* – COMPITCH) to represent it in further discussions. COMPITCH was created in 1994 in the context of the encouragement of traditional medicine, and has been an ally of pro-Zapatista social organizations. It is a loosely knit council comprising some 1,100 men and women healers from 11 organizations, including OMIECH. COMPITCH in turn belongs to a national-level organization of Indian traditional healers which consists of 43 organizations from 17 states (Nigh 2002).

COMPITCH requested support from the Rural Advancement Foundation International (RAFI, which subsequently changed its name to Action Group on Erosion, Technology and Concentration, or ETC Group), an international NGO with a history of advocating

---

[5] The Rural Advancement Foundation International (RAFI), the project's main international NGO opponent, also described the project as having 'honourable intentions' (RAFI 1999b).

[6] Following the Declaration of the Alma-Ata International Conference on Primary Health Care in 1978, the Mexican government's *Instituto Nacional Indigenista* (National Indigenous Institute) created OMIECH, through which local healers and midwives were officially recognized as part of national health systems. However, medical staff in Mexico's national public health service never quite recognized this status. When Dr Brent Berlin and Dr Elois Ann Berlin started PROCOMITH (*Programa de Colaboración sobre Medicina Indígena Tradicional y Herbolaria* – Collaborative Programme in Traditional Indigenous Herbal Medicine) (Berlin et al. 1990) to promote ethnobotanical research in the highlands of Chiapas, they concluded collaboration agreements with the official health service and were thus seen by OMIECH as enemies.

[7] According to Nigh (2002), members of COMPITCH, including OMIECH, had been developing their own proposals for research and possible commercial development of Maya medicine for many years, but these initiatives were not included in the priorities of the ICBG.

against biopiracy and opposing bioprospecting. After consulting with RAFI and local advisers, COMPITCH concluded that the project was potentially damaging to Maya peoples' interests and asked Maya community authorities not to sign agreements until everyone had been adequately informed of their implications and proper legal and regulatory frameworks were in place (Nigh 2002). In September 1999 an article in a newspaper in the Chiapas town of San Cristóbal de las Casas reported the concerns of Maya groups (UGA 1999). In December 1999 the story was reported internationally by RAFI, who characterized the Maya ICBG as 'biopiracy', thus focusing world attention on the issue (RAFI 1999a).

Four main areas of concern emerged. The first related to prior informed consent procedures used in the Maya ICBG. RAFI questioned the validity of individual community agreements. There was also controversy over whether scientific plant collection which had occurred before the Maya ICBG started was included in, or required, these consent agreements (RAFI 2000).[8]

Opponents also questioned the quality and completeness of the information provided to communities, arguing that Maya ICBG prior informed consent procedures minimized information about patents and other bioprospecting risks and misled potential participants (RAFI 1999b). This raised a third concern about cultural erosion due to the introduction of 'the concepts of marketing, privatizing and individualizing knowledge and resources that were previously owned collectively and freely exchanged' (RAFI 2000) in a context where 'communities reject both intellectual property itself as well as the process of sharing benefits through PROMAYA' (Ceceña 2000).

Finally, opponents argued that a 25% share of 1% of total pharmaceutical profits was an unacceptable return. Of this amount, the whole indigenous population would receive the same share as one of the research partner institutions (RAFI 2000).

Subsequently an acrimonious national and international debate developed around the Maya ICBG amid questions of proprietary versus public knowledge as well as issues concerning the legal and social legitimacy of local communities' control over their biological resources. Maya communities that had indicated their willingness to participate in the project, lacking any umbrella organization to represent them as a whole, were unable, or chose not, to become involved. An evolving network of local, national and international indigenous organizations and NGOs openly joined opposition to the project (Berlin and Berlin 2004).

The Mexican Department of the Environment, Natural Resources and Fishing (*Secretaría de Medio Ambiente, Recursos Naturales y Pesca* – SEMARNAP) attempted to mediate what it saw as a conflict between COMPITCH and the scientists, but little progress was made in a series of talks through 2000, which were riddled with problems generated internally or by the local and national political

---

[8] Lauren Naville writes: 'Berlin admits that the plants sent to UGA prior to the ICBG-Maya project created confusion and decreased the trust between both parties. He offered to clear up the situation by sending people to check the plants at UGA to prove that they were not being used for bioprospecting research. ECOSUR also agreed to stop any type of collections' (Naville 2004).

context. SEMARNAP was perceived as an illegitimate social actor by some of the stakeholders, and general proposals, specific activities and/or resolution timeframes presented by either side were mostly seen as unacceptable by the other. For example, when the ICBG proposed a fast-track plan for consulting communities jointly, COMPITCH did not accept, and instead proposed a long-term unilateral consultation of their members and communities, as well as a nationwide discussion of a legal framework for bioprospecting. The indigenous organizations continued to insist that the government establish clear regulations, and the environmental ministry continued to delay issuing these (Nigh 2002).

Throughout this process, some Maya ICBG members continued to rely on the possibility of obtaining government permits for bioprospecting collection, based on the agreements issued by hamlets. Most ECOSUR members of the ICBG insisted on re-establishing dialogue with COMPITCH and including more indigenous organizations in order to bring a more plural and representative spectrum of indigenous partners to the negotiating table. Both strategies failed.

In September 2000 SEMARNAP denied ECOSUR's permit application for biological assays, on the basis that according to current legislation, agreement at the municipal community level and not only at the *paraje* community level was required (ECOSUR-San Cristóbal n.d.; Naville 2004).

Amid increasing concern about the future of the project, ECOSUR declared a moratorium on the Maya ICBG in a statement published in a popular national newspaper. The local researchers said that no further attempts would be made to begin bioprospecting until (1) administrative and legal mechanisms for bioprospecting in Indian territories and procedures for obtaining prior informed consent and final authorization were in place, and (2) Indian communities and organizations had established a formal authoritative representative body to represent and protect their interests (Barrios and Espinosa 2000). ECOSUR's decision reflected its recognition that the fourth member of the project should be a self-organized, widely representative indigenous organization responsible for facilitating information provision and discussion as well as conducting the informed consent process.

Following further attempts to modify the project, ECOSUR announced a suspension of its activities with the Maya ICBG in October 2001 on the grounds that appropriate conditions were not in place for the project to be seen as both legal and legitimate. It argued that continuing involvement could be a further source of conflict within and among communities, which would harm ECOSUR's other long-term outreach efforts and could hamper its own capacity to participate in national discussions directed at developing bioprospecting regulations. Without a local research partner, the funders terminated the project grant (Rosenthal 2006).

Bioprospecting projects are frequently beset by turmoil. Often analysis of such difficulties focuses on conflicts of interest, but another way to understand the problems is to analyse the diverse actors' views and assumptions with attention to the particular cultural context (Naville 2004). Looking for an explanation of failure in the other parties is not always helpful. For example, Dr Berlin has claimed that rather than simply focusing public attention on the project's shortcomings through their campaigns, the NGOs themselves are actually responsible for the failure of the

Maya ICBG (Berlin and Berlin 2003), while Rosenthal – director of the ICBG Program at NIH (Berlin and Berlin 2004) with 'firsthand knowledge of the facts' (Berlin and Berlin 2006) – argues that 'the political, cultural, and governance context in which the members of the Maya ICBG chose to erect the project may have doomed it from the beginning' (Rosenthal 2006). Neither approach accepts much responsibility for the project's failure on the part of the research community.

17.8 Comparison with San Case

The effect of NGOs and other organized social networks has been significant in both this case and that of the San. The South African-based NGO Biowatch first exposed the story of the agreement between South Africa's Council for Scientific and Industrial Research (CSIR) and the British company Phytopharm on the world stage, initiating international interest in the story (Wynberg 2004). This contributed to the creation of conditions in which, with the active assistance of a San NGO, the South African San Institute, successful negotiations could take place. The Maya ICBG controversy was significantly influenced by the involvement of NGOs, which in this case exposed the general mistrust between Mexican institutions and the indigenous people in a way that demonstrated the power of coalitions between social actors when they are united by the same assumptions, claims and motivations (Naville 2004).

Both cases played out in a domestic legal and policy vacuum. When the San concluded their first benefit-sharing agreement with the CSIR in 2003, the South African Biodiversity Act was yet to be promulgated, and when they further negotiated with commercial *Hoodia* growers in 2006, regulations regarding benefit-sharing agreements were not yet in force. The San learned that it was possible to negotiate binding rights despite this lack of an enabling domestic legal environment, with no policy or law yet in place. Roger Chennells described the achievement as 'evidence of the power and ability of negotiating parties to meet each other, to establish rules of engagement, to commit to act in good faith, and to attempt to strike a balance encompassing the long term requirements of divergent parties' (2007; see also, Wynberg and Chennells, Chapter 6).

In Mexico, bioprospecting regulations remain incomplete and imprecise, and do not fully protect the interests of indigenous people nor consider their claim for autonomy. This lack of a proper legal system exacerbated the conflict over the Maya ICBG, in which each social actor interpreted the law according to its own interests in the post-CBD but pre-Bonn Guidelines (CBD 2002) environment. Legislation requires social legitimacy which can only be acquired through consultation; and indigenous people, social organizations and greater segments of society at large need to be informed and consulted about bioprospecting to develop this legal framework successfully.

Questions about the legitimacy of processes and decisions emerge as fundamental in both of these cases. In the case of the Maya ICBG they were of more significance to the opposition than the lack of a legal framework. Everyone agreed that the indigenous Maya people had to be considered owners of the resources if legitimacy was to be ensured. But there was fundamental disagreement about the way prior informed consent and decision-making procedures should be undertaken. This is markedly different from the *Hoodia* case, in which the P57 patent, although clearly dependent on San traditional knowledge, was developed without any active engagement with the San peoples – an approach which was originally defended on the basis that they were 'extinct' (Chennells 2007).

However despite the Maya ICBG's 'elaborate informed-consent protocol' (Rosenthal 2006), the failure of the project was basically due to the lack of an appropriate prior informed consent process built on trust and adequate representation. Problems of trust and communication were inherent in the bioprospecting model created by the project. In the absence of a comprehensive, representative sociopolitical organization in Chiapas to which prior informed consent requests could be directed, it looked to the researchers as though they were confronted with an insurmountable obstacle when the legitimacy of permissions received from agreeable communities was contested. (Indeed, their legal status to make these agreements was eventually denied by SEMARNAP's decision.) As we have seen, some of the researchers at ECOSUR understood that a project of this kind needed to be built upon the gradual establishment of trust and collaboration among the stakeholders.

We believe that no single actor should bear the total responsibility for what happened, but that none is devoid of it. Both ECOSUR and the UGA team were too optimistic about their ability to build trust and insufficiently sensitive regarding the level and type of communication and trust that were necessary for such a complex endeavor to succeed.

The San peoples have experienced an appalling history of genocidal oppression, and it is out of this history that they have recently started to build representational structures with which to negotiate with a united voice. These structures are not without their complications, but have been central to the success of the *Hoodia* benefit-sharing negotiations. Likewise the Maya people and their cultural traditions suffered at the hands of colonialism. The Maya ICBG project was planned during the Zapatista revolution and amid the failure of the government to honour the 1996 San Andrés Accords, which would have strengthened the legitimate status of the community, in particular in relation to natural resources, and would also have privileged indigenous communities with regard to obtaining the benefits from those natural resources (Berlin and Berlin 2003). This situation inevitably affected basic conditions for trust.

Since the 1994 Zapatista rebellion, a plethora of decision-making processes and grassroots organizations have flourished, so that 'the Chiapas Maya have become increasingly capable of making their decisions felt' (Simonelli 2006). However, the Maya ICBG chose not to include existing broadly representative organizations such as COMPITCH as active partners in their research. Despite the fact that to be recognized by COMPITCH, a healer had to have community backing (Ayora Diaz 2002), 'later Maya ICBG investigators argued that COMPITCH was not 'representative' of Maya Indians as a whole' (Nigh 2002).

Rosenthal maintains that the most fundamental factor in the outcome of such a project 'is the existence of an established, credible, and politically representative governance system of the indigenous communities involved' (2006) and that the Maya ICBG failed because of a lack of such structures. However, other field anthropologists in Chiapas dispute the notion that such structures do not exist (Hunn 2006), and Shane Greene has asked why COMPITCH, for example, was not deemed to hold such status, when it 'sounds a lot like a Western-style professional association to me' (Greene 2006).

The question of Maya identity and self-representation through forms that are 'credible' to outside bioprospectors is therefore an ongoing issue. But the concept of 'the Maya people' itself is a relatively recent, imported ethnographic category (Nigh 2002). This parallels the experience of the groups whom the Harvard Kalahari Research Group in 1961 named the San peoples, a term that has since been recognized as the only known one to include all their peoples. In 1996 San leaders formed the Working Group of Indigenous Minorities in Southern Africa (WIMSA) as a San umbrella organization to unite and represent the interests of San communities across Botswana, Namibia and South Africa. The 1998 WIMSA General Assembly confirmed that San culture and heritage was a collective asset, an assertion that became very significant later, when they came to negotiate their rights in the *Hoodia* case, according to which benefits are to be shared among San peoples across South Africa, Botswana and Namibia (Chennells 2007).[9] The pan-Mayan identity which is currently under construction in Chiapas faces similar challenges. In relation to the Maya ICBG, COMPITCH has also raised the issue of 'collective resources', which 'do not belong to one, two or even 50 communities in the Highlands of Chiapas, but may belong to the entire area inhabited by the Maya, well beyond the Mexican borders' (RAFI 2000).

In addition to posing problems for the 'national sovereignty' application of the CBD, this indigenous approach to territory which crosses political borders challenges Western intellectual property concepts. The Berlins have more recently suggested: 'Knowledge that extends across many local communities, throughout a region or country, or across national boundaries must necessarily be deemed to be knowledge in the public domain' (Berlin and Berlin 2004). While agreeing that it is essential to acquire the consent of local communities to the collection of such resources, they argue that the use of such publicly accessible information does not infringe any legal right and is not subject to any intellectual property protection.

This approach would seem to undermine the southern African San peoples' hard-won *Hoodia* benefit-sharing agreement, which was negotiated on legal grounds, as well as ethical imperatives, after an 'access and benefit-sharing' story in which everything that could go wrong did. It is a position that seems to take an unnecessarily narrow view of what constitutes a 'legal right' in the context of the growing international consensus on the extent of collectively held intellectual property

---

[9] Interestingly, WIMSA avoided using 'San' in their title as the name was 'strongly disapproved of by certain southern African Governments' (Chennells 2007).

rights (WIPO 2008a, 2008b, 2008c). This emphasis on the lack of legal rights would be an unfortunate legacy of the Maya ICBG, which was intended to be a model of good practice.

The current global context regarding the conservation of genetic resources, intellectual property and trade is having an impact on relationships between people, land and natural resources all over the world. Indigenous peoples are particularly affected by these changes, through which intellectual property rights and conflicts assume increasing economic importance. However, 'indigenous people don't necessarily believe that they have interests in common with anthropologists' (Brush 2002) or any other researchers. Edward Fischer (2002) puts it rather more strongly:

> More than just opposing the ICBG project, COMPITCH and the other Maya opposition were making a statement ... about the new power relations that hold in this globalized world, decrying epistemological violence to their culture in the name of scientific progress.

One positive impact of the Maya ICGB in Chiapas is that almost all of the communities are aware of the controversy and now insist on researchers explaining and answering many more questions about the potential risks and benefits of proposed projects (Nigh 2002; Simonelli 2006). This parallels lessons learned and applied by the San. Mathambo Ngakaeaja, a WIMSA delegate to an international conference in 1997, stated that until recently 'the San have been treated as objects of research. ... they have not been involved in the research agendas of the academics, and their own needs and aspirations have been ignored'. He went on to announce WIMSA's policy that no research would take place on the San in the future without the researcher completing a comprehensive research contract with the San leadership (Chennells 2007).[10] Likewise, in response to the situation in post-Maya ICBG Chiapas, the design and conduct of research is being rethought, with a move away from 'extractive research' which provides information to outsiders, towards 'enriching research', which is more empowering for local people (Nigh 2002).

Benefit sharing can easily become controversial due to conflicting assumptions about how and to what extent different people, or groups of people, should benefit from the potential royalties, and who should make these decisions. Issues of power relationships always complicate benefit sharing. These two cases are structurally similar in that both involved a for-profit partner which appeared to be small and therefore risk-taking, whereas, in the event, both enterprises seem linked to much larger operations apparently able to make large amounts of money from the development of pharmaceutical products. In the *Hoodia* case, Phytopharm was able to sublicense the P57 patent to international drug manufacturer Pfizer for development and exploitation, and later to sublicense commercialization rights to Unilever for development into a food product (Chennells 2007). In the Maya ICBG case, MNL was presented as a small enterprise, but acted as a technological intermediary to multinational pharmaceutical companies including Glaxo (RAFI 2000). Like the

---

[10] 'This assertion was angrily received by some social scientists, accustomed to unilateral research on hapless San communities' (Chennells 2007).

San peoples, the Maya stood to receive a very small proportion of any profit that might come from the development of pharmaceutical products.

Nevertheless, it was possible that this could amount to a large amount of money flowing into the community. Wynberg has raised concerns about the 'potentially divisive impact' this can have in an indigenous community, and draws attention to

> the fraught questions of administering the funds, of determining beneficiaries and specific benefits across geographical boundaries and within different communities, and of minimizing the social and economic impacts and conflicts that could arise with the introduction of large sums of money into poor communities (Wynberg 2004).

Rosenthal has expressed concern about the way that research and development partnerships 'alter the power dynamics and membership of local political organizations', which 'raises other important ethical concerns regarding the origins of social change in a traditional society' (Rosenthal 2006). Chennells describes the specific challenges faced by the San beneficiaries, both within and outside of their own communities, as they face 'unenviable, but historic tasks' in '[f]inding the uncharted path representing the divergent aspirations of their communities' (Chennells 2007). It is clear that these windfall benefits, whether they are ever actually realized or not, are never ethically neutral, so the transparent, full and free prior informed consent of communities to accept the risk of going along this path is absolutely essential.

If more acceptable prior informed consent processes had been conducted in Chiapas, further challenges would inevitably have emerged linked to issues of trust, defining fair benefits and clarifying the diverse needs and interests of the stakeholders. If a dialogue had been successfully initiated among the external partners and the Maya people, then there would have been a better chance of consolidating a legitimate body of representation to carry this programme forward, and a truly collaborative, enriching research project might have resulted. This is the real loss of the failure of the Maya ICBG.

References

Ayora Diaz, S. I. (2002). Comment, in Nigh, R., 'Maya medicine in the biological gaze: bioprospecting research as herbal fetishism'. *Current Anthropology, 43*(3), 464–465.

Barrios, L. E. G., & Espinosa, M. G. (2000). ECOSUR y el proyecto de bioprospección ICBG-Maya en Chiapas. A los pueblos y comunidades indígenas, a todos los miembros de COMPITCH, a la opinión pública. *La Jornada*, Mexico City. http://chiapas.laneta.org/noticias/icbgmaya.htm. Accessed 12 November 2008.

Berlin, B., & Berlin, E. A. (2002). Comment, in Nigh, R., 'Maya medicine in the biological gaze: bioprospecting research as herbal fetishism'. *Current Anthropology, 43*(3), 466–467.

Berlin, B., & Berlin, E. A. (2003). NGOs and the process of prior informed consent in bioprospecting research: the Maya ICBG project in Chiapas, Mexico. *International Social Science Journal, 55*(178), 629–638. http://unesdoc.unesco.org/ulis/cgi-bin/ulis.pl?req=2&mt=100&mt_p=%3C&by=2&sc1=1&look=default&sc2=1&lin=1&mode=e&futf8=1&gp=1&text=Maya+ICBG+Mexico&text_p=inc. Accessed 30 June 2008.

Berlin, B., & Berlin, E. A. (2004). Community autonomy and the Maya ICBG Project in Chiapas, Mexico: how a bioprospecting project that should have succeeded failed. *Human Organization*,

Winter 2004. http://findarticles.com/p/articles/mi_qa3800/is_200401/ai_n9466403. Accessed 27 November 2008.

Berlin, B., & Berlin, E. A. (2006). Comment, in Rosenthal J. P., 'Politics, culture, and governance in the development of prior informed consent in indigenous communities', *Current Anthropology, 47*(1), 129–130.

Berlin, B., Berlin, E. A., Breedlove, D. E., Duncan, T., Astorga, V. M. J., Laughlin, R. M., et al. (1990). *La herbolaria médica tzeltal-tzotzil en los altos de Chiapas.* Chiapas, Mexico: Gobienos del Estado de Chiapas, Consejo Estatal de Fomento a la Investigación y Difusión de la Cultura, DIF-Chiapas, and Instituto Chiapenco de Cultura, Tuxtla Gutienez.

Berlin, B., Berlin, E. A., Ugalde, J. C. F., Barrios, L. G., Puett, D., Nash, R., et al. (1999). The Maya ICBG: drug discovery, medical ethnobiology, and alternative forms of economic development in the highland Maya region of Chiapas, Mexico. *Pharmaceutical Biology,* 37(Suppl.), 1–19.

Brush, S. B. (2002). Comment, in Nigh, R., 'Maya medicine in the biological gaze: bioprospecting research as herbal fetishism', *Current Anthropology, 43*(3), 467–468.

CBD (1992). *Convention on Biological Diversity.* Quebec: Secretariat of the Convention on Biological Diversity. www.cbd.int/convention/convention.shtml. Accessed 18 June 2008.

CBD (2002). *Bonn guidelines on access to genetic resources and fair and equitable sharing of the benefits arising out of their utilization.* Quebec: Secretariat of the Convention on Biological Diversity. www.biodiv.org/doc/publications/cbd-bonn-gdls-en.pdf. Accessed 27 June 2008.

Ceceña, A. E. (2000). *¿Biopiratería o desarrollo sustentable?* Chiapas, México: Era-IIE. http://membres.lycos.fr/revistachiapas/No9/ch9cecena.html. Accessed 11 November 2008.

Chennells, R. (2007). *San Hoodia case: a report for GenBenefit.* http://www.uclan.ac.uk/old/facs/health/ethics/staff/projects/GenBenefit/docs/cases/San_Case.pdf. Accessed 7 December 2008.

Field, L. (2006). Comment, in Rosenthal, J. P., 'Politics, culture, and governance in the development of prior informed consent in indigenous communities'. *Current Anthropology, 47*(1), 131.

Fischer, E. F. (2002) Comment, in Nigh. R., 'Maya medicine in the biological gaze: bioprospecting research as herbal fetishism'. *Current Anthropology, 43*(3), 470–471.

Greene, S. (2006) Comment, in Rosenthal, J. P., 'Politics, culture, and governance in the development of prior informed consent in indigenous communities'. *Current Anthropology, 47*(1), 131–132.

Hunn, E. S. (2006) Comment, in Rosenthal, J. P., 'Politics, culture, and governance in the development of prior informed consent in indigenous communities'. *Current Anthropology, 47*(1), 133–134.

ISE (2006) *Code of ethics.* International Society of Ethnobiology. http://ise.arts.ubc.ca/global_coalition/ethics.php. Accessed 27 November 2008.

Naville, L. (2004). *The experts, the heroes, and the indigenous people: the story of the ICBG-Maya bioprospecting project in Chiapas, Mexico.* Ås, Norway: Noragric, Agricultural University of Norway. www.umb.no/noragric/publications/msctheses/2004/2004_ds_Lauren_Naville.pdf. Accessed 15 October 2008.

Nigh, R. (2002). Maya medicine in the biological gaze: bioprospecting research as herbal fetishism. *Current Anthropology, 43*(3), 451–477.

RAFI (1999a). Biopiracy project in Chiapas, Mexico denounced by Mayan Indigenous Groups. Rural Advancement Foundation International. www.etcgroup.org/en/materials/publications.html?pub_id=348. Accessed 8 May 2008.

RAFI (1999b). Messages from the Chiapas 'bioprospecting' dispute. Rural Advancement Foundation International. www.etcgroup.org/en/materials/publications.html?id=344. Accessed 18 June 2008.

RAFI (2000). Stop biopiracy in Mexico – Indigenous people's organizations from Chiapas demand immediate moratorium, Mexican government says no to bioprospecting permits. Rural Advancement Foundation International. www.etcgroup.org/en/materials/publications.html?pub_id=304. Accessed 18 June 2008.

Rosenthal, J. P. (2006). Politics, culture, and governance in the development of prior informed consent in indigenous communities. *Current Anthropology, 47*(1), 119–142.

Sánchez, C. (1999). *Los pueblos indígenas: del indigenismo a la autonomia.* Mexico City: Siglo Veintiuno, Editores.

SEMARNAP (1997). *Ley general del equilibrio ecologico y la proteccion al ambiente*, Secretaria de Medio Ambiente, Recursos naturales Y Pesca, Mexico City, quoted in Nigh, R. (2002). Maya medicine in the biological gaze: bioprospecting research as herbal fetishism. *Current Anthropology, 43*(3), 451–477.

Simonelli, J. (2006). Comment, in Rosenthal, J. P., 'Politics, culture, and governance in the development of prior informed consent in indigenous communities'. *Current Anthropology, 47*(1), 135–136.

Simonelli, J., & Earle, D. (2003). Meeting resistance: autonomy, development, and "informed permission" in Chiapas, Mexico. *Qualitative Inquiry, 9*(1), 74–89.

UGA (1999). *Drug discovery and biodiversity project among highland Maya subject of misunderstanding, according to UGA professor*, news release, University of Georgia, GA, 6th December. www.uga.edu/news/newsbureau/releases/1999releases/berlin_maya.html. Accessed 15 May 2008.

WIPO (2008a). *Recognition of traditional knowledge within the patent system*. WIPO Intergovernmental Committee on Intellectual Property and Genetic Resources, Traditional Knowledge and Folklore, Thirteenth Session, Geneva, 13–17th October. www.wipo.int/edocs/mdocs/tk/en/wipo_grtkf_ic_13/wipo_grtkf_ic_13_7.doc. Accessed 20 November 2008.

WIPO (2008b). *Genetic resources: list of options*. WIPO Intergovernmental Committee on Intellectual Property and Genetic Resources, Traditional Knowledge and Folklore, Thirteenth Session, Geneva, 13–17th October. www.wipo.int/edocs/mdocs/tk/en/wipo_grtkf_ic_13/wipo_grtkf_ic_13_8_a.doc. Accessed 21 November 2008.

WIPO (2008c). *Genetic resources: factual update of international developments*. WIPO Intergovernmental Committee on Intellectual Property and Genetic Resources, Traditional Knowledge and Folklore, Thirteenth Session, Geneva, 13–17th October. www.wipo.int/edocs/mdocs/tk/en/wipo_grtkf_ic_13/wipo_grtkf_ic_13_8_b.doc. Accessed 21 November 2008.

Wynberg, R. (2004). Rhetoric, realism and benefit sharing: use of traditional knowledge of Hoodia species in the development of an appetite suppressant. *The Journal of World Intellectual Property, 7*(6), 851–876.

ECOSUR-San Cristóbal (n.d.). *El ICBG Maya en Los Altos de Chiapas*. Project archives of the General Director.

Part IV
Conclusions and Recommendations

To share in our knowledge you need to understand us, you need to understand how we relate to ... [each other]. You need to understand how we communicate, and the only way to do that is to develop a relationship, to grow the relationship and [then] the knowledge will be shared on the basis of ... [the relationship]. ... My biggest advice would be, please, ... do ... not ... just focus on the economic gains, because for indigenous people the most important thing is the relationship.

Jack Beetson

Aboriginal teacher and UN 'Unsung Hero of the 20th Century'

Presentation to a San-!Khoba project workshop, Kalk Bay, South Africa, June 2006.

Chapter 18
Conclusions and Recommendations: Towards Best Practice for Community Consent and Benefit Sharing

Rachel Wynberg, Roger Chennells, and Doris Schroeder

18.1 Introduction

The adoption of the Convention on Biological Diversity (CBD) is one of the great policy success stories of the twentieth century. One hundred and ninety-one parties have signed this broad and forward-thinking convention after exceptionally wide processes of consultation. Only Andorra, the Holy See, Somalia and notably the United States are not party to the CBD. Yet the treaty's full implementation is hindered by unresolved practical matters. This collection of papers has illuminated some of the most crucial issues facing both policymakers and practitioners, and these are drawn together in the following conclusions and recommendations.

18.2 Land: The Foundation of Indigenous Peoples' Rights

Indigenous peoples regard knowledge as being closely tied to and associated with land. Traditional knowledge encapsulates peoples' spiritual experiences and relationships to land, lying at the very heart of their cultures. As a number of authors

---

R. Wynberg (✉)
Environmental Evaluation Unit, University of Cape Town, Private Bag X3, Rondebosch, 7701, Cape Town, South Africa
e-mail: rachel@iafrica.com

R. Chennells
Chennells Albertyn: Attorneys, Notaries and Conveyancers, 44 Alexander Street, Stellenbosch, South Africa
e-mail: scarlin@iafrica.com

D. Schroeder
UCLAN, Centre for Professional Ethics, Brook, 317, Preston, PR1 2HE, United Kingdom
e-mail: dschroeder@uclan.ac.uk

in this book have shown, indigenous peoples are currently waging struggles for self-determination, which include rights to ancestral lands and the associated resources that were lost on a vast scale during the colonial expansion of the past centuries. The brutal annexation of lands which had belonged to indigenous peoples for thousands of years was supported by the widely held principle of *terra nullius*, which stated that land governed by indigenous peoples in a 'foreign' manner (in the opinion of the colonizing state) was not 'owned' in a recognized sense, and could therefore lawfully be seized and occupied.

Ancestral lands are more to indigenous peoples than simply the geographical source of their historical livelihoods and identity. As Jon Altman explains, Aboriginal Australians conceive of land as being animate or 'sentient', and this cosmology is common to and shared by many indigenous peoples worldwide. The San of southern Africa have the same reverence for nature, and suffer the same powerlessness and dislocation, as the Kani of India, the indigenous tribes of the Philippines and other indigenous peoples of the world that have been removed from their ancestral lands. Knowledge of the medicinal and spiritual values of plants, and of the associated remedies, myths and rituals, become meaningless when people are removed from their traditional lands.

Saskia Vermeylen, recounting the colonial and post-colonial process by which the San in Namibia were alienated from their lands, refers to the slogan, 'Kill the tribe to build the nation,' which underpinned colonial attempts to homogenize sociocultural differences among ethnic groups. The replacement of traditional land tenure rights by other, Western forms of landholding inevitably alienated indigenous peoples from their livelihoods and further disturbed the ancient relationship between them and their lands. Aboriginal title has, however, grown in acceptance and importance as a legal principle, affirming the inalienable right of indigenous people who are able to show, in place of a formal system of governance, unbroken occupation since time immemorial of particular ancestral lands.

The restoration of land and resource rights to indigenous peoples over the past decades has been closely associated with global attempts to reform the related injustices of the past that are recounted in this book. Theft by nations and corporations of traditional knowledge, or of the products of biodiversity, is now similarly acknowledged to be unfair and warranting corrective attention. Legal victories such as cases of the Australian Mabo (Mabo 1992), the South African Richtersveld (Alexkor 2003) and the Botswana Central Kalahari Game reserve (Roy Sesana 2002) all support the cautious conclusion that indigenous peoples are successfully reclaiming their ancient rights to land.

> Land is the foundation on which indigenous peoples preserve and protect their traditional knowledge. Equitable benefit sharing for indigenous peoples should therefore be linked to broader initiatives to secure their rights to the resources, knowledge and land that have been alienated over centuries.

18.3 Justice and the Convention on Biological Diversity

Throughout colonial history, plants and other biological resources were taken from their habitats by foreign botanists and collected for display in European botanical gardens, use in commercially focused research or straightforward sale. Based on the assumption that resources from the wild belonged to the common heritage of humankind, Southern biodiversity was depleted to promote Northern interests. No benefits were shared with local communities and countries of origin, but instead the 'common heritage of humankind' principle was interpreted as a free-for-all on a first-come-first-served basis. As a result, the 'open access to resources' approach was considered highly exploitative.

Yet, independent of context, the common heritage principle is not in itself unfair. For instance, the 1982 United Nations Convention on the Law of the Sea established that resources falling under the agreement belonged to humanity as a whole and should not become the property of any state. Instead their use should *benefit humankind as a whole*. This interpretation of the common heritage principle might even promote overall social utility best, as it allows freely available resources to be used for the benefit of all. Such open access to natural resources could lead to rapid medical progress and new treatments accessible to everyone. Alas, this is not the case at the beginning of the twenty first century and it never has been. Open access in the past has led to countries feeling increasingly suspicious of any research to develop their genetic resources for new medicines, crops or other purposes within an international economic order characterized by enormous disparity.

A third of all human deaths are from avoidable, poverty-related causes, such as a lack of access to vaccines, medicines or food. Poverty-related deaths occur mostly in countries with high biodiversity that used to be under colonial rule. For Northern companies and researchers today to insist on the common heritage of humankind principle would simply validate a first-come-first-served approach that greatly favours the affluent and powerful over the poor. As Vandana Shiva (1991) notes, as long as the North does not open the doors to its industrial products (e.g. vaccines) based on open access policies, it is unacceptable for it to insist on seeing developing country resources as a common good.

As Doris Schroeder argues, the CBD is founded on a spirit of justice, redressing past injustices by creating a social rule that affirms the sovereignty of nation states over their resources. Its effectiveness will be judged by the extent to which wealth and, perhaps more significantly, non-monetary benefits such as technology transfer are shared equitably between bioprospectors, local communities and developing nations. The San-*Hoodia* case could have been the most exciting and advanced of its kind, with millions of dollars flowing into a highly marginalized community ravaged by poverty-related diseases such as tuberculosis and without adequate access to education. Besides the potential difficulties of coping with significant inflowing money, the vagaries of markets and the risks associated with natural product development mean that aspirations of development facilitated through the CBD are largely unrealistic. This recalls Altman's warning that benefit sharing can only ever be a partial solution for development: despite over 300 agreements between mining companies and indigenous communities throughout Australia, indigenous peoples remain economically marginalized and socially disadvantaged.

As is noted below in the section on managing expectations, it is important to distinguish between what benefit sharing *can* achieve and what it *cannot* do. It cannot remedy the ills of an international economic order that is pitched against the interests of the poor (e.g. by pricing essential drugs out of their range in order to provide incentives to pharmaceutical companies). But it can make some progress on redressing past injustices by forcing those who access biodiversity to negotiate mutually agreed terms with providers.

> While benefit sharing rightly aims for justice in exchange (e.g. traditional knowledge holders must be consulted with and compensated for the use of their resources), it cannot take over the role of governments. Governments have a legal duty to ensure that their citizens can live in dignity and, importantly, have their basic needs met. Basic human rights must continue to be secured for all citizens by their nation states. Benefit sharing is no quick fix for deep-seated international injustice and never can be.

18.4 Commodification

Since ancient times, philosophers have argued that some entities are commodifiable – that is, exchangeable for money – while others are not. Most famously, Immanuel Kant (1996) asserted that a human being is

> exalted above any price; for as a person … he is not to be valued merely as a means to the ends of others or even to his own ends, but as an end in himself, that is, he possesses a dignity (an absolute inner worth) …

Determining which entities or actions are beyond any price is still contentious in the twenty first century. Kidneys? Surrogate motherhood? Embryonic stem cells? For some, traditional knowledge belongs to the group of entities that should never be traded or commodified. As Vermeylen notes in Chapter 10, the commercial use of indigenous knowledge or heritage by outsiders has been called 'sacrilege' or 'defamation'. It has even been argued that fair and equitable benefit sharing is *impossible* under 'the prevailing [CBD] paradigm of privatisation and commodification of nature and knowledge' (Sridhar et al. 2008) and that benefit sharing is dead (Sharma 2005). By contrast, others have pointed out that trading traditional knowledge is an appropriate means to mitigate serious poverty or that it can be a liberating act.

Research for this book undertaken by Vermeylen among San populations in Botswana, South Africa and Namibia showed no strong rejection of or support for

18 Conclusions and Recommendations: Towards Best Practice for Community

the commodification of traditional knowledge. But splitting responses along gender lines did give a clearer picture. Women respondents noted a slight preference for the sharing of knowledge for financial reward over the two alternative scenarios (refusal to share knowledge and the sharing of knowledge with legal protection). When asked to justify their preference for commodification, some noted that they wanted their children to shake off their parents' impoverished and stigmatized identity through education, which was costly. On the other hand, the majority of male respondents favoured the sharing of knowledge in exchange for legal protection. When asked to justify this choice, some argued that gaining legal rights over their knowledge and other assets (e.g. natural resources and land) was essential to restoring their human dignity.

The San-*Hoodia* case illustrates the difficulties the San faced when deciding whether or not to commodify their knowledge. The route of signing a benefit sharing agreement involved an implicit acceptance of the patent obtained by the Council for Scientific and Industrial Research (CSIR), and thus of the commodification of San knowledge. Yet, as San representatives noted in interviews undertaken for Chapter 12 by Rachel Wynberg, Doris Schroeder, Samantha Williams and Saskia Vermeylen, the pressure to reach this decision, combined with the complexity of the issues, suggest that sufficient time and capacity could well have led to another decision.

> Knowledge has been traded for millennia without being considered non-commodifiable. It is not for policymakers, academics, activists, lawyers or other outsiders to decide whether traditional knowledge should be commodified in particular circumstances or not. This decision has to rest with those directly concerned, but they must be assured of sufficient time to gather information, and build capacity and knowledge, in order to be able to act appropriately and independently. The pros and cons of commodification are local choices that cannot be made universally, and communities that categorically reject the possibility of sharing their knowledge need to be sure that they will be heard and respected. Hence it is paramount to take seriously and strengthen the CBD's provision for prior informed consent.

18.5 Prior Informed Consent

According to CBD decision V/16, access to traditional knowledge is subject to formal prior informed consent requirements.

> Access to the traditional knowledge, innovations and practices of indigenous and local communities should be subject to prior informed consent or prior informed approval from the holders of such knowledge, innovations and practices.

Realizing the full potential of prior informed consent to end the exploitative use of biological resources and traditional knowledge is an enormous challenge.

As Rachel Wynberg and Roger Chennells describe in their rendition of the San-*Hoodia* case, no effort was made to obtain consent from the San before using their knowledge of the appetite suppressant qualities of *Hoodia*. Understandably, this created serious dissonance between the parties from the start and could well have led to a challenge of the patent. One of the most important steps in building relationships between the CSIR and the San after the patent application was the former's open and unreserved acknowledgement that the San's knowledge had been used in the research process. Though a benefit sharing agreement was reached in this case, such an acknowledgement should never be seen as an acceptable substitute for *prior* informed consent.

Yet, as the chapter by Graham Dutfield and that of Dafna Feinholz-Klip, Luis García Barrios and Julie Cook Lucas show, even the best-laid schemes of mice and men often go awry in the context of obtaining consent. Using cases from Peru and Mexico, the authors demonstrate that even the most sincere and painstakingly planned efforts to obtain prior informed consent can fail and lead to unforeseen complications.

Some progress can be made in developing and improving prior informed consent procedures through learning from the medical context, as Schroeder shows. For instance, the practice of involving an intermediary between researcher and research participant ensures that information is conveyed in a relatively neutral manner. But the analogies between obtaining consent for a medical procedure and obtaining prior informed consent to access traditional knowledge fail in other areas, as both Dutfield and Schroeder explain. In particular, and as elaborated below, the identification of who can legitimately give consent – hardly an issue in the medical context – can seem an insurmountable obstacle in the context of the CBD.

As Jack Beetson eloquently appeals on the last page of this book, sustained efforts need to be made by those desiring access to traditional knowledge to build long-term relationships that minimize the potential for exclusion and misunderstandings.

> The principle of prior informed consent is not negotiable, but significant flexibility is required in its attainment, recognizing that circumstances vary from case to case and community to community. All parties need to approach the complex and challenging consent process with a willingness to adapt to circumstances and focus on building relationships over time. Obtaining consent is not a quick, one-off process as in the medical context, but rather an iterative, progressive process, which benefits significantly from collaboration with local intermediaries and support organizations. In the words of Mason Durie (2008), deputy vice-chancellor at Massey University, New Zealand, and a Māori, 'an encounter [with indigenous peoples] is more likely to have a good outcome if mutual benefits are on the agenda, agreement is reached about the terms, and … a commitment to a long term relationship is made'.

18.6 Identification of Traditional Knowledge Holders and Relevant Authorities

Many of the difficulties faced by those wanting to engage in access and benefit sharing negotiations revolve around two fundamental assumptions: firstly, that there are traditional knowledge holders who are identifiable, organized, coherent and able to enter into negotiations, and, secondly, that it is straightforward and tenable to identify authorities through which agreements can be processed and ratified. As several authors highlight, neither of these assumptions is necessarily true, and, indeed, they constitute the exception rather than the rule.

In the first instance, and as described earlier, there is often a lack of clarity with regard to the rights that traditional knowledge holders have to land and resources. Vermeylen, Chennells and Altman all emphasize that without rights, benefits are extremely difficult to secure, yet such rights are often poorly defined, legally absent or, in the case of the Philippines, legally recognized but nonetheless undermined by the state.

The situation is made all the more complex by the difficulties of identifying beneficiaries and traditional knowledge holders. In fact, simply setting out to do so raises more questions than answers. If the knowledge resides in a small sector of a community (e.g. among traditional healers, individuals or particular families) for example, rather than in the community as a whole, should the community as a whole benefit or rather the knowledge holders themselves? Such decisions require sensitive and careful deliberation. Altman, for example, describes how tensions have resulted from Aboriginal peoples in Australia receiving payments from mining for group benefit as against payments being distributed to individuals and families for income supplementation.

Similarly, if the knowledge is held widely across a number of communities, must an applicant for a permit get the agreement of all those communities? What protocols should apply where knowledge or resources straddle political boundaries, as is the case for *Hoodia*? What if the knowledge is held by indigenous communities, as described by Kelly Bannister in the case of the Pacific yew in Canada, but is identified through other means, in that case by a random screening programme of the National Cancer Institute? What if the community is not formally organized? What if the very definition of 'community' is contested or, as Feinholz Klip et al. show for the Chiapas community, continually changing? How is representation determined under such circumstances and how, as Vermeylen asks pointedly, can one meaningfully insist on single-voice representation to reflect these diversities and complexities?

The *Hoodia* case sheds some light on these questions, highlighting in particular the incremental and iterative manner in which solutions must be found, and the importance of defining clear roles and responsibilities for different categories of beneficiaries. However, frustrations remain for those attempting to comply with the CBD and navigate a path through national regulatory mazes. This has resulted in companies often adopting a hands-off approach to the use of traditional knowledge.

Not only are traditional knowledge holders difficult to identify, but so too are the authorities vested with the power to vet and approve bioprospecting applications. Wynberg explains the myriad of permitting procedures in place for *Hoodia* alone, each with its own set of decision-makers, in some cases located in different government departments at local, provincial and national levels. Despite repeated calls for a 'one-stop shop' permit system, governments the world over find it difficult to move away from a silo approach to resource management. This makes the cross-cutting regulatory arena of access and benefit sharing especially challenging to negotiate.

> Beneficiaries of bioprospecting initiatives, and traditional knowledge holders in particular, should be identified through processes that are incremental, iterative and socially astute. Principles should be developed in collaboration with the wider community to guide the process of identifying the correct beneficiaries. Clear roles and responsibilities should be developed for different categories of beneficiaries. Governments should act to establish simple and clear information channels to inform bioprospectors and researchers about the authorities that need to be consulted to obtain permits, and the procedures to be followed.

18.7 Capacity Development

Capacity development is increasingly under the spotlight as a prerequisite to fair benefit sharing. This centrality is confirmed by a number of authors in this book who point towards the need to build capacity not only among indigenous peoples and local communities, but also among supporting NGOs, the research fraternity that seeks to access biodiversity and traditional knowledge, and the government bodies that aim to regulate research and bioprospecting activities.

Indigenous peoples and local communities require strengthened capacity on a number of fronts and, as the case studies from Mexico, Australia and the Philippines show, the earlier this investment is made, the better. Indigenous peoples and local communities need the know-how, knowledge and organizational capacity both to engage in the prior informed consent process in a manner that is substantive and meaningful, and then to enter into negotiations and decision-making processes to develop equitable benefit sharing agreements. In reality, few communities are equipped to do this and, as the San-*Hoodia* case has shown, advocacy NGOs and lawyers have to step into the void. Although there is a need to provide continued support to NGOs playing this role, the long-term goal of building local institutional capacity among indigenous groups is vital if communities are to become equal partners in negotiations and decision-making.

Organizational capacity is needed not only to engage in informed consent processes and negotiations, but also to set in place efficient administrative systems, management structures and financial procedures. However, as Bannister remarks regarding Aboriginal communities in Canada, while this is vital, it often takes second place to other priorities such as health, education, housing and employment. Capacity development for access and benefit sharing thus needs to be dovetailed with wider developmental initiatives to build institutional and administrative capacity within the community as a whole.

Roger Chennells, Victoria Haraseb and Mathambo Ngakaeaja paint a vivid picture of San-owned NGOs in southern Africa, and of the dangers of putting untrained leaders in positions of control over complex organizations with large, multi-donor projects. In the San-*Hoodia* case a trust was established which over time developed that capacity to manage itself and make decisions about benefit sharing. The early setting up of this structure, prior to the inflow of significant sums of money, was an important forward-looking step, irrespective of the uncertainty it now faces about the future of *Hoodia*.

Feinholz-Klip et al. and Bannister also deal with the critical need to build capacity among researchers. As Bannister notes, it is not always clear what level of understanding Western-trained scientists have about the nature of traditional knowledge – a point well demonstrated by the Chiapas case, which was beset by problems of mistrust between Mexican researchers and indigenous people. Ironically, however, the controversy served as a lesson in itself, and communities in the mountains of Chiapas, like the San, now insist on researchers explaining the costs and benefits of proposed projects.

Lastly, the capacity needs of governments in implementing access and benefit sharing are often overlooked. Wynberg notes the frustrations of the *Hoodia* industry and the San alike in attempting to engage the South African government about implications for the San of the new bioprospecting, access and benefit sharing regulations and requirements. The cross-cutting and controversial nature of access and benefit sharing, combined with the complexity of the issues they enfold, makes management and regulation enormously challenging.

> Strong efforts are required to build the capacity of indigenous institutions to engage in the consent process, to negotiate with potential partners, and to receive and distribute funds from benefit sharing. Stable, robust and representative institutions are needed to ensure equitable benefit sharing. Sufficient time, financial support and advice are essential elements of building this capacity. Efforts should also be made to develop the capacity of NGOs to support indigenous communities, of researchers to facilitate communication, and of governments to drive the implementation of laws. The earlier capacity-building investment is undertaken, the better.

18.8 Managing the Expectations of Bioprospecting

Expectations are often raised about the development and economic opportunities bioprospecting can bring, especially in marginalized and poor communities. As the San-*Hoodia* case has shown, reasonable expectations of income from benefit sharing can range from millions of dollars one day to very little when licence holders suddenly abandon research the next day.

Expectations of bioprospecting results, in terms of both product outcome and potential financial benefits, need to be managed carefully, as they are often unrealistic. The chapter by Rachel Wynberg and Sarah Laird describes how few understand the complexities of natural product cycles, the vagaries of markets, and the risks and costs involved in product development. Moreover, most bioprospecting activities do not yield commercial products, many do not use traditional knowledge, and typically the main benefits lie in the discovery phase and are of a scientific and technological nature. The San-*Hoodia* case is one of the few exceptions where indigenous peoples have been involved in receiving direct financial benefits arising from the use of their knowledge, and it is important that this experience not be regarded as the norm.

Several of the cases in this book emphasize regular communication between parties, information sharing and participatory processes as critical elements of a strategy to manage expectations and realize positive outcomes. Sachin Chaturvedi, for example, notes the key role played by scientists in facilitating information sharing and the development of an agreement with the Kani to commercialize Jeevani. Feinholz-Klip et al. in contrast, describe how the Mayan ICBG project floundered because of the lack of a participatory process to engage beneficiary communities about the specific details in the development of the benefit sharing agreement. Bannister identifies a more conscious way of engaging with Aboriginal communities in Canada, and makes the point that biodiversity studies today are subject to higher levels of scrutiny than in the past and are governed by different ethical expectations relating to Aboriginal rights and interests.

Disappointments can also be mitigated by exploring the diversity of opportunities associated with a particular product, or – if industry partners exist – by starting new ventures involving different plants. The San-*Hoodia* case has shown the merit of proactively using existing institutional structures that are set up to administer benefit sharing for a variety of related purposes – in this case to negotiate new agreements with *Hoodia* growers and to enter discussions about using San traditional knowledge of other plants. This approach can serve as insurance against the most serious disappointments.

> Communities involved in bioprospecting initiatives need to be fully and honestly informed about the benefits they are expected to derive from their engagement in the development of any benefit-sharing agreements. Information sharing from researchers and industry should include ongoing feedback about progress with research, development and commercialization. Communities should be active partners from the beginning. To minimize the risk associated with bioprospecting, communities should proactively use the institutions they set up and the knowledge gained to pursue new ventures.

18.9 Intercultural Encounters and Governance

When bioprospectors negotiate access and benefit sharing agreements with representatives of indigenous communities, two very different worlds meet. Such encounters provide practical challenges that cannot always be resolved according to the letter of the law. For instance, during negotiations bioprospectors are likely to concentrate on expediency and fast decision-making through a small group of highly educated professionals. By contrast, decisions within indigenous communities are more likely to emerge over long periods of time involving consultation with and seeking consensus by large groups. The chapter on decision-making in the San-*Hoodia* case by Wynberg et al. provides insights into the tensions that arise when formal Western institutions and expectations meet the norms and practices of indigenous peoples and local communities.

The clash over decision-making procedures and speeds often leaves the traditional knowledge holders worse off, as the San-*Hoodia* case has shown. When benefit sharing negotiations started between the San and the bioprospecting partner (the CSIR), it was a race against time to form a body that could legally represent the San according to Western governance principles. As a result, it would appear that the decision-making abilities of the San were compromised by the need for urgent resolution on the part of the CSIR and its commercial partners.

There is no obvious solution to this dilemma. Investors are unlikely to allow years for community consultation processes and are more likely to drop a particular research lead than risk the time and money required to adhere to local protocols and indulge continued indecision. On the other hand, the imposition of rapid decision-making processes on traditional communities is both ethically unacceptable and unfeasible in practice. In the San-*Hoodia* case, the tensions about time frames were aggravated by the lack of money to fund meetings, obtain advice and hone negotiating skills, all vital components for effective decision-making under such circumstances. One could ask where the responsibility lies for securing these components. In South Africa, the Biodiversity Act (Act 10 of 2004) locates support for consultations firmly with the government. This may be a partial solution in some cases, but requires governments to be committed and honest, and to have the capacity and knowledge to ensure that benefit sharing agreements are negotiated on an equal footing. In addition it requires governments to be free of the considerable pressure that is all too often applied by powerful resource extractive corporations.

> Efforts by national governments to ensure that their citizens are not disadvantaged in benefit-sharing negotiations are both commendable and necessary to overcome inbuilt power imbalances based on Western governance expectations. However, alternative approaches may be necessary where governments are not legitimate or are not believed to be capable of representing the interests of their citizens. It is important to strike a balance between fulfilling the expectations of Western governance and respecting the norms and practices of indigenous communities. In helping maintain that balance, NGOs and local support organizations can play an important bridging role. However, flexibility on procedures and governance expectations must not be used to justify corruption, nepotism or theft.

18.10 Policies and Laws for Indigenous Peoples, Access and Benefit sharing

The policy lens of *Hoodia* has revealed a number of broader lessons about the development and implementation of access and benefit sharing. It has been shown, for example, that the laws and policies governing access and benefit sharing, conservation, traditional knowledge and trade are vastly different across the countries that regulate the use and trade of *Hoodia*. Ideally, common regional policies should govern strategic resources such as *Hoodia*, but in practice, the complexity and diversity of legal and institutional mechanisms across countries mean that governments have found it difficult to fully streamline policies. Some steps have been put in place by southern Africa countries to collaborate more strongly on *Hoodia* poaching and trade and the transport of illegally harvested material, but the more slippery political issues of benefit sharing and indigenous peoples' rights remain disconnected and incoherent between countries.

In part this is due to the cross-cutting nature of access and benefit sharing, which includes multiple jurisdictions and many different stakeholder groups. Additionally, many southern African countries lack legal and policy frameworks that regulate the way in which resources and knowledge should be obtained, the types of benefit sharing agreements that should be developed and the role of the state in implementing access and benefit sharing. Although South Africa now has detailed regulations that spell out these requirements, these were not in place at the time of negotiations between the San, the CSIR and the Southern African *Hoodia* Growers Association. Although this policy vacuum did not impede the development of legally binding agreements, it was restrictive in failing to provide the overall framework within which contracts should be conceptualized, crafted and administered.

Many of these findings are echoed in other cases. Feinholz-Klip et al. describe how the lack of a proper legal system in Mexico exacerbated the conflicts over the Maya ICBG, and how different social actors interpreted draft laws according to their own interests. Altman finds similar shortcomings, suggesting that benefit sharing agreements in Australia have been poor instruments for the delivery of constructive development because of the absence of a cogent policy framework to ensure they generate positive outcomes. The lack of specificity in policies determining free and prior informed consent is underscored as a major weakness of Philippine law, with the chapter by Fatima Alvarez-Castillo and Rosa Cordillera A. Castillo advocating specificity in the description of the type of information to be disclosed, the language to be used and the potential risks and adverse impacts of activities. Mechanisms to ensure compliance and accountability, and to revoke permission if need be, are considered especially important.

An emerging concern is the multiplicity of legal systems that exist, often within one geographic area, with state and customary laws operating alongside each other. While this is not always problematic, Altman's descriptions of developing benefit sharing agreements among Aboriginal communities in Australia suggest that it creates an additional layer of tension in an already fragile context. An important step has been taken through Canadian research ethics policies, which now recognize that overarching national guidelines must intercalate in meaningful and effective ways with community-level guidelines and research protocols articulated by Aboriginal communities. As Bannister remarks, 'The full implications of attempting to integrate these parallel processes and share decision-making power in research have yet to be understood, but the principle is a significant step towards decolonizing research and an important learning opportunity for all involved.'

> When indigenous traditional knowledge holders reside in several countries and biological resources are shared across national borders, it is important that governments cooperate successfully and communicate effectively with each other. Cooperation should be especially encouraged in the development of common policy approaches towards trade and benefit-sharing, and in pursuing joint strategies to promote and protect indigenous knowledge and local industries (for example, the development of geographic indications). Access and benefit-sharing laws and policies will only be effective if they have social legitimacy and through consultation reflect the viewpoints of indigenous peoples, local communities and society at large. Laws and policies should be as specific as possible with regard to the procedures to obtain free and prior informed consent, and the nature of benefits expected. State laws and policies must recognize and work in meaningful ways with indigenous and customary approaches.

18.11 The Law is Not Enough

The benefit sharing provisions contained in the CBD followed years of social activism aimed at reforming the way in which nations managed and used the planet's increasingly threatened biodiversity. As part of a broader framework designed to further equity and human rights, the CBD and the domestic laws that followed its ratification by countries were widely praised as heralding a new era in which the benefits flowing from genetic resources would be fairly shared among those who had given access to these biological resources or traditional knowledge. A range of legal reforms to recognize the rights of indigenous peoples has accompanied these initiatives, along with more progressive legislation aimed at protecting the biodiversity upon which indigenous peoples rely. But has this been enough?

Altman provides some answers to this question in his description of the succession of mining-related laws enacted in Australia following the abandonment of forced assimilation, and the acceptance of self-determination as a central theme of indigenous affairs policy. The mining industry is an important generator of wealth in Australia, and his historical analysis of successive attempts to reform the laws and temper the combined power of the mining houses and the state over Aboriginal landowners provides a fascinating – albeit depressing – account of the difficulties faced by marginalized communities attempting to assert their rights to benefits. As the complex array of laws has evolved, so have the various players developed strategies to manage and often circumvent the original intentions. Indigenous peoples, previously excluded from participating in mining wealth, were forced to use existing laws creatively in order to secure their rightful share of the proceeds of extracted minerals. As an example, the legal right to veto mineral exploration and production on lands granted to Aboriginal peoples, rather than vesting the more valuable mineral rights in Aboriginal hands, developed over time into a *de facto* property right of considerable value. Similarly the right to negotiate contained in the Native Title Act, while even less powerful than a right of veto, has been used creatively to strive for a fairer share of mineral wealth.

The Castillos likewise recount the conspiracy in the Philippines between large resource-extractive corporations and the state, both ostensibly committed to laws and principles of justice, and describe the way in which these agencies pit themselves against relatively powerless indigenous peoples. Internationally approved legal requirements such as the duty to obtain free, prior and informed consent are routinely and fraudulently circumvented by these powerful players, using a range of means. The domestic laws of the Philippines, internationally acclaimed for their express protection of the rights of indigenous peoples to ancestral lands and resources, have proved virtually useless in the face of mining laws supported by a government and corporate alliance. The power differential between the protagonists remains formidable, and the vastly different cosmologies and cultures of indigenous peoples pose an additional challenge to indigenous peoples' advocates and lawyers attempting to turn back the tide. Confronting injustices perpetuated by influential resource-extractive companies requires an understanding of the inherent power imbalance between the forces at play, and secure mechanisms to enable indigenous peoples to engage on a more equal footing.

Chennells et al. traverse the range of indignities visited upon the San peoples of southern Africa despite the plethora of laws and regulations that exist to prevent injustice. For centuries the San, along with indigenous peoples worldwide, suffered the loss of land, intellectual property and culture under the benign and patronizing guise of 'development'. When the San finally decided in 1996 to assert their own power, and in particular to claim the rights that had been slipping away from them for so long, a remarkable series of successes followed. This required workshops to help them understand their rights and the development of strategies to convert those paper or theoretical rights into tangible outcomes. It was to be expected that San empowerment efforts would not be welcomed by certain opposing interests. The San message of hope is that international covenants and progressive domestic laws are welcome, but indigenous peoples need to actively claim and assert their rights in order to bring about the fairer dispensation sought by the law.

> Progressive laws to enforce benefit sharing and recognize the rights of indigenous peoples are necessary but, on their own, insufficient to ensure equity and secure human rights. Experience has shown that states often collude with industrial and corporate interests to the detriment of their own indigenous citizens. Indigenous peoples must therefore become organized and empowered, if necessary in alliance with NGOs, in order to counter the asymmetry of power and ensure that their own rights are asserted and secured.

18.12 Conclusion

This book has told the story of one of the most remarkable bioprospecting initiatives to emerge from post-CBD deliberations. It has done so not only with an in-depth account of the San-*Hoodia* case, but also by introducing important new concepts about benefit sharing and bringing in comparative material from other countries to illuminate broader trends about indigenous peoples, consent and benefit sharing. While much of the focus has been on bioprospecting and natural product development, the book also yields significant lessons about informed consent and benefit sharing from the health sciences and sectors such as mining.

As has been shown, a remarkably consistent suite of issues emerges from these diverse experiences. First, it is clear that too much has been made of the potential of bioprospecting, access and benefit sharing to resolve poverty, address global injustices and deal with the complex set of issues associated with the commodification of traditional knowledge, the patenting of life and the rapacious nature of many companies. While these are critical issues to debate and resolve, this will not happen through benefit sharing agreements or successful bioprospecting initiatives.

Second, it is vital to recognize the interface between indigenous approaches and processes and those adopted by governments, researchers and companies, and, where appropriate, integrate them as comprehensively as possible. They may be in the form of laws and policies, consultative processes, institutional arrangements, research protocols, or company practices. It is especially vital to make linkages to the rights of indigenous peoples to land, knowledge and natural resources if benefit sharing is to emerge and succeed as a strategy to empower indigenous peoples.

A third common theme to emerge is the importance of legal clarity and specificity in informed consent and benefit sharing procedures. Here there is something of a conundrum. On the one hand, there is a need for flexibility and adaptability in such procedures, given the range of circumstances that have been described. On the other hand, specificity is important to ensure the provision of adequate information, and to enable compliance and accountability. A delicate balance between these needs should be struck by legislators setting up access and benefit sharing systems.

Finally, the importance of communication, cooperation and consultation has been stressed throughout the book. This is crucial not only for indigenous peoples – who need to be active partners from the beginning of access and benefit sharing initiatives, and to receive clear and honest information about projected benefits and risks – but also for governments, which have a responsibility to establish simple and clear information channels, to cooperate among themselves where resources or knowledge crosses national borders, and to establish transparent and fair consultation processes for the development of access and benefit sharing policies.

The question remains: will the enormous efforts and funds that have been expended over decades to achieve the conservation of biological diversity, the sustainable use of its components and the fair and equitable sharing of benefits be rewarded? Although the jury is still out, there is significant evidence that collective efforts can influence the verdict positively.

References

Alexkor limited and another v Richtersveld community and others (2003). 12 BCLR 1202 CC.
Durie, M. (2008). Bioethics in research: the ethics of indigeneity. Ninth Global Forum on Bioethics in Research, 3–5 December. http://gfbr9.hrc.govt.nz/presentations/Mason%20Durie%20-%20The%20Ethics%20of%20Indigeneity.doc. Accessed 16 January 2009.
Kant, I. (1996). *Metaphysics of morals* (trans: Gregor, M.). Cambridge: Cambridge University Press.
Mabo v Queensland (1992). 175 CLR1; 66 ALJR 408 (Australia).
Roy Sesana and others v the Attorney General of Botswana (2002). Misca 52 (Botswana).
Sharma, D. (2005). Selling biodiversity: benefit sharing is a dead concept. In B. Burrows (Ed.), *The catch*. Washington, DC: Edmonds Institute.
Shiva, V. (1991). *The violence of the green revolution*. London: Zed Books.
Sridhar, R., Usha, S., & Wolff, K. (2008). Commodification of nature and knowledge: the TBGRI-Kani deal in Kerala. Paper presented at the National Conference on Traditional Knowledge Systems, Intellectual Property Rights and their Relevance for Sustainable Development, Delhi, 24–26 November.

Index

Please note: Page numbers in *italics* refer to tables and figures; page numbers in **bold** refer to examples in boxes.

A

Aboriginal cosmology, 288
Aboriginal Land Rights (Northern Territory) Act, 291, 292, 295
Aboriginals Benefit Trust Account (ABTA), 291–292
Aboriginals Benefits Trust Fund, 290
Aboriginal title, 152–154
　claims, 158–159
　and the San, 154–156
　sui generis proprietary interest, 152, 152n12
ABS, *see* access and benefit sharing
ABTA, *see* Aboriginals Benefit Trust Account
access and benefit sharing (ABS), 261–262
　agreements, 77–79
　Business Management Committee (BMC), 263, 269
　laws and policies, *130–132*
　legislation, 116
　perceptions, 77–79
　policies and laws, 346–347
　policy framework for Canada, 306
　regional coherence, 136–137
　regulation in Southern Africa, 79, 128–129
　regulations for Hoodia trade, 137
　San and Kani agreements between stakeholders, 269
Action Aid, 101, 266
Action Group on Erosion, Technology and Concentration (ETC Group), 322–323
adequate comprehension, 40
　of disclosed information, 45
Ad-Hoc Open-Ended Working Group on Access and Benefit Sharing, 73
African Commission on Human and Peoples' Rights, 172
African Union's Model Law:
　Protection of the Rights of Local Communities, Farmers and Breeders, 101
　Regulation of Access to Biological Resources, 101
Aglubang Mining Corporation (subsidiary of Mindex), 274
Agreement Treaties and Negotiated Settlements (Australia), 293
Aguaruna people (Amazonian), 61–63, 115, 166
AICRPE, *see* All India Coordinated Research Project on Ethnobiology
Akwé: Kon Voluntary Guidelines, 43
ALAMIN, *see* Alliance Against Mining
alcohol, 169, 170, 171
Alexkor Limited diamond mining, 153
Alliance Against Mining (ALAMIN), 275
All India Coordinated Research Project on Ethnobiology (AICRPE), 263, 265
altruism model of research, 23
Alzheimer's patients, 36
Amazon region, 14, 15
America, first peoples, 170
American ginseng species, 305
American Herbal Products Association, 112
ancestral land, 336
Ancestral Land of All Mangyan, 274
Andriesvale settlement, 196, 202, 203–204
apartheid, 91, 91n2, **98**, 104, 154, 243
Arnold, Chief John (!Kung), 183

Arogyappacha (*Trichopus zeylanicas* subsp. *travancoricus*), 8, 115, 263; *see also* Jeevani
art, 217
 and crafts, 223–224
Artemisia plant, 14
Arya Vaidya Pharmacy (AVP) (India), 263
 good manufacturing practice (GMP), 266
 licence fee and royalty, 263
Australasia
 first peoples, 170
Australia, 9, 145, 285–286
 Aboriginal and Torres Strait Islanders, 152–153
 Aboriginal art, 223
 Arnhem Land reserve, 289
 background of indigenous people, 287–289
 benefit-sharing agreements, 289–291,294–295
 Commonwealth, 289, 289n3, 290
 land rights, 288
 Mabo v Queensland case, 152–153
 miners and indigenous people, 289–291
 Mining Ordinance, 290
 payment under benefit-sharing agreements, 295–296
 Tiwi artists, 224
 totemic ancestors, 288
 utilization of benefit-sharing agreements, 296
Australian Mining Industry Council, 291
autonomous decisions, 36
autonomy of individuals, 234
avian flu, 70

B

Bacopa moneri (brahmi, memory-enhancing drug), 267
Basel Convention on the Control of Transboundary Movements of Hazardous Wastes and Their Disposal, 59
Batswana (Botswana), 173
beneficence, 65
beneficiaries
 bioprospecting initiatives, 342
 identification of, 104, 299, 341, 342
beneficiary communities, 233, 296
benefit sharing, 20–24, 89, 338
 decision-making and distribution of benefits, 114–116
 decision-making processes, 8, 231
 negotiations, 7, 345
 policies and regional differences, 116–117

progressive laws, 348
provisions of CBD, 73, 232
benefit-sharing agreements, 5, 6, 11, 37, 44, 82, 285
 with CSIR, 114
 initiating talks, 100–102
 issues of concern, 103–107
 memorandum of understanding, 102–103
 with SAHGA, *111*, 112–114
 with San-CSIR-Phytopharm-Unilever, 108
Berne Convention, 214
biochemicals and derivatives, 5, 69
biodiversity, 3, 4, 11, 22, 23, 28, 41, 70–72, 81, 82, 89, 107, 108, 112, 118, 128, 129, 215, 232, 241, 265, 272, 276, 404, 316, 317, 336, 344, 347
 Canadian policy, 303, 306, 307
 capacity development, 342
 commercial use, 74–77
 conservation, 128
 of Himalayas, 265n5
 human research ethics, 309
 justice and convention, 337–338
 legislation, 300
 products, 272
 research, 303
 threats to, 72, 110, 137
 traditional knowledge of, 224–225, 308
Biodiversity, Access and Benefit-Sharing Regulations, 114
Biodiversity Act (South Africa), 45, 81n5, 106, 112, 113, 129, 136
biodiversity-related fields, 308–311
biological resources, 5
 transactions, 53
BioMed Pharmaceuticals, 110
biopiracy, 4, 5, 9, 11, 20, 22, 48n6, 53, 54, 55–58, 70, 76, 78, 83, 90, 101, 115, 217, 218, 266, 279, 308, 316, 323
 allegations, 216, 216n4
 Madagascar, 64
 patent-related version, 58
 of traditional knowledge, 57
bioprospecting, 4, 44, 45n14, 69–70, 77, 81
 ABS Regulations for permit, *135*
 and benefit-sharing agreements, 232
 or biotrade, 136
 cases, 166
 communities involved in initiatives, 344
 through history, 71–72
 and indigenous communities, 46, 231
 managing expectations, 343–344
 regulatory frameworks, 72–73
 ventures, 307–308

Index 353

biotechnology, 5
biotrade, 81
Biowatch South Africa, 101, 266, 325
Blouberg settlement, 196, 201
Bok, Gert (chairperson of CPA), 202
Bonn Guidelines, 39, 45, 73
    Access to Genetic Resources and Fair and Equitable Sharing of the Benefits Arising out of their Utilization, 101, 215–216
    disclosure requirements, 41–42
Boswellia plant (frankincense), 71
botanical medicine companies, 74–75
Botswana, 6, 7, 93, 104, 129, 189
    assessment study, 178
    Khwedom council, 182
    nature conservation laws, 137
    political representation, 181–182
Brazil, 79–80
Bristol-Myers Squibb (pharmaceutical company), 305
Bulun Bulun v R and T Textiles (Australian case), 223–224
Burma, 20

C

Canada, 9–10, 145, 303
    Aboriginal peoples, 304, 304n1
    comparison with San-*Hoodia* case, 304–307
    first peoples, 170
    national ethics policy for research, 309–311
    Salishan people, 305
    Tsimshian people, 305
Canadian Institutes of Health Research (CIHR), 309
    Guidelines for Health Research Involving Aboriginal Peoples, 310–311
capacity, 8, 92, 93, 99, 171, 223, 246, 310, 312, 339, 342–343, 345
    to account for expenditure, 107
    agronomic, 96
    to consent, 35–36
    constraints, 232, 239
    decision-making, 30, 36
    development, 232, 342–343, 248
    to distribute funds, 233
    to document violations, 278
    of extraction of product, 117
    indigenous institutions, 343
    indigenous representative bodies, 299
    law enforcement, 138
    management, 251, 300, 316
    marketing, 267
    national discussions, 324
    to negotiate, 241
    organizational and technical, 116, 242
    rule of law, 148
    San board members, 176
    San individuals, 178
capacity-building, 9, 194, 244, 247, 248, 286, 297, 301, 303, 316, 339
    Canadian government, 313
    Maya collaborators, 317
    programme, 73
    science and technology, 78
Caribbean, 64
cattle fund, 173
CBD, *see* Convention of Biological Diversity
Central Drug Research Institute (India), 267
Central Kalahari Game Reserve, 154, 169, 182
Centre for Research Information Action in Africa–Southern African Development and Consulting (CRIAA SA-DC), 196, 201n8
Century zinc mine (Australia), 293
cereal crop (*Eragrostis tef*), 82, 82n6
chemical compounds, 81
Chiapas (Mexico), 10, 166, 315–317, 319, 321, 322, 322n6, 323, 327–329
    community, 318, 318n1, 341, 343
    socio-political organization, 326
Chiapas Council of Traditional Indigenous Doctors and Midwives (COMPITCH), 322–327
CIHR, *see* Canadian Institutes of Health Research
CIPR, *see* Commission on Intellectual Property Rights
CITES, *see* Convention on International Trade in Endangered Species
coercion, 46–47
collection and use, 57
collective action, 281
collective trauma
    societal problems, 170n1, 170–171
colonial botanical accounts, 95
colonial invasions, 92, 143
colonialism, 46, 143
    exploitation, 41
    legacy, 40–41
Columbus, Christopher, 71
Comalco bauxite mining company (Australia), 289, 293

commercialization, 196, 197
    cultural perspective, 203–205
    of knowledge, 203
    of trance dance, 205
Commission on Intellectual Property Rights (CIPR), 212, 217
commitment to fairness, 37
commodification, 338–339
    of medical knowledge, *199*, 199–203
    scenario survey, 195–197
    of traditional knowledge, 7, 193–195
common heritage, 20
    of humankind, 14, 15, 23, 24
    principle, 13n1
    versus national sovereignty, 13–15
common property (medicinal field plants), 151
communal land tenure ('native reserves'), 149–150
Communal Property Associations Act (SA), 180n14, 185
community
    decision-making processes, 39
    deliberation, 32
Community Property Association (CPA), 202, 251
COMPITCH, *see* Chiapas Council of Traditional Indigenous Doctors and Midwives
Confederación de Nacionalidades Amazónicas del Perú (CONAP), 62, 63
Conference of the Parties to the Convention on Biological Diversity Fifth Ordinary Meeting (Nairobi), 59–60
conflict
    of interest, 29–30
    management system, 238
    resolution among the San, **236–237**
Consejo Aguaruna Huambisa (CAH), 61–62, 63
consensus in band societies, 235, 235n1
consent, 33
    and cultural differences, **35**
    ethical basis for traditional knowledge, 39
    ethical basis in medical field, 36, 38
    and health care, 33, 35, 36
    legitimate authority, 45
    stakeholders, 42
    and traditional knowledge, 36-39
constituencies
    linguistic and geographical, 181
contract law, 226
Convention on Biological Diversity (CBD), 3, 9, 11, 28, 116n12, 215, 232, 252, 266, 335
    access and benefit sharing, 54

Ad Hoc Inter-Sessional Working Group on Article, 8(j), 41
bioprospecting, 70, 72
Canada, 306
ethical foundation, 3
genetic resources, 72–73, 136
Gove case (Australia), 291
international treaties, 128
justice, 337–338
legislation, 21
objectives, 12
policy process, 69
prior informed consent, 27
Convention on International Trade in Endangered Species (CITES), 128, 137–139
copyright law, 214–214
    clothing manufacturer in Australia, 223
cosmopolitan human rights, 19
Costa Rica, 80
    Biodiversity Law, 81n5
Council for Scientific and Industrial Research (CSIR) (SA), 6, 41, 90, 325
    benefit-sharing agreement, 100–109, 248
    benefit-sharing model, 266
    licensing agreement with Phytopharm, 95, 96, 101
    negotiations, 266
    patent application, 233
    patenting without consent, 129
    payments to San-*Hoodia* Trust, *110*, 114
    protection of traditional knowledge, 225
CPA, *see* Community Property Association
Crew Development of Canada, 275
Crew Minerals Philippines Inc., 275
CRIAA SA-DC, *see* Centre for Research Information Action in Africa–Southern African
Development and Consulting cross-country cooperation, 251–252
CSIR, *see* Council for Scientific and Industrial Research
cultural dispossession and alienation, 171
culture
    conceptions of, 157–158
customary land rights, 150
customary law and communal areas, 149–152

D

Damara (minority group, Namibia), 94, 103, 104, 105, 115
dam building, 28, 37, 272

Index 355

dance, 217
　cultural heritage, 221–222
decision-making, 231
　Aboriginal people, 296–298
　communal challenge, 247
　democratic and traditional, 238
　and democratic model, 233–235
　San-*Hoodia* benefit-sharing agreements, 239
　traditional, among the San, 235, 238, 239
Declaration of Belém, 265
Declaration on the Rights of Indigenous Peoples, 144n2, 178, 178n12
Denmark, 80
Department of Arts and Culture (SA), 136–137
Department of Land Affairs (SA), 181, 185
Department of Science and Technology (SA), 266
Department of Trade and Industry (SA), 225
Department of Environmental Affairs (SA), 137
de Vattel, Emeric, 149
Didima Rock Art Centre (KwaZulu-Natal), 222–223
disclosure
　of origin, 80
　proposal, 79–80
distributive justice, 18
　and benefit sharing, 20–24
　and justice in exchange, 18–20
Dixey, Richard, 96, 101
D'Kar community (Botswana), 173
　Bokamoso Trust, 175n8
　D'Kar Trust, 175n8
　Komku Trust, 175n8
Dobe area, 196, 198
donor-funded projects, 73
donor funding, access to, 170
drug discovery and development, 74

E
Earth Summit in Rio de Janeiro, Brazil, 12, 28, 70
Echinacea species, 305
economic benefit, 7
economic status, 7
egalitarian values, 170
El Colgegio de la Frontera Sur (ECOSUR), 317, 320, 321, 326
Eli Lilly (pharmaceutical company), 63–64
Empire Exhibition (Johannesburg), 172
empowerment, local, 296
equality, 234

equity and effectiveness, 285, 293–294
Eragrostis tef (cereal crop), 82, 82n6
essentialist conception of culture, 157–158
Ethiopia, 82
Ethiopian Agricultural Research Organization, 82n6
Ethiopian Institute of Biodiversity Conservation, 82n6
ethnobiological information, 65
ethnocide, 279
Europe, 71
Evolvulus alsinoides species, 263
exclusive property rights, 14–15
Ezemvelo KZN Wildlife, 223

F
farm labour communities
　Omaheke region of Namibia, 170
FDA, *see* Food and Drug Administration
Federal/Provincial/Territorial Task Group (Canada), 306
Federal Trade Commission (FTC), 111, 112
fee simple land, 149, 149n7
First Nations, 92, 92n4
First People of the Kalahari (FPK), 182
folklore, 211, 222
Food and Drug Administration (FDA), 96, 111, 112
FPIC, *see* free and prior informed consent
free and prior informed consent (FPIC), 46–47, 272–273
　Indigenous Peoples Rights Act (IRPA), 276–278
　legal framework, 276–278
　Mining Act, 275–276
　mining companies, 276n7
　political problematique, 280
FRELIMO (Mozambique), 149
FTC, *see* Federal Trade Commission
full disclosure, 40, 43
　of relevant information, 45
funds, receiving and disbursing, 245–247

G
Gagudju Association (Australia), 296
Geingob, Prime Minister Hage (Namibia), 146
gender inequality, 200
genetic material, 81
　certificate or 'passport', 80–81
genetic resources, 56, 71, 75
　access to, 28, 41–42

genocidal predations on San, 92
'genome mining', 72
Germany, 14
    colonialist practices, 146, 147
germplasm, 14, 15, 16
glaucoma treatment, 14, 15
GlaxoSmithKline pharmaceutical company, 64, 328
Global Bio-Collection Society, 47
globalization, 5, 79
global distributive justice, 4
global trade regime, 272
goldenseal species, 305
'grand bargain', 72

H
Hai//om San, 147, 158
Hambukushu people, 198
Harvard Kalahari Research group, 327
Hasluck, Paul (Minister for Territories, Australia), 289–290
    legacy, 291
hazardous materials, 28
H. Currorii (known as *sekopane*), 94
    diabetes remedy, 94–95
health care, 31
    of San, 170
Hereros (Namibia), 147, 197, 198
H5N1 virus, 70
Hi-Tech Pharmaceuticals, 110
Hoodia gordonii (succulent plant) extracts, 96
    Aloe Hoodia, 110
    flower, 120
    safe (GRAS) status, 96–97
    Lipodrene, 110
    Trimphetamine, 110
Hoodia plant, 3
    advertisements for products, 194
    appetite-suppressing properties, 78, 89, 94
    benefit-sharing agreement, 194–195
    booms and busts, 109–112
    Case, 38, 44–45, 81–82
    commercial development, 95–97, *98–100*
    conservation, trade and ABS, 137–139
    distribution and occurrence, 105, *105*
    explanation of, **119–120**
    molecular structure, 264
    ownership of genetic resources, 139
    patenting of active ingredients, 179, 218
    policy frameworks, 6
    research for commercial application, 95–97
    traditional use and knowledge, 93–95

Hoodia Value Chain Based on Trade of Raw Material, *111*
'Hottentots' (Khoe peoples), 93–94
Human Genome Project, 79
    Ethics Committee Statement on Benefit Sharing, 13
Human guinea pigs, **34**
hunter-gatherers, 91, 167
    world view, 7, 165, 169
hunting parties, 171

I
ICBG, *see* International Cooperative Biodiversity Group
ICC, *see* International Chamber of Commerce
ICESCR, *see* International Covenant on Economic, Social and Cultural Rights
IKS, *see* indigenous knowledge systems
India, 20–21, 79–80
Indians (Brazil), 14
indigeneity, idealization of, 157–159
Indigenous and Tribal Peoples in Independent Countries, 218
indigenous children, 171
indigenous communities
    disclosure elements, 42–43
indigenous knowledge, 101, 106, 109, 217, 218, 218n8, 227, 261, 262
indigenous peoples
    agreements, 144, 144n1
    democratic work for rights, 282
    foundation of rights, 335–336
    lands and resources, 272
    militarization of community areas, 278
    rights of self-determination, 27, 28, 37
    socio-economic status, 293
    United Nations definition, 137
Indigenous Peoples of Africa Coordinating Committee (IPACC), 178
individual property (small agricultural plots), 151
Indonesia, 70
informed consent, 4, 27, 233
    health care, 39–40
    international guidelines, 30, *30*
    medical context, 29–30
    procedures, 27
    roles, *30*
    traditional knowledge, 40–43
Institutional Framework for Bioprospecting and ABS, 267–269
    Botswana, **134**
    Namibia, **134**
    South Africa, **133–134**

Index

intellectual piracy, 56, 58
intellectual property rights (IPRs), 5, 7,
    55–56, 78
  breach of and use of law, 219–220
  disclosure mechanism, 79
  Hoodia species, 95
  international and national, 213–217
  legal system, 225
  and the media, 220–221
  modified laws, 79
  music and dance, 221–222
  protection, 128, 261
  research, 219–220
  system, 211–213
intercultural encounters and governance, 345
International Biodiversity Group, 115
International Chamber of Commerce (ICC),
    80n4
International Cooperative Biodiversity Group
    (ICBG), 10, 62, 63, 166, 315–317
  Drug Discovery and Biodiversity among
    the Maya of Mexico, 316–317
  informed-consent protocol, 326–329
  Mexican and Peruvian cases, 167
International Covenant on Civil and Political
    Rights, 214
International Covenant on Economic, Social
    and Cultural Rights (ICESCR),
    37, 214
International Labour Organisation Convention,
    54, 54n1, 101, 147, 218
International Society of Ethnobiology (ISE), 265
  Code of Ethics, 60, 321
  disclosure requirements, 42
International Treaty on Plant Genetic Resources
    for Food and Agriculture, 128
IPACC, *see* Indigenous Peoples of Africa
    Coordinating Committee
IPRs, *see* intellectual property rights
ISE, *see* International Society of Ethnobiology

J
Jeevani (*Trichopus zeylanicus*), 78, 115, 263
Johannesburg Plan of Implementation, 73
Ju/'hoansi of Nyae Nyae (Namibia), 165, 173,
    198–199, 203
justice
  concepts of, 15–20
  domains of, *16*
justice in exchange, 4, 18–19
  and benefit sharing, 20–24
Ju/wa Bushman Development Foundation
    (JBDF), 173

Ju/wa Farmers Union, 173

K
Kakadu National Park, 294
Kakadu Region Social Impact Study
    (Australia), 294
Kani, Mottu (head of Kani tribe), 263
Kani tribe (Kerala, India), 8–9, 78, 115, 262
  involvement of state government, 266
  Jeevani (herbal drug), 78
  overview of case, 263
Kalahari, 156
  debate, 91
  Game Reserve, 172
Kerala Kani Samudaya Kshema Trust, 263,
    263n1, 267–269
Kerala State Drugs and Pharmaceuticals Ltd.,
    267
KFO, *see* Kuru Family of Organizations
Kgalagadi Transfrontier Park, 198
Khoe-speaking peoples, 105
Khoi-speaking peoples, 94
≠Khomani San, 180, 244
  attitudes to commodification, 202
  collapse of formal structures, 184–185
  land claims, 157, 186, 198, 248
!Kung San, 183
'Khwedom' council (Botswana), 182
Khwe San community, 148, 180, 198,
    203, 244
  land claims, 156–157, 248
  knowledge, 7, 14
  trading of, 339
Kuru (in D'Kar, Botswana), 173, 173n6
  Art Project, 223
  Dance festival, 222
  development organization, 174–175
  Development Trust (Botswana), 7, 165, 173
Kuru Family of Organisations (KFO), 93,
    175–177, 185
  organogram, *175*
Kwangali tribe (Namibia), 147

L
land, 7
  claims, 6
  de jure rights, 197
  equitable allocation, 149
  legal rights, 144
  restoration and resource rights, 336
  rights of San, 169
  tenure security, 150

language
  Afrikaans, 244
  of patients, 40
leadership, 182–184
Legal Assistance Centre (Windhoek), 169
legal competency, 35–36
Letloa Trust (Botswana), 176–177, 196, 198
Lewis, Walter, 61
liberalized global trade regime, 272
licence
  agreement, 62
  and benefit-sharing agreements, 97
  fees, 62
Like-Minded Megadiverse Countries, 73, 73n2
local community governance, lack of success, 250–251
Locke, John, 144, 149, 158
logging, 28, 37, 272

M

Mabo and Delgamuukwe (Canadian) case, 153, 153n14
Mabo v Queensland (Australian) case, 219
Madagascar, 63–64
Magu, Selina (chairperson of Letloa Trust), 177
malaria vaccine study site, 43
  community permission, *44*
Mangyan organizations (Philippines), 274–275
manufacturing practices, 261
market capitalism, 23
Maya civilization (Chiapas, Mexico), 166
  political context, 317–318
Maya
  International Cooperative Biodiversity Group (ICBG), 10, 315–317
  controversy, 325
  intellectual property rights and benefit sharing, 321–322
  permit application, 319–320
  prior informed consent, 320
Mbukushu tribe (Namibia), 147, 157
medical practice, modern, 27, 28
medicines, traditional, 217, 218
  Plathis (Kani tribal healers), 264
Merck Pharmaceuticals, 14, 15
Mexican Department of the Environment, Natural Resources and Fishing (SEMARNAP), 323–324, 326
Mexican national regulations, 319–320
Mexico, 166
  bioprospecting regulations, 325
  comparison with San Case, 325–329
  intellectual property rights and benefit sharing, 321–322
  opposition to project, 322–325
Milirrpum and others v Nabalco and the Commonwealth, 290
militarization of community areas, 278
Millennium Development Goals, 212
Mindex Resources Development Inc. (Norway), 274–275
mineral resources and biodiversity, 272
Minerals Council of Australia, 286, 299
mining, 28, 37, 271–272, *see also* Australia; Philippines
  bauxite, 289–290
  corporations, 286
  fair benefit sharing and compensation, 279
  freedom to decide, 279
  indigenous leverage in agreements, 291–293
  resistance to foreign investment, 278
  risks or adverse impacts, 279
Ministry of Land Resettlement and Rehabilitation (Namibia), 146
misappropriation, 56n2
Mohamed, Pops and Gcubi family of musicians, 221
Molecular Nature Limited (MNL) (Wales), 317
Molopo Lodge (Kalahari), 186, 251–252
Molopo (San-!Khoba) Declaration, **252**
  methods and approach, 253–256
Mozambique, 149
Museo de Historia Natural de la Universidad San Marcos, 61
music, 211

N

Nabalco mine (Australia), 290
Nama (minority group, Namibia), 94, 103, 104, 105, 115
Namibia, 6, 7, 93, 94, 104, 115–116, 129, 145
  assessment study, 178
  land law, 148
  land policy, 151
  land reform, 146–148
  nature conservation laws, 137
  redistribution of land, 148–149
Namibian independence struggle, 146
Namibian San Council, 181
National Commission of Indigenous Peoples (Philippines), 274

Index

National Conference on Land Reform and the Land Question, 147
National Environmental Management Biodiversity Act, 129, 226n20, 232, 248
National Institutes of Health (NIH), 61, 316
National Land Policy (Namibia), 147, 147n6
National Native Title Tribunal (Australia), 292
National Science Foundation, 61
national sovereignty, 20, 23
Native Title Act (Australia), 292
native title law and native title rights, 300
natural law, 21
natural products, 74, 75, 80, 82, 317
natural rights, 20
  versus social utility, 16–18
Natural Sciences and Engineering Research Council of Canada (NSERC), 309
Natureceuticals group, 96
Nazi experiments, **34**
Ndebele (Namibia), 155
neem tree (*Azadirachta indica*), 20–21
negotiations
  lack of resources for San team, 248
  relationship, 232–234
  San and CSIR, 239–243
  tension of time frames, 247–248
neuropsychological tests, 36
New Xade (Botswana), 169
New Zealand, 145
NGOs, *see* non-governmental organizations
Ngubane, Ben, 107
Ngurratjuta Aboriginal Corporation, 296
NIH, *see* National Institutes of Health
non-capacity to consent, 36
non-governmental organizations (NGOs), 7, 93, 167
  development and support, 171–177
  and donors, 186–187
North America, 71
Northern Territory (Australia), 289
  Aboriginal challenge in Supreme Court, 290
  Mining and Aboriginals Ordinances, 290
Norway, 80
Nuremberg Code, 28
Nyae Nyae Conservancy, 173
  Farmers Cooperative, 173

O

Omaheke San in Namibia, 158
  farm labour communities, 170
  musicians, 221
  North and South, 148
Omatako community, 196, 196n2

open access (grazing areas), 151
Organización Central de Comunidades Aguarunas del Alto Marañon (OCCAAM), 61, 62
organizational structures, 184–186
Organization of Indigenous Physicians of the State of Chiapas (OMIECH), 322
Organization of United Mangyan Alangan Inc. (SANAMA), 274
Organization of United Mangyan Tadyawan Inc. (KAMTI), 274

P

Pacific yew tree (*Taxus brevifolia*), 305
paclitaxel anti-cancer agent, 305
Paris Convention, 214
patenting, 7, 16, 57–58, 129, 193, 218, 225, 320
  gene sequences, 79
  Hoodia, 102, 179, 272
  life forms and natural products, 80, 241, 349
  'spurious inventions', 216n4
patents, 14, 55, 214
  applications, 47
patients
  autonomy of, 33–34
  language, 40
Permanent Forum for Indigenous Issues, 178
Peru, 166
  International Cooperative Biodiversity Groups Project, 61–63, 115
Peruvian Medicinal Plant Sources of New Pharmaceuticals, 61
P57 programme, 95–96
Pfizer (pharmaceutical company), 3, 41, 95–96, 248, 328
Pharmaceutical Cooperation (I.M.) Kerala Ltd., 267
Phillipines, 9, 47, 280, *see also* Mangyan organizations
  Alangan community and mining, **273–274**
  human rights violations, 273–274
  Kabilogan people, 274–275
  large-scale mining, 275–276
  mining activities, 271–273
  Tadyawan community, 274
phytomedicines, 95
Phytopharm (British company), 95–97, 101, 129, 325
PIC, *see* prior informed consent
Piper longum species, 263
piracy, 57
Pitjantjatjara people (Australia), 220

plant(s), 217
  collection expeditions, 71
  cultivation and harvesting, 261
  transfers, 71
  variety protection, 55
polyherbal drug, 263, 263n2
popular sovereignty, 234
post-informed refusal, 44
poverty, 41, 170
  and death, 337
primordialist conception of culture, 157–158, 169
prior informed consent (PIC), 5, 15, 27, 43–45, 53, 54, 166, 232, 339–340
  accessing traditional knowledge, 31–33
  definition, 60
  concept, 58–60
  international guidelines, 32
  legitimacy vs legality, 319–320
  in practice, 61–64
  roles, 31
profit-sharing, 296
Promotion of Intellectual Property Rights of the Highland Maya of Chiapas, Mexico (PROMAYA), 317, 318, 321
property
  individualization, 150
  rights, 22
Proteus Initiative, The, 176–177
Pushpangadan, Dr. P. (head of AICRPE), 265

Q
Queen Hatshepsut (Egypt), 71

R
rainforests, vanishing 'medicinal riches', 72
Ranger uranium mine agreement (Australia), 294
Regional Conference on Development Programmes for Africa's San Populations, 177–178
Regional Research Laboratory (RRL), 265
Rehoboth Baster appeal, 154–155
repressive authenticity, 295, 295n10
research
  altruism model, 23
  ethics for biodiversity policy, 309–311
  ethics guidelines, 10
researchers
  agents of appropriation, 307–308
  national code of conduct, 308
resettlement, 28
res nullius principle, 144, 144n3, 146, 150

resources
  legal rights, 144
  misappropriation of, 22
Re Southern Rhodesia claims, 155
Richtersveld
  Land Claims Court, 153
  Restitution of Land Rights Act, 153, 153n13
  Richtersveld Community and Others v Alexkor and Another, 153
rights of self-determination, 4
rituals, 217
rock art, San, 106, 214–215
  and cultural symbols, 222–223
  Didima Rock Art Centre, 222–223
  Ukahlamba-Drakensberg Park, 222
rosy periwinkle (*Catharanthus roseus*), 63–64
  anti-cancer alkaloids, 63–64
  moral obligations question, 65
royalties, 7, 112–113, 114, 221, 242, 244
  ad valorem, 294
  quantum-based, 294
  statutory, 292
Roy Sesana and Others v the Attorney General of Botswana, 154
RRL, *see* Regional Research Laboratory

S
SAHGA, *see* Southern African *Hoodia* Growers Association
San, 6, 91n1, 91–93
  institutions prior to modernity, 167–168
  land and resource rights, 145
  modern institutions, 172–173
  population figures, 91
  poverty, 41, 170
  protecting culture and knowledge, 106–107
  representative organizations, 165–166, 177–182
  royalties from *Hoodia* product, 78
  socio-economic and political situation, 145
  stakeholders, 107
  traditional authorities, 147
San communities
  Andriesvale-Witdraai (SA), 198
  Blouberg and Vergenoeg (Namibia), 197–198
  Shaikarawe and Dobe area (Botswana), 198–199
San-*Hoodia* benefit-sharing case, 4, 6, 233
  comparative implications, 298–301
  comparative insights, 279–280
San-*Hoodia* Benefit-Sharing Trust, 107n9, 107–109, 114, 117, 185, 267, 300
  clarity on roles and responsibilities, 249–250

decision-making in allocation of funds, *250*
establishment of, 243–245
expenditure of milestone payments, *245*
lack of resources for, 248–249
Sanscapes (music CD), 221
    royalties, 221
San Trust, *see* San-*Hoodia* Benefit-Sharing Trust
SASI, *see* South African San Institute
scientists, 264–266
Second International Indigenous Forum on Biodiversity (SAIIC), 42
self-determination, rights to, 27, 37, 279, 296, 347
self-rule (*autonomia*), 33, 35
SEMARNAP, *see* Mexican Department of the Environment, Natural Resources and Fishing
semi-nomadic groups, 167
Sexual Offences (Amendment) Act, 33
Shaikarawe settlements (Okavango Delta), 196, 198
social faultlines, 273, 273n2
Social Sciences and Humanities Research Council of Canada (SSHRC), 309
social utility, 17–18, 20
Solomon, Maui, 60
South Africa, 6, 7, 80, 93, 104, 129
    assessment study, 178
    benefit-sharing agreement, 112
    colonial practices, 146, 147
    nature conservation laws, 137
    role of, 266
South African Biodiversity Act, 8, 325
South African Council for Scientific and Industrial Research (CSIR), 261, 264–266, 279
South African Defence Force, 95
South African Human Rights Commission (SAHRC), 185
South African National Gallery, 93
    Miscast exhibition, 93, 171
South African San Council, 102, 106–107, 112, 179–180, 202, 249
    memorandum of understanding with CSIR, 102–103
    building trust, 104
    holders of traditional knowledge and beneficiaries, 104–106
    Hoodia case, 167
    opposition to CSIR's secret registration, 182
    organogram, *180*
South African San Institute (SASI), 102, 181, 196, 261, 325
South American vine (*Banisteriopsis caapi*), 78
    religious and healing ceremonies, 78

Southern African *Hoodia* Growers Association (SAHGA), 6
    benefit-sharing agreement, *111*, 112–115, 117–118
South West Africa People's Organization (SWAPO), 198
Sri Lanka, 20
Stapelia gordoni (now *H. gordonii*), 93
state, role of, 266
Sustainably Harvested Devil's Claw Project (SHDC), 201
SWAPO, *see* South West Africa People's Organization
symbols, 217

T

Tanzania, 149
    Ujamaa policy, 149
tenure niches, 151–152
terra nullius doctrine, 144, 144n3, 146, 158
    atrocities, 170, 170n2, 218
territoriality of the San, 156
Thoma, Axel (former WIMSA chief executive), 179
Tiwi artists (Australia), 224
Topnaar (minority group, Namibia), 94, 103, 115
trade, 5
    in biological resources, 81
    control of, 117
    and markets, 117–118
    trademarks, 214, 215
Trade-Related Aspects of Intellectual Property Rights (TRIPS), 72–80
    Agreement of the World Trade Organization, 128
    submission to, 80n3
traditional art, 211
traditional cultures, 217
traditional knowledge, 5, 6, 21–22, 23–24, 38
    access to, 15, 28
    of biodiversity, 224–225
    commodification of, 7
    consent, 36–39
    documentation, 308
    identification of holders and authorities, 341–342
    informed consent, 40
    prior informed consent, 53
    protection, 128, 306
    protection and just reparation, 78
    regulating the protection and commercial use, 75–77
    vulnerability, 217–219

traditional lands of San, 167
traditional leaders or chiefs, 181
trance
    dance, 205, 221
    music, 221–222
Tribal Authorities Act (Namibia), 183
Trichocaulon piliferum, 93, 93n6
Trichopus zeylanicus subsp. *travancoricus* (Arogyppacha or Jeevani), 8, 115, 263
TRIPS, *see* Trade-Related Aspects of Intellectual Property Rights
Tropical Botanic Garden and Research Institute (TBGRI), 261, 263, 265
    negotiation, 266–267
Tsumkwe, East and West, 147, 197
Tswana (Botswana), 198
Tuskegee study, **34**, 40

U

Ukahlamba-Drakensberg Park (KwaZulu-Natal), 222
UNESCO, *see* United Nations Educational, Scientific and Cultural Organization
unfair free-riding, 56n2
Unilever food manufacturer, 3, 6, 41, 96, 97, 328
United Nations (UN)
    Agreement Governing the Activities of States on the Moon and Other Celestial Bodies, 13
    Convention on the Law of the Sea, 13, 337
    Declaration on the Rights of Indigenous Peoples, 128, 29, 46–47, 137, 215, 215n3, 216
    Educational, Scientific and Cultural Organization (UNESCO), 217
    institutions, 144, 144n1
    International Decade of the World's Indigenous People, 178
    Universal Declaration on Human Rights, 20, 22, 213
    Working Group on Indigenous Populations (Geneva), 178
United States (US)
    Department of Agriculture, 61, 316
    Federal Trade Commission (FTC), 111, 112
    Food and Drug Administration (FDA), 96, 111, 112
    National Cancer Institute, 305
    National Institutes of Health (NIH), 61, 316

Universidad Peruana Cayteno Heredia, 61
University of Bamako (Mali) Faculty of Medicine, 43
University of Georgia, USA (UGA) research fund, 321
University of Maryland (US) School of Medicine, 43
University of Western Ontario, 64, 65

V

Vaalbooi, Petrus (chairperson of SA San Council), 106–107, 223
veld food collection, 197
Vergenoeg settlement, 196, 201
voluntariness of decisions, 45, 46

W

weaponry, superior, 170
Wechsler Intelligence Scale, 36
welfare liberal human rights, 19
West Caprivi, 156
Western Sahara case, 155, 155n16
WHO, *see* World Health Organization
wild plants, 16
WIMSA, *see* Working Group of Indigenous Minorities in Southern Africa
WIPO, *see* World Intellectual Property Organization
Withania somnifera (Ashwagandha), 263
Working Group of Indigenous Minorities in Southern Africa (WIMSA), 7, 93, 102, 106, 166, 186, 196, 266, 327
    benefit-sharing agreement, 112–114, 239
    development assistance, 261
    general assembly, 179
    media and research contract, 220
    San-owned representative organization, 178
    terms of reference for leaders, 183
    workshops with San leaders, 221
workshop (Upington, SA), 244
World Bank, 45, 272
World Conservation Union, 45
World Health Organization (WHO), 70, 266
World Heritage Site, 222
World Intellectual Property Organization (WIPO), 212, 214
    global issues, 217

World Summit on Sustainable
    Development, 73
World Trade Organization
    (WTO), 72
    free trade policy, 272

X
!Xun San, 244, 248
    community, 180

and Khwe San Art and Culture Project,
    224

Z
Zambia
    assessment study, 178
Zapatista rebellion (Mexico), 326
Zimbabwe, 149
    assessment study, 178